D1132670

Introduction to Stochastic Models

Second Edition

Roe Goodman
Rutgers University

DOVER PUBLICATIONS, INC.
Mineola, New York

Copyright

Bibliographical Note

This Dover edition, first published in 2006, is a revised republication of the work originally published by Benjamin/Cummings Publishing Company, Inc., Menlo Park, California, 1988. For this second edition, the author has added more than fifty exercises and several more examples of random variables in Chapters 1–4, as well as minor revisions and corrections throughout the book.

International Standard Book Number: 0-486-45037-6

Manufactured in the United States of America
Dover Publications, Inc., 31 East 2nd Street, Mineola, N.Y. 11501

Preface to the First Edition

This book introduces the mathematical theory of probability and stochastic processes at the level of a junior-senior undergraduate course for students in mathematics, computer science, and engineering. This is an immense and rapidly-growing subject with applications to physical and biological sciences, engineering, and computer science. We approach it using both *computer simulations* and *mathematical models* of random events that occur in real-world examples. To make the models accessible to students with minimal technical backgrounds, we have chosen examples that can be described without specific knowledge of the underlying physical laws governing their behavior. This includes systems such as computer facilities, telephone switchboards, and industrial production and distribution facilities.

Most of the systems we study are described in terms of their state at a particular moment. The state of such a system changes randomly as time goes on. When time is measured in discrete steps, Markov chains furnish an appropriate class of models. Systems with a continuous time parameter, such as queueing systems, are modeled using birth–death processes and renewal processes. In both cases we set up stochastic models and obtain information about the long-term behavior of the systems.

Many powerful mathematical tools have been developed for solving problems in probability and stochastic processes. In this book we deliberately keep the mathematical techniques at an elementary level and stress a probabilistic point of view in carrying out the mathematical analysis. The minimum prerequisite is one year of calculus and some elementary linear algebra (matrix multiplication, solution of systems of linear equations). Occasionally we use some basic multivariable calculus (multiple integrals). We make a careful analysis of the mathematical assumptions used in setting up the stochastic models. We also emphasize the use of computer-generated simulations of random events—both for its practical value in situations where analytic solutions are unobtainable and as an aid to understanding the sample paths of a stochastic process. After finishing this book, a student can proceed to further applications of probability in operations research and computer systems analysis, or to a graduate-level course on probability and stochastic processes.

The book is organized to fit either a one- or two-semester course and to allow considerable flexibility in choice of topics. Chapters 1–5 constitute a self-contained introduction to the basic notions of probability theory and random variables, and can serve for a one-quarter or one-semester introductory course. We have emphasized some of the classic probability problems (the problem of the points, the mismatch problem, gambler's ruin), the basic families of discrete and continuous random variables, and the fundamental limit theorems for sums of random variables. We focus on random variables and expectations, rather than using the combinatorial approach to probability, since many students encounter combinatorics in other courses at this level.

A course in stochastic processes for students who have already had one quarter or one semester of probability can begin with a quick review of random variables and conditional expectation from Chapters 3–5, proceed to a detailed study of Markov chains in Chapter 6, develop the Poisson process in Chapter 7, and finish with a selection of topics from Chapters 8 and 9. An alternate course emphasizing queueing theory and operations research might review material from Chapters 3–5, omitting Chapter

6, and cover all of Chapters 7-12. (Although there are many important links between discrete-time and continuous-time Markov chains, we have written Chapters 7–12 to be independent of Chapter 6.)

In the exposition we have tried to be as logically complete as possible. When a rigorous argument is beyond the elementary level of the book, we often give a heuristic argument to illustrate the probabilistic ideas involved, and we stress the "sample function" point of view. Examples are used throughout to motivate the development of general results and to illustrate how these results are applied. Numerical evidence and detailed solutions of special cases are given to make the harder limit theorems plausible; for a course stressing applications these can serve as a substitute for the analytic proofs (for example, in Sections 6.6, 6.7, and 6.9 and in the treatment of the Kolmogorov equations in Section 9.5).

In general, we follow a spiral approach. The classic gambler's ruin is studied first in Chapter 2, as an illustration of repeated trials. It appears again in Chapter 6 as a random walk with absorbing barriers, and it serves as a prototype for the class of absorbing Markov chains. The sum of a random number of random variables is first introduced in Chapter 5; this notion returns in Chapter 8 for the compound Poisson process and in Chapter 12 in the setting of random stopping rules and Wald's identity. The Kolmogorov differential equations for transition probabilities are first derived for the Poisson process in Chapter 7, and then are obtained for a general birth–death process in Chapter 9. Exponential (memoryless) models are emphasized in Chapters 7–11. Then in Chapter 12, we introduce the ideas of renewal theory to treat more general queueing systems, and we compare the results of these investigations with those obtained for exponential queues in the earlier chapters.

We do not treat the basic problem of fitting theoretical models to real-world data. To do so would have required developing a significant amount of statistical theory, which can be found in many sources (see the references). We also accept uncritically the notion of a random-number generator, for the same reason.

The text includes worked examples in each section and exercises at the end of each chapter. There are numerical solutions for the exercises at the back of the book. Suggestions for further reading in stochastic processes, simulation, and various applications are also given there.

Many of the models in the text can be treated by simulation, as indicated in the exercises. Some instructors will prefer to use a special simulation language such as GPSS for this purpose (see the references). Some sample queue simulations in GPSS are described in Chapter 11. Others may want students to write their own programs in a general-purpose language.

This book developed from several courses in probability, statistics, and stochastic models taught by the author at Rutgers University. The enthusiastic reaction of the students in these courses encouraged me to organize the material into book form. I would like to express special thanks to Denise Feder, Margaret A. Readdy, and Elizabeth W. Rutman for many helpful criticisms and careful proofreading of several drafts of the manuscript. Their help was invaluable. I would like to thank my Rutgers collegues Terrence Butler and Bertram Walsh for teaching courses based on versions of this book, and Kenneth Kaplan, David Rorhlich, and Eugene Speer for sharing their views on teaching probability. I appreciate the help of Frank Klatil and his assistants in the Rutgers Graphics Laboratory in preparing the graphs. My thanks also go to my editor Craig Bartholomew, the copy editor Steven Gray, and the anonymous reviewers; their suggestions aided me greatly in revising earlier drafts of the book. The final work on the manuscript was supported by the Rutgers University Faculty Academic Study Program.

ROE GOODMAN
JUNE, 1987

Preface to the Second Edition

This edition follows closely the first edition. The main change is the addition in chapters 1–4 of more than 50 exercises and several more examples of random variables, in order to make the first half of the book more useful as an introductory probability text. Minor revisions to the exposition and corrections to a few of the proofs were made throughout the book. The goal, as stated in the preface to the first edition, remains the same: to introduce students to an interesting collection of stochastic models while requiring only modest mathematical prerequisites.

The continued interest of students and teachers in the book, as evidenced by inquiries from various parts of the world over the last decade, has been the motivation for preparing this edition.

ROE GOODMAN
JUNE, 2005

Contents

Chapter 1

Sample Spaces

In this opening chapter, we look at some examples of real-world experiments that have two common characteristics: there are many possible outcomes, and the actual outcome that occurs in a particular performance (or trial) of the experiment depends on random influences. To set up a stochastic model for these experiments, we introduce the fundamental concepts of **sample spaces**, **events**, and **random variables**. These will provide a mathematical framework for studying the outcomes of an experiment. In Chapter 2 we will set up the remaining apparatus of mathematical probability for measuring the randomness in an experiment.

1.1 Experiments with Random Outcomes

Example 1.1 Tossing a Coin

The fundamental experiment with a random outcome is the **coin toss**. This term applies to any experiment that can have either of two possible outcomes—call them *success* and *failure* (or *heads* and *tails*). Because of chance, the outcome that actually occurs can't be predicted in advance with certainty, although we may have some idea as to the likelihood of success. For example, consider an industrial production system, where the experiment consists of manufacturing an item such as an integrated circuit chip. Here a *success* means that the item meets the quality-control specifications. Another example occurs in public-health medicine, when a vaccine is administered to a group of people. Here a *success* means that the vaccinated individual acquires the desired immunity. In both of these examples, chance effects influence the outcome, and the likelihood of success in a single trial can be estimated on the basis of past experience. ■

Example 1.2 Wheel of Fortune

If an experiment has n possible outcomes, then a trial of the experiment corresponds conceptually to spinning a wheel of fortune that has n different sectors of various widths and an immovable pointer. Each possible outcome of the experiment corresponds to a particular sector. The outcome that occurs in the experiment corresponds to the sector that the pointer occupies when the wheel stops spinning. Highly likely outcomes correspond to wide sectors; unlikely outcomes correspond to narrow sectors. (Of course, this is only a thought-experiment description of the real-world situation, but it is conceptually useful.) This model is adaptable to rolling a die, dealing a hand of bridge, or playing any other such game of chance. For instance, a fair die would correspond to a wheel with six equal sectors. ■

Example 1.3 Replacing a Light Bulb

Consider an electric lighting fixture. We install a 1000-hour light bulb in the fixture and turn the bulb on and off randomly. Eventually the bulb burns out, after t hours of use; the actual value of t depends on random factors such as the manufacturing process of the bulb, power-line fluctuations during periods when the bulb is turned on, and the number of on/off cycles that the bulb is subjected to. If our model ignores any other features of the operating environment, however, the outcome of the experiment is the single real number $t > 0$. Alternately, we could set up a more complicated stochastic model for this experiment. For example, if we also included the number n of on–off cycles that occurred during the lifetime of the bulb, an outcome would consist of the pair (t, n). ■

Example 1.4 Repairing Machines

Assume that a factory contains three identical machines, and that two people are available to service the machines. Each machine operates for a random length of time, breaks down, and then needs servicing. The service takes a variable (random) length of time and may not be done immediately if both service people are busy fixing another machines. In this case, the machine remains idle until a repair person is available. Suppose the experiment consists of observing the repair process in the factory at some particular moment. The outcome of the experiment consists of data concerning the state of the repair process. For example, if we only observe m, the number of machines that are not working at the time of observation (either because they are being repaired or waiting to be repaired), the outcome of the experiment is the single number m. We could also include as part of the outcome other data such as the length of time x_i that machine i has operated since its last repair and the length of time y_i that machine i has been down just prior to the observation. If we included these additional data, an outcome of the experiment would be the seven numbers $(m, x_1, x_2, x_3, y_1, y_2, y_3)$. We can think of these numbers as a single point in a seven-dimensional space. ■

Example 1.5 Repeated Trials

The real-world experiments that can be modeled by using probability theory are repeatable. For example, instead of tossing a coin just once, we can toss it n times. A single outcome of such a sequence of tosses could be described symbolically by a string of n binary digits $10011\ldots10$, where $1 =$ success and $0 =$ failure in each trial. Similarly in Example 1.3, if we collected data for the lifetimes of three bulbs, then an outcome would consist of the point (x_1, x_2, x_3) in three-dimensional space, where x_i is the lifetime of the ith bulb installed. ■

1.2 Sample Spaces, Events, and Random Variables

As a first step in setting up stochastic models for real-world examples of the type described in Section 1.1, we focus on the data-gathering aspect: the experiment is performed, and then the outcome is recorded in some systematic way. For example, if a coin is tossed three times, the outcome is a sequence such as HTH (which could also be coded as the binary number 101, of course). We define the **sample space** S for an experiment to be the set of all possible outcomes. We refer to a particular outcome as a *point* in the sample space. For the example of three tosses of a coin, S consists of the eight points $\{HHH, HHT, \ldots, TTT\}$.

A typical **event** that we observe in an experiment can be described by a verbal statement of a logical true-or-false type, such as

$$\begin{aligned} A &= \text{"The light bulb has burned for more than 400 hours"} \\ &\quad \text{(Example 1.3)} \\ B &= \text{"Machines 1 and 2 are operating and machine 3 is being repaired"} \\ &\quad \text{(Example 1.4)} \\ C &= \text{"At least one result of heads occurred in three coin tosses"} \\ &\quad \text{(Example 1.5)} \end{aligned}$$

Suppose that we perform an experiment and obtain the outcome ω. If the statement defining an event E is true for this particular outcome, we write $\omega \in E$ ("ω is a point of E"), and say that the event E occurred in the experiment. For example, $\omega = HTT$ is a point of C in the preceding example.

Some events logically imply other events. For example, if

$$D = \text{"The light bulb burned for more than 600 hours"}$$

(in Example 1.3), then the truth of D implies the truth of A, and we write $D \subset A$. We can combine statements (events) using the following logical operations: **and** — denoted symbolically by $\cap$; **or** (non-exclusive) — denoted symbolically by $\cup$, and **not** — denoted symbolically by a superscript c (complement). Thus, for the events above,

$$\begin{aligned} A^c &= \text{"The light bulb lasted at most 400 hours"} \\ B &= B_1 \cap B_2 \cap B_3^c, \quad \text{where } B_k = \text{"machine } k \text{ is operating"} \\ C &= C_1 \cup C_2 \cup C_3, \quad \text{where } C_k = \text{"heads showed on toss } k\text{"} \end{aligned}$$

If an event E is described by some verbal statement (or proposition) P as in the examples above, we shall indicate this by using the following set-theoretic notation:

$$E = \{\omega \in S : \omega \text{ satisfies } P\}$$

Given this correspondence, the logical operations $\cup$, $\cap$, and $P \to P^c$ on propositions correspond to set union, set intersection, and set complementation in the sample space. These operations can be visualized by means of Venn diagrams, as shown in Figure 1.1.

If E corresponds to a proposition P, and F corresponds to a proposition Q, then the logical relation

"P and Q are mutually exclusive"

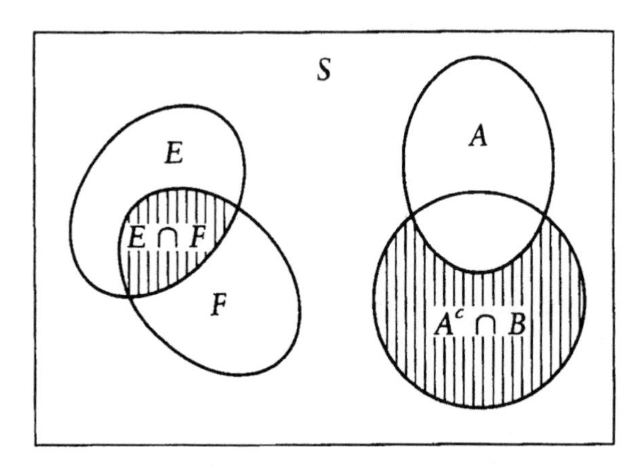

Figure 1.1: Venn Diagrams

corresponds to the set-theoretic relation

$$E \cap F = \emptyset \quad \text{(empty set)}$$

For example, if the experiment consists of rolling a die, then the propositions

$$P = \text{“an even number of points show on the die”}$$

and

$$Q = \text{“an odd number of points show on the die”}$$

are *mutually exclusive*: no matter what the outcome ω of the experiment, ω cannot satisfy both P and Q. The relation

$$\text{“}P \text{ implies } Q\text{” (if } P \text{ is true, then } Q \text{ is true)}$$

corresponds to the set-inclusion relation

$$E \subset F \quad \text{(every outcome satisfying } P \text{ also satisfies } Q\text{).}$$

For example, in the same die-rolling experiment, take

$$P = \text{“the die showed three points”}$$
$$Q = \text{“an odd number of points showed on the die”}$$

By relating the set operations to logical operations on propositions (or by means of informal geometric reasoning using Venn diagrams), we can verify the following laws of *Boolean algebra*:

(commutative) $E \cup F = F \cup E, \quad E \cap F = F \cap E$

(associative) $(E \cup F) \cup G = E \cup (F \cup G), \quad (E \cap F) \cap G = E \cap (F \cap G)$

(distributive) $(E \cup F) \cap G = (E \cap G) \cup (F \cap G), \quad (E \cap F) \cup G = (E \cup G) \cap (F \cup G)$

for any events E, F and G. Notice that each of these laws consists of a pair of equations related by interchanging **and** with **or**. This symmetry is a consequence of **DeMorgan's Laws**:

$$\begin{aligned} (E^c)^c &= E \\ (E \cup F)^c &= E^c \cap F^c \\ (E \cap F)^c &= E^c \cup F^c \end{aligned}$$

In most of the models we shall study, the observed events will be described in terms of *numerical measurements*, such as

$E =$ "The second light bulb installed has burned for more than 300 hours"

It is natural to consider these measurements themselves as the principal object of study. For example, in connection with event E, we can let X_n denote the *total* lifetime of the nth bulb installed. Then E corresponds to the inequality

$$X_2 > 300 \text{ hours}$$

and we shall write $E = \{X_2 > 300\}$. With this notation, we then have

$$E^c = \{X_2 \leq 300\}$$

Other inequalities involving the quantities X_n similarly correspond to observed events in the repeated experiment of replacing a light bulb.

Even in experiments for which the data are not necessarily numerical, it is often convenient to introduce numerical measurements. For example, in connection with repeated coin tosses, we can set $X_k = 1$ if a result of heads appears on the kth toss, and otherwise set $X_k = 0$. Then event "At least one result of heads occurred in the first three tosses" happens if and only if the inequality

$$X_1 + X_2 + X_3 \geq 1$$

is true.

In terms of the sample-space formulation of a probability model, numerically measured quantities such as X_n can be viewed as ordinary *functions*: for every outcome ω of the experiment (= one point of the sample space), there is a specific numerical value—say $X_n(\omega)$. The notation previously introduced is thus an abbreviated version of the set-theoretic notation:

$$\{X_2 > 300\} = \{\omega \in S : X_2(\omega) > 300\}$$

Numerical functions defined on the sample space of an experiment are traditionally called *random variables*.

Exercises

1. Suppose that an experiment consists of rolling a six-sided die twice and observing the number on its top face each time.

a. Describe the sample space for this experiment. How many possible outcomes are there?

b. Consider the events

$$E = \text{An even number showed on at least one roll}$$

$$F = \text{An odd number showed on at least one roll}$$

List the outcomes in $E \cap F$, $E \cup F$, E^c, $E^c \cap F$, and $E^c \cup F$, and give a verbal description of each of these events.

2. Let E, F, and G be three events in a sample space. Find Boolean expressions in terms of these events for the following:

   a. At least one of the events occurs.

   b. None of the events occurs.

   c. Exactly one of the events occurs.

   d. At most one of the events occurs.

3. Prove DeMorgan's Law: $(A \cup B)^c = A^c \cap B^c$.

4. If E is an event, then the *indicator* function for E is the random variable I_E defined by the rule:

$$I_E = \begin{cases} 1, & \text{when } E \text{ occurs} \\ 0, & \text{when } E \text{ does not occur} \end{cases}$$

   a. For any two events E and F, prove that the indicator function of $E \cap F$ is the product $I_E \cdot I_F$.

   b. For any two *mutually exclusive* events E and F, prove that the indicator function of $E \cup F$ is the sum $I_E + I_F$.

   c. Prove that the indicator function of E^c is $1 - I_E$.

   d. Find a formula for the indicator function of $E \cup F$ when E and F are not mutually exclusive. (HINT: Consider the set $(E \cup F)^c$.)

5. A public opinion poll is gathering information about the following characteristics of the people in a certain population:

   M: subscribes to a newsmagazine

   N: gets a daily newspaper

   T: watches news on television

   A person is picked at random from the population. Give Boolean expressions in terms of M, N, T for the following events:

   a. the person either gets a daily newspaper or subscribes to a newsmagazine (or both), but does not watch television news.

   b. the person watches television news and either gets a daily newspaper or else subscribes to a newsmagazine (but not both).

   c. the person does not satisfy descriptions (a) or (b).

6. Urn A contains three balls numbered 1–3. Urn B contains four balls numbered 4–7. The even-numbered balls are red and the odd-numbered balls are white. The experiment consists of picking one ball at random from each urn.

   a. Describe the sample space for this experiment. How many possible outcomes are there?

   b. Let $E = \{$ exactly one ball picked is red$\}$. List the outcomes in E.

   c. Let $F = \{$ at least one ball picked is red$\}$. List the outcomes in F.

   d. Let $G = \{$ both balls picked are the same color$\}$. List the outcomes in G.

Chapter 2

Probabilities

In this chapter, we develop the mathematical theory of probability. First we explore the intuitive notions of repeated experiment and relative frequency in the context of computer-generated random simulations. This provides motivation for adopting the axioms of probability theory. After introducing these axioms, we develop some of their consequences and obtain exact solutions for some of the examples already studied through the use of simulation methods. We then examine the key notions of conditional probability and stochastic independence, and we describe the stochastic model for repeated independent trials of an experiment.

2.1 Relative Frequency of Events

In everyday conversation, a statement such as "The probability that Team X will beat Team Y today is 1/3," usually is based on some relative frequency information. For example, the person making the statement might know that Team X won four out of its last twelve games against Team Y. Similarly, an insurance company estimates the probability that a person aged 25 will survive to age 26 by using past information about the proportion of all people (in a large population) of this age who lived at least one more year. This requires gathering statistical data from real life and compiling so-called mortality tables. In the historical development of the theory of probability, another important source of empirical data was various gambling games. Much of the early theory of probability was devoted to obtaining a theoretical explanation for the observed frequencies of the outcomes in these games.

The widespread availability of computers gives us a convenient and flexible tool for generating random data. Instead of actually shuffling and dealing cards or tossing a coin, we can let the computer carry out the experiments by using a random number generator. Simulating real-world experiments via a computer is a useful tool that finds many applications far beyond simple gambling games.

Example 1.1 Rolling a Die

One way to simulate the rolling of a fair six-sided die is to use a table of random numbers. Table 4 at the back of this book is an example of such a table; in it each entry is a five-digit random number. To set up a correspondence between these numbers and the outcomes of tossing a die, we divide the numbers from 0 to 99999 into six essentially equal ranges. The correspondence is then as follows:

RANGE	FACE OF DIE SHOWING
00001–16666	1
16667–33333	2
33334–50000	3
50001–66666	4
66667–83333	5
83334–99999	6

From the first five entries in column 1 of Table 4, for example, we get five simulated tosses with outcomes as follows: $10480 \to 1$, $22368 \to 2$, $24130 \to 2$, $42167 \to 3$, and $37570 \to 3$. Continuing in this way, we can use each column of the table to simulate the experiment of tossing a die fifty times. For column 1, we obtain the simulation data in Table 2.1. Can we conclude from these data that the simulated die is really fair? Face 1 seems to be highly favored.

Table 2.1 Simulation of 50 rolls from Column 1 of Random Number Table

Face showing	1	2	3	4	5	6
Occurrences	16	8	5	6	5	10

Let's continue with the experiment by using column 2 of Table 4 to generate another fifty simulated tosses. This time we obtain the simulation data in Table 2.2. In this simulated sequence of rolls, face 1 is no longer favored, but face 4 seems to have a slight edge.

Table 2.2 Simulation of 50 rolls from Column 2 of Random Number Table

Face showing	1	2	3	4	5	6
Occurrences	9	9	6	12	6	8

We can merge the simulations from columns 1 and 2 to obtain Table 2.3. In this table we have used the **relative frequencies** (number of occurrences/number of samples). If we denote by $f_1, f_2, \ldots$ the frequency that face 1, face 2, ... occurred, then the numbers f_i satisfy the conditions

$$0 \leq f_i \leq 1, \quad f_1 + f_2 + \cdots + f_6 = 1$$

Table 2.3 Simulation of 100 rolls of a Fair Die

Face showing	1	2	3	4	5	6
Frequency	0.25	0.17	0.11	0.18	0.11	0.18

For a fair die, we would expect that in the long run, the relative frequencies f_i would approach 1/6 since each of the six faces should be equally likely to show. The data for 100 rolls given in Table 2.3 seem to be tending in this direction, but they aren't altogether conclusive.

We shall call each real or simulated roll of the die a **random trial**. To simulate long sequences of random trials, we turn to the computer. The random number key on a hand calculator (or the RND function in a computer language such as BASIC) returns a quasi-random number X between 0 and 1. Repeated pressing of the key (or calls to the RND function) generates a random sequence of numbers. We set

$$Y = \text{INT}(6X + 1) \tag{2.1}$$

where INT denotes the integer part of a real number. Since $1 < 6X + 1 < 7$, we see that the possible values of Y are the integers 1 to 6. Also, $Y = 1$ if and only if $1 < 6X + 1 < 2$. This, in turn, is equivalent to $0 < X < 1/6$. Thus, if we generate a long sequence of values of Y by formula (2.1) using repeated calls to the random number generator, then we expect that about 1/6 of the Y values will be 1. Likewise, we would expect about 1/6 of the values of Y will be 2, and so forth. Table 2.4 gives a sample output from such a simulation program.

Table 2.4 Sample Frequencies from Simulation Program

Face showing	1	2	3	4	5	6
500 rolls	0.188	0.178	0.158	0.148	0.172	0.156
1000 rolls	0.165	0.156	0.179	0.178	0.161	0.161
10,000 rolls	0.166	0.159	0.170	0.167	0.166	0.172

The data in this table suggest that as the number of rolls increases, the frequencies will tend to cluster around the value $1/6$. ■

Example 2.2 Waiting Game

In Example 2.1, we could already predict the outcome of the computer experiment. The only unknown element was how many simulated rolls of the die needed in order for the proportion of 1's to be close to $\frac{1}{6}$. (We will study this more closely in Section 4.6, in connection with the Law of Large Numbers, but for now we'll consider things from an experimental point of view.) It's easy to ask questions about the frequency of other outcomes in this experiment, whose answers are not immediately obvious. For example, suppose that we play the Waiting Game: we roll the die until 1 appears for the first time. What are the relative frequencies of the number of rolls needed to get a 1? A simple extension of the program used to generate the simulation data in Example 2.1 furnished the graphs in Figure 2.1 when this game was simulated 10 times and 1000 times. This figure shows a plot of the *cumulative frequency distribution* $F(n)$ —that is, the proportion of games that required $\leq n$ rolls to get a 1. Notice how the distribution for 1000 games seems to follow a smooth curve. Later, in Example 3.4, we will obtain the theoretical frequency distribution of this game, which confirms the simulation results. ■

Example 2.3 Meeting for Lunch

Two friends who have unpredictable lunch hours try to meet at a particular restaurant whenever possible. They each arrive at a random time between noon and 1:00 P.M.,

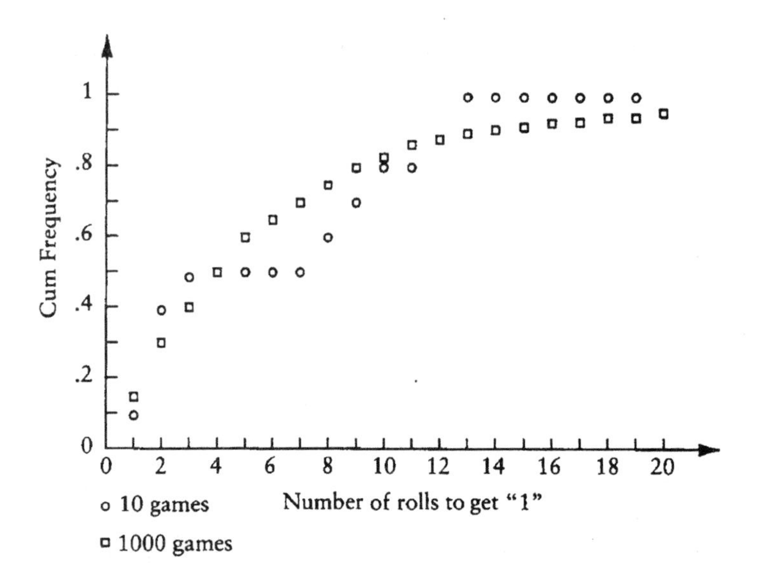

Figure 2.1: Simulation of Waiting Game

and each waits for the other for at most 15 minutes. The restaurant closes its doors at 1:00 P.M. What is the proportion of days on which the friends actually eat together at this restaurant?

Solution by Simulation Let X and Y be the returns from two successive calls to a random number generator. By the description of the randomness in the problem, it is reasonable to take these values as representing the arrival times of the two people (measured in decimal parts of an hour beginning at $0 = 12$ noon). The friends will have lunch together at the restaurant when

$$|X - Y| < 0.25 \tag{2.2}$$

Table 2.5 shows some computer-generated simulation data for this problem.

Table 2.5 Frequency of Meeting for Lunch

Days simulated	50	100	200	400	1000	2000	4000
Frequencies	0.46	0.47	0.395	0.445	0.427	0.436	0.442

The simulation frequencies seem to be tending toward a value near 0.44. This is somewhat puzzling, since the obvious guess for the limiting frequency is 0.5 ("The two people wait 30 minutes between them. This is half of the one-hour period, so the probability ought to be 0.5.") We shall obtain the exact analytical solution in Section 2.5; it confirms the simulation data. ■

Example 2.4 The Birthday Problem

Suppose that there are K people in a room. What are the chances that at least two of them have the same birthday?

Solution by Simulation This is a classic problem in elementary probability, and it can easily be studied experimentally via the computer. Just as in the case of rolling a die, we generate a random sequence $B(I)$, with $I = 1, \ldots, K$, of possible birthdays for the K people in the room by setting

$$B(I) = \text{INT}(365X + 1) \tag{2.3}$$

where X is generated by successive calls to the random number generator (we ignore February 29). We can then test whether

$$B(I) = B(J) \tag{2.4}$$

for $I = 2$ to K and for $J = 1$ to $I - 1$. If equation (2.4) is true for some pair (I, J), then in our simulated room with K people, there is a match—a pair with the same birthday. We now generate a large number N of simulated rooms with K people by repeating this procedure, and we calculate the proportion of rooms that contain a match. In this case, there are two parameters, K and N; for fixed K, the relative frequency of a match stabilizes when N gets large, as shown in Table 2.6.

Table 2.6 Simulation of the Birthday Problem ($K = 25$ people)

Number of rooms simulated	10	50	100	500	1000
Frequency of match	0.50	0.58	0.53	0.585	0.569

The simulation data suggest that in a room with 25 people the chance of a birthday match is greater than $\frac{1}{2}$ (we shall verify this surprising prediction using probability theory later in the chapter). ■

Example 2.5 Mixed-up Mail

Suppose that N form letters and N envelopes are printed from a computerized mailing list. A careless clerk then inserts the letters into the envelopes in a completely random way, ignoring the addresses. What is the probability that not one of the addresses on the envelopes matches the inside address on the letter it contains?

Table 2.7 Simulation of the Mismatch Problem (10 letters and envelopes)

Number of trials simulated	10	50	100	500	1000
Frequency of no match	0.60	0.32	0.35	0.366	0.379

Solution by Simulation If the addresses are numbered $1, 2, \ldots, N$, then the pairing up of letters and envelopes is determined by a permutation of $1, 2, \ldots, N$. We can generate a random permutation of the integers $1, 2, \ldots, N$ as follows: First we generate N successive random numbers $x_1, x_2, \ldots, x_N$ in the interval $(0, 1)$ using the RND function. Next, we use a sorting routine to rearrange these numbers in increasing order:

$$x_{j_1} < x_{j_2} < \ldots < x_{j_N}$$

The random permutation is thus the sequence $j_1, j_2, \ldots, j_N$. Now we check to see whether there are any fixed points in this permutation—that is, whether there are any indices k such that $j_k = k$. If there is at least one fixed point, then the permutation corresponds to a match between an outside address and an inside address. Finally, we repeat this procedure a large number of times and record the proportion of trials in which there was no match. Table 2.7 gives a sample output from this program.

Running the simulation for different values of N (the number of letters) we find that the frequency of the event "no match" seems to tend toward a limiting value (see Table 2.8).

Table 2.8 Simulation of the Mismatch Problem (15 letters and envelopes)

Number of trials simulated	10	50	100	500	1000
Frequency of no match	0.40	0.36	0.35	0.35	0.355

The data in Table 2.8 are surprising; it would seem reasonable that, with a large number of letters, a *complete* mismatch would be very unlikely. The simulation indicates that, to the contrary, the chance of a complete mismatch is about 0.36. We shall confirm these simulation results by theoretical analysis in Section 2.2; the limiting probability (as $N \to \infty$) is $e^{-1} \approx 0.368$. ■

Empirical data concerning random events, whether gathered from real-world experience or created on a computer by a random number generator, are usually complicated and often of limited accuracy. A mathematical model of the situation can reveal hidden patterns and regularity, and can provide estimates for the size of the random fluctuations in the data. In constructing a stochastic model, however, we need to use a purely mathematical notion of probability. By treating probability as a branch of theoretical mathematics, we gain the advantage of having an unambiguous basis for calculations—one that, in principle, can be as accurate as needed.

Here the computer can play another important role in the modeling process, carrying out the analytical calculations (sums, integrals, matrix multiplications, solutions of differential equations) that occur in the model. Nonetheless, real-world data still play the vital roles of suggesting reasonable assumptions to build into models and of validating predictions of models. Even when a mathematical model is too complicated to be solved exactly, it can be combined with computer simulations to obtain some information about the behavior of the real-world system.

2.2 Axioms of Probability

The calculation of relative frequencies (proportions) in Section 2.1 serves as motivation for the axioms of probability. If E is an observable event in an experiment, then we define the relative frequency as

$$f(E) = \text{proportion of times that } E \text{ occurs in } N \text{ repetitions of the experiment}$$

Here the experiment may be carried out in real life or on a computer. Clearly, we have

$$0 \leq f(E) \leq 1$$

If S denotes the event "The experiment had some outcome," then S always occurs, so that

$$f(S) = 1$$

Furthermore, if E and F are mutually exclusive events, then

$$f(E \cup F) = f(E) + f(F)$$

For example, if 40 people in a sample of 100 people had brown eyes and 30 other people in the sample had blue eyes, then the relative frequency of the event "brown or blue eyes" was $0.4 + 0.3$ (assuming that no one in the sample had one brown eye and one blue eye).

What is the relation between these relative frequencies $f(E)$ and a stochastic model for the experiment? Observe that the numbers $f(E)$ refer to a *particular* sequence of N repetitions of the experiment and can only be calculated after the experiments are performed. Another sequence of repetitions (either in the real world or on a computer) or a different value of N would generally yield a different value of $f(E)$. A stochastic model of the experiment, on the other hand, should offer a unique number (probability) that can be calculated for each event in the experiment by a mathematical algorithm. These can then be interpreted as theoretical limiting frequencies. (We shall amplify on this point later in connection with the Law of Large Numbers.)

Now let's consider in more detail the conditions that any algorithm for calculating exact probabilities should satisfy. Suppose that S is a **sample space** —a set consisting of all the possible outcomes of a particular experiment. We shall say that we have a *probability model* (or *stochastic model*) for the experiment if, for every observable event $E \subset S$, we have a rule for calculating a number $P(E)$ (the probability that E will occur) that satisfies the following conditions:

(i) Normalization Axiom: $0 \leq P(E) \leq 1$ for every event E, and $P(S) = 1$.

(ii) Additivity Axiom: If E and F are mutually exclusive events, then

$$P(E \cup F) = P(E) + P(F)$$

Both of these axioms are modeled on the corresponding properties of the relative frequency function $f(E)$. Some immediate consequences are

$$P(E^c) = 1 - P(E) \tag{2.5}$$

$$P(E_1 \cup \cdots \cup E_n) = \sum_{i=1}^{n} P(E_i) \tag{2.6}$$

if the events $E_1, \ldots, E_n$ are mutually exclusive; furthermore,

$$P(E) \leq P(F) \quad \text{if } E \subset F \tag{2.7}$$

For example, we can prove property (2.7) by first writing $F = E \cup G$, where $G = E^c \cap F$. Since E and G are disjoint, we can then calculate $P(F)$ by using the Additivity Axiom:

$$P(F) = P(E) + P(G)$$

But $P(G) \geq 0$ by the Normalization Axiom, so we can drop it from this last equation to obtain property (2.7).

A third axiom is indispensable in a theoretical treatment of unlimited repetitions of an experiment:

(iii) Continuity Axiom: If $A_1 \subset A_2 \subset A_3 \subset \cdots$ is an increasing family of events, and $A = \bigcup_i A_i$, then

$$P(A) = \lim_{i \to \infty} P(A_i)$$

Notice that, by property (2.7), the sequence of numbers $p_i = P(A_i)$ is monotone increasing and bounded above by 1; hence it has a limit (the smallest real number r that satisfies $r \geq p_i$ for all i). We can also think of the set A in the Continuity Axiom as being the limit of the sets A_i (it is the smallest set that contains all of the sets A_i). This axiom thus asserts that Prob (Limit) = Limit (Prob) in this situation.

An important consequence of the Continuity Axiom is that the additivity property (2.6) is also valid for an infinite collection of mutually exclusive events $E_1, E_2, \ldots$ in the following form:

$$P\left(\bigcup_{i=1}^{\infty} E_i\right) = \sum_{i=1}^{\infty} P(E_i). \tag{2.8}$$

To prove property (2.8), we form the increasing family of events $A_1 = E_1$, $A_2 = E_1 \cup E_2$, $A_3 = E_1 \cup E_2 \cup E_3$, and so on, whose limit (union) is the same as the union of all of the events $\{E_i\}$. Note that, by property (2.6), we have

$$P(A_k) = \sum_{i=1}^{k} P(E_i)$$

for each finite integer k. Now we let $k \to \infty$ and use the Continuity Axiom to obtain property (2.8).

Example 2.6 Finite Sample Space

Suppose that the experiment can be completely described by a finite set S of possible outcomes, which we enumerate as $\omega_1, \omega_2, \ldots, \omega_n$. If we have a probability model for the experiment, the numbers

$$p(\omega) = P\{\omega\} \tag{2.9}$$

satisfy the properties

$$0 \leq p(\omega) \leq 1 \quad \text{and} \quad \sum_{\omega \in S} p(\omega) = 1 \tag{2.10}$$

by the Normalization Axiom and formula (2.6). Conversely, if we take any function p defined on S and satisfying properties (2.10), we may use formula (2.9) to define a probability model for the experiment in an obvious way. For any event E, we define $P(E)$ by adding up the probabilities of all of the individual outcomes in E:

$$P(E) = \sum_{\omega \in E} p(\omega)$$

With this definition, the Normalization and Additivity axioms obviously hold, and the Continuity Axiom is automatically satisfied. (Since S is finite, the sequence of sets A_i doesn't get any bigger after a certain index i, so the numbers $P(A_i)$ are eventually constant for large i.)

For the experiment of rolling a single die, where the possible outcomes are the numbers 1, ..., 6 showing on the upturned face, assigning probability $\frac{1}{6}$ to each outcome gives a model for a balanced die. On the other hand, we could take the observed frequencies f_i from a real (or computer-simulated) sequence of rolls, and *define* a probability model by setting $p(i) = f_i$. For example, take the relative frequency data in Table 2.3 regarding the simulation of 100 rolls of a fair die in Example 2.1. This model would give an exact fit to the given experimental data, but might or might not accurately predict the outcomes of another sequence of rolls. (Compare the simulation of 100 rolls in Table 2.3 with the computer simulation of 10,000 rolls in Table 2.4.) ■

Equally Likely Outcomes

Many classic problems of elementary probability involve such experiments as picking a card from a deck or drawing three balls from an urn "at random". In these situations we have a finite sample space S. The phrase "at random" is interpreted to mean that, for any outcome ω,

$$p(\omega) = \frac{1}{|S|} \qquad \text{(equally likely outcomes)}$$

Here we use the notation $|E|$ to denote the number of elements in a set E. In these models, $P(E) = |E|/|S|$ for any event E, so calculating probabilities reduces to the problem of counting the number of outcomes in which a particular event occurs.

In counting the outcomes in finite sample spaces, we shall often need the **binomial coefficient:**

$$\binom{n}{k} = \frac{n!}{k!(n-k)!} = \frac{n(n-1)\cdots(n-k+1)}{k(k-1)\cdots 2\cdot 1}$$

This is the number of possible ways to pick a set of k elements out of a set of n elements, where n and k are nonnegative integers. We will use the convention that

$$\binom{n}{k} = 0 \qquad \text{if } k > n \text{ or if } k < 0$$

This convention is consistent with the combinatorial interpretation of the binomial coefficients.

Example 2.7 Birthday Problem Probabilities

In the birthday problem (Example 2.4), a natural choice of S is the set of all ordered lists of the birthdays of k people. Thus, $|S| = (365)^k$ (if we ignore leap years). Let E be the event that at least two people in the room have the same birthday. Of course, if $k > 365$, then $E = S$ and $P(E) = 1$. If $k \leq 365$, then the complementary event

$$E^c = \text{"Every birthday is distinct"}$$

is true for $365 \cdot 364 \cdots (365 - k + 1)$ different lists—since there are 365 choices for the birthday of the first person, 364 choices among the remaining days for the birthday of the second person, and so on. Thus,

$$P(E^c) = \frac{|E^c|}{|S|} = \frac{365 \cdot 364 \cdots (365 - k + 1)}{365 \cdot 365 \cdots 365}$$

(k factors in numerator and denominator), and $P(E) = 1 - P(E^c)$ by formula (2.5). Numerical calculations with this formula confirm the simulation results: this probability is surprisingly large even for k of moderate size. For example, when $k = 15$, then $P(E) = 0.25$. When $k = 30$, then $P(E) = 0.71$. ■

Example 2.8 Poker Probabilities

In five-card poker, there are

$$\binom{52}{5} = 2,598,960$$

different subsets of five cards that a player can draw from a deck of fifty-two cards. A *straight* consists of five cards with denominations in order (where the ace can be either low or high), but the cards not all of the same suit. Let E be the event "A player draws a straight." To calculate $P(E)$, we must count the number of possible straight hands. One way to do this is to enumerate the straights by listing the low card first. Thus if

$$E_i = \{\text{All straights with low card } i\}$$

then the possible values of i are 1 (= ace), 2, ..., 10, because of the convention about aces. Clearly, the sets E_i are mutually disjoint, and

$$E = \bigcup_{i=1}^{10} E_i$$

Now fix i. To specify an element of E_i, we only have to give the suits of the five cards, since the index i determines the denominations. There are 4 choices for each card, giving a total of 4^5 choices. However, we must exclude the four cases where the five cards are all in the same suit. Hence E_i consists of $4^5 - 4$ elements. It follows from the Additivity Axiom that

$$P(E) = \sum_{i=1}^{10} P(E_i) = 10(4^5 - 4) \Big/ \binom{52}{5}$$

Numerical calculations yield the value 0.0039 for $P(E)$. ■

Inclusion-Exclusion Formula

In Example 2.8, we saw how the Additivity Axiom can simplify the calculation of probabilities. However, this axiom only applies to *mutually exclusive* events. When events E and F are not mutually exclusive, we can calculate $P(E \cup F)$ by including both E and F, and then excluding the overlap $E \cap F$ (see Figure 2.2):

$$P(E \cup F) = P(E) + P(F) - P(E \cap F) \tag{2.11}$$

To prove formula (2.11), split up E, F, and $E \cup F$ using the three disjoint sets $E \cap F^c$, $F \cap E^c$, and $E \cap F$. Then $P(E \cup F)$ is the sum of the probabilities of these three sets. On the other hand,

$$P(E) = P(E \cap F) + P(E \cap F^c)$$

and

$$P(F) = P(F \cap E) + P(F \cap E^c)$$

Adding these two equations, we get

$$P(E) + P(F) = P(E \cap F^c) + P(F \cap E^c) + 2P(E \cap F)$$

The right side of this last equation equals $P(E \cup F) + P(E \cap F)$. Rearranging terms, we obtain formula (2.11).

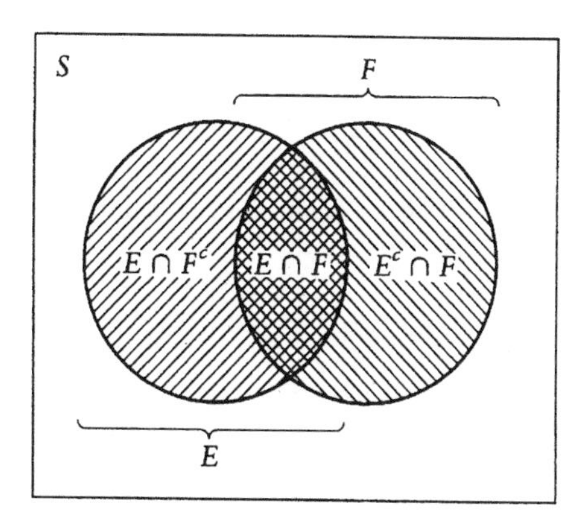

Figure 2.2: Inclusion-Exclusion

The general version of the inclusion-exclusion formula for n sets is

$$\begin{aligned} P\Big(\bigcup_{i=1}^{N} E_i\Big) &= \sum_{i=1}^{N} P(E_i) - \sum_{i_1<i_2} P(E_{i_1} \cap E_{i_2}) + \sum_{i_1<i_2<i_3} P(E_{i_1} \cap E_{i_2} \cap E_{i_3}) \\ &\quad - \cdots + (-1)^{N+1} P(E_1 \cap E_2 \cap \cdots \cap E_N) \end{aligned} \tag{2.12}$$

This formidable-looking formula is relatively easy to describe verbally. We compute the probabilities of all possible combinations of the events E_i, taken k at a time. We multiply each probability involving k events by $(-1)^{k+1}$. Finally, we add up the terms for $k = 1$ through N. The result is formula (2.12).

Proof of General Inclusion-Exclusion Formula We know the formula is true when $N = 2$. We make the induction hypothesis that it is true when $N = n$. If we have $n + 1$ sets $E_1, \ldots, E_{n+1}$, then we can write their union as

$$\bigcup_{i=1}^{n+1} E_i = E \cup E_{n+1}$$

where $E = E_1 \cup \cdots \cup E_n$. Now set $F = E_{n+1}$, and apply formula (2.11):

$$\begin{aligned} P\Big(\bigcup_{i=1}^{n+1} E_i\Big) &= P(E) + P(E_{n+1}) - P(E \cap E_{n+1}) \\ &= P\Big(\bigcup_{i=1}^{n} E_i\Big) + P(E_{n+1}) - P\Big(\bigcup_{i=1}^{n} (E_i \cap E_{n+1})\Big) \end{aligned}$$

Finally, we may apply the induction hypothesis to calculate the probabilities in this last formula, since each union has n sets. Applying formula (2.12) to

$$P\Big(\bigcup_{i=1}^{n} E_i\Big)$$

we get all of the terms *excluding* E_{n+1}. Using the inclusion–exclusion formula on the n sets $E_1 \cap E_{n+1}, E_2 \cap E_{n+1}, \ldots, E_n \cap E_{n+1}$ to calculate

$$-P\Big(\bigcup_{i=1}^{n}(E_i \cap E_{n+1})\Big)$$

we get a sum of terms *including* E_{n+1} and k other events E_i. Because of the minus sign, each such term is multiplied by $(-1)^{k+1}$. Finally, adding the term $P(E_{n+1})$ to those already calculated, we obtain the inclusion–exclusion formula for the case of $n + 1$ sets. This completes the inductive proof.

Example 2.9 Mixed-up Mail

Consider Example 2.5 again. We take the sample space to be the set of all $N!$ permutations of $1, 2, \ldots, N$, as suggested by the simulation solution. Let A be the event "No letter is in a correct envelope." The complementary event A^c is "At least one letter is in its correct envelope." Thus, if we let E_i be the event "The letter for person i is in the correct envelope," then

$$A^c = \bigcup_{i=1}^{N} E_i \tag{2.13}$$

The events E_i on the right side of formula (2.13) are not mutually exclusive, so the sum of their probabilities does not equal $P(A^c)$. With the inclusion-exclusion formula (2.12) available, however, the solution to the mismatch problem is straightforward. For any given set of k integers—say $1, 2, \ldots, k$—there are $(N - k)!$ permutations of $1, 2, \ldots, N$ that leave these particular integers fixed. Hence

$$P(E_{i_1} \cap \ldots \cap E_{i_k}) = \frac{(N-k)!}{N!}$$

There are $\binom{N}{k}$ different ways to choose a set of k integers out of N, so the sum of the terms in formula (2.12) involving k sets is

$$(-1)^{k+1}\binom{N}{k}\frac{(N-k)!}{N!}$$

This term simplifies to $(-1)^{k+1}/k!$. Adding up all of these terms, we find that

$$P(A^c) = 1 - \frac{1}{2!} + \frac{1}{3!} - \cdots + \frac{(-1)^{N+1}}{N!}$$

Hence, the probability of a complete mismatch is

$$1 - P(A^c) = 1 - 1 + \frac{1}{2!} - \frac{1}{3!} + \cdots + \frac{(-1)^N}{N!} \tag{2.14}$$

This formula confirms something that is already evident from the simulation solution of the problem: as soon as N reaches moderate size, this probability is essentially independent of N. This is because the right side of formula (2.14) is simply the Taylor polynomial of degree N for e^x evaluated at $x = -1$, and the successive terms in the polynomial go to zero very rapidly. Hence the probability is approximately $e^{-1} = 0.36788\ldots$. ■

Geometric Probabilities

Let R be some region in the plane $\mathbb{R}^2$ that has a finite nonzero area. We can set up a probability model for picking a point at random in R as follows. We take the sample space $S = R$, since the outcome of the experiment is a point in R. If E is a subset of R, then we define

$$P(E) = \frac{\text{Area }(E)}{\text{Area }(R)} \tag{2.15}$$

Because of the denominator in formula (2.15) and the geometric properties of area, it is clear that this definition of probabilities satisfies the Normalization and Additivity axioms. (An analytical verification of the Continuity axiom in this case requires a more sophisticated definition of area and is beyond the scope of this book.) Equation (2.15) can obviously be extended to regions R in the n-dimensional space $\mathbb{R}^n$, with area replaced by n-dimensional volume.

Example 2.10 Random Points in a Circle

Let R be the square $\{(x, y) : -1 \le x \le 1, \quad -1 \le y \le 1\}$, and let E be the unit circle $\{(x, y) : x^2 + y^2 \le 1\}$. Then by formula (2.15), the probability that a random chosen point in R lies in E is $\pi/4$. This gives us a way of estimating the number $\pi = 3.14159265\ldots$ by simulation. We generate a large number of points (X, Y) at random in R on a computer, by setting

$$X = 2 * \text{RND} - 1 \quad \text{and } Y = 2 * \text{RND} - 1$$

(where each appearance of the variable RND generates a new random number). The Law of Large Numbers—which we will prove in Section 4.6—predicts that the proportion of these points satisfying the inequality $X^2 + Y^2 \le 1$ will be approximately $\pi/4$. Multiplying this proportion by 4, we obtain an estimate for π. Table 2.9 gives some estimates obtained by this simulation method.

Table 2.9 Calculation of π by Generating Random Points in a Square

Number of points	50	100	400	1600	3200	6400
Estimate for π	3.04	2.88	3.21	3.162	3.109	3.139

The data in Table 2.9 illustrate one drawback of simulation methods: They give results of rather low accuracy. (The simulated frequencies of an event converge quite slowly toward the theoretical probability of the event, as the number of trials increases.) ■

In Example 2.10, individual points of the sample space have probability *zero* of being picked. (In fact, the same is true for any set of area 0, such as a smooth curve.)

While this may seem paradoxical at first, it is completely consistent with the limitations of physical measurement or computer computation: we can only measure the location of a point to a finite degree of accuracy (that is, to finite number of decimal places). The probability that a point lies inside a square of side $1/n$ is positive, and we can pin down the location of the point as precisely as we wish by making n large. Of course, the corresponding probability becomes very small.

Example 2.11 Random Chords

Suppose that we draw a chord at random in a circle of radius 1. What is the probability that the chord is longer than the side of an equilateral triangle inscribed in the circle?

This problem seems to be well-posed, but in fact the phrase "at random" is ambiguous, and there are several different reasonable answers, depending on the particular stochastic model chosen for the experiment. Let r be the distance from the midpoint of the random chord to the center of the circle. By elementary geometry (see Figure 2.3), the event

$$E = \text{"The random chord is longer than the side of an inscribed equilateral triangle"}$$

occurs if and only if $r < \frac{1}{2}$. We shall calculate $P(E)$ in three different models.

Model 1: Random Radius Suppose we generate a random chord by the following mechanism: we pick a random radius in the circle, pick a point at random on this radius, and draw the chord through this point perpendicular to the radius (see Figure 2.3). In this model, the distance r is picked at random on the interval $0 < r < 1$, using geometric probabilities on this interval. Thus, $P(E) = P\{r < \frac{1}{2}\} = \frac{1}{2}$.

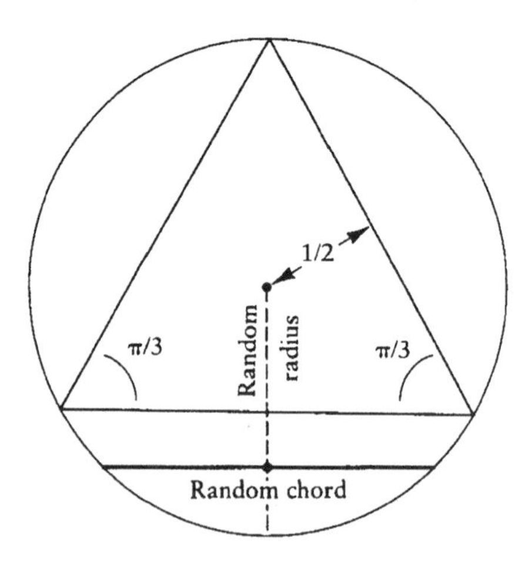

Figure 2.3: Random Chords: Model 1

Model 2: Random Point on Circumference Next, suppose that we generate a random chord by the following mechanism: we label the points on the circumference of the circle by reference to the polar coordinate θ, we fix one point—say $\theta = 0$—on the circumference of the circle, we pick another point with polar coordinate θ at random on the circumference, and we draw the chord connecting these two points. In this model, the angle θ is picked at random, uniformly between 0 and 2π. From Figure 2.4, we see that if

$$2\pi/3 < \theta < 4\pi/3 \tag{2.16}$$

then the random chord will be longer than the side of the equilateral triangle. Since this range of values of θ covers one-third of the total range $0 < \pi < 2\pi$, it follows that $P(E) = \frac{1}{3}$ in this model.

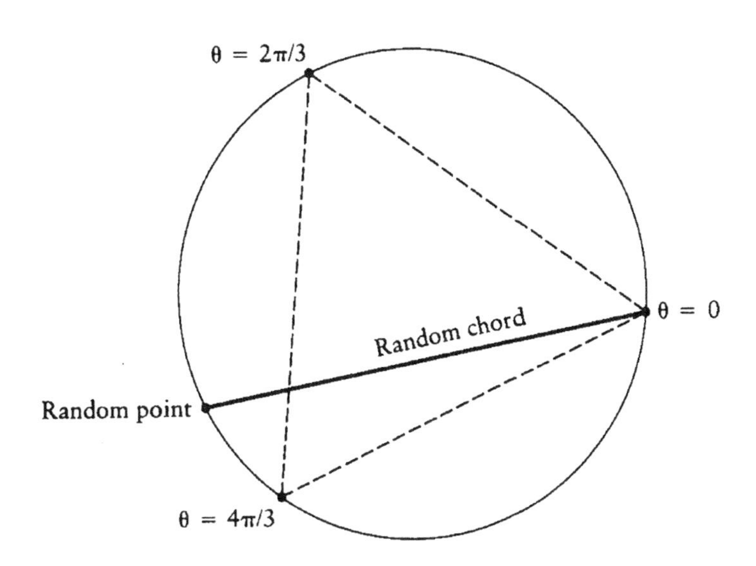

Figure 2.4: Random Chords: Model 2

Model 3: Random Point in Circle Finally, suppose that we generate a random chord by the following mechanism: we pick a random point $Q = (x, y)$ in the unit circle (using geometric probabilities as in Example 2.10), we draw the radius through Q, and we draw the chord through Q perpendicular to the radius. In this model, the distance $r < \frac{1}{2}$ if and only if the random point is inside the concentric circle of radius $\frac{1}{2}$. Since this smaller circle has one-fourth the area of the unit circle, it follows from the definition of geometric probabilities that $P(E) = \frac{1}{4}$.

Solution by Simulation We can run a computer simulation of each of these models very simply. For model 1, we set R = RND, generate N sample values of R, and count the proportion of values that are less than $\frac{1}{2}$. For model 2 we set

$$\text{THETA} = 2 * \text{PI} * \text{RND}, \qquad \text{where PI} = 3.1415\ldots$$

generate N sample values of THETA, and count the proportion that satisfy inequalities (2.16). (Notice that 2*PI*RND gives random numbers between 0 and 2π.) For model

3, we generate random points (X, Y) in the circle by means of a two-step procedure. Set

$$X = 2 * \text{RND} - 1 \quad Y = 2 * \text{RND} - 1$$

which gives a random point $Q = (X, Y)$ in the square $-1 < x < 1, -1 < y < 1$. Generate N points by this formula. Let $N(1)$ be the number of points that lie inside the circle of radius 1, and let $N(1/2)$ be the number of points that fall inside the circle of radius $\frac{1}{2}$. The estimate for $P(E)$ is then the ratio $N(\frac{1}{2})/N(1)$. Table 2.10 gives the results of a particular simulation. Notice how the simulation results for the three models approach the theoretical probabilities ($\frac{1}{2}$, $\frac{1}{3}$, and $\frac{1}{4}$, respectively) as N gets larger.

Table 2.10 Simulation of Random Chord Models

N	MODEL 1	MODEL 2	MODEL 3
20	0.550	0.250	0.250
40	0.700	0.325	0.200
80	0.550	0.300	0.225
160	0.538	0.388	0.256
640	0.522	0.334	0.242
1280	0.496	0.328	0.260

Which is the correct model for drawing a random chord? Each of these three models is consistent with the axioms of probability. These axioms, however, don't prescribe the numerical values of probabilities; they merely give a consistent way of carrying out calculations. If we were trying to set up a stochastic model for random chords in some real-world situation, such as the random track made by a cosmic ray traversing a circular detection chamber, it is quite possible that none of these models would be suitable. ■

2.3 Conditional Probability and Bayes Formula

Suppose that we perform N repetitions (trials) of an experiment of the type described in Section 2.1. For each possible event E associated with the experiment, we let $n(E)$ be the number of times that E actually occurs in the sequence of trials. Thus, the relative frequency of occurrence of E is $f(E) = n(E)/N$. Now we pick a particular event B that has occurred at least once in the sequence of N trials. For any other event A, we can calculate the *conditional* relative frequency of A, given B, by restricting attention to the trials in which B occurred. Since there are $n(B)$ such trials, this relative frequency is

$$f(A \mid B) = \frac{n(A \cap B)}{n(B)}$$

Dividing numerator and denominator by N, we can also write this formula as

$$f(A \mid B) = \frac{f(A \cap B)}{f(B)} \tag{2.17}$$

It is also instructive to write equation (2.17) in multiplicative form:

$$f(A \cap B) = f(A \mid B)f(B) \tag{2.18}$$

This shows that the frequency of A and B together can be calculated in stages: in the first stage, we may assume that B always occurred in the experiment, and we check whether A occurred; in the second stage, we check the whole set of experiments to determine the frequency of occurrence of B. Observe that by including only the trials in which B occurred in the calculation of $f(A \mid B)$, we have in effect reduced the sample space to events that are certain to include B.

Example 2.12 Simulation of Geometric Probabilities

Suppose S is the square $-1 \leq x \leq 1, -1 \leq y \leq 1$ in the (x, y) plane, and that the experiment consists of picking a point at random in S. Let B be the event "The point lies in the unit circle $x^2 + y^2 \leq 1$." We already described in Example 2.10 a random number generator can be used to simulate any number of trials of this experiment. If A is a subset of S, we will identify A with the event "The randomly-picked point lies in A." Thus, we can calculate the conditional relative frequencies $f(A \mid B)$ by formula (2.17). These conditional relative frequencies obviously correspond to picking a point at random in B. This provides an easy way of simulating geometric probabilities relative to a nonrectangular region such as B: we generate a sequence of points randomly in a rectangular region containing B, using a random number generator for each coordinate, and then simply discard the trials in which the point doesn't fall in B. We already used this method in the simulation solution to random chord model 3 (Example 2.11). ■

Now suppose that we have defined a probability model for our experiment, and that B is an event for which $P(B) > 0$. With formula equation (2.17) as our guide, we define the **conditional probability of A, given B** to be

$$P(A \mid B) = \frac{P(A \cap B)}{P(B)} \tag{2.19}$$

Just as for the conditional relative frequency, we can write formula (2.19) in multiplicative form

$$P(A \cap B) = P(A \mid B)P(B) \tag{2.20}$$

and the same remarks made concerning calculation in stages apply here. This version of the formula also leads us to define $P(A \mid B) = 0$ when $P(B) = 0$. This is consistent with (2.20), since $P(A \cap B) = 0$ in this case. The left side of (2.20) is symmetric in A and B, so it follows that

$$P(A \mid B)P(B) = P(B \mid A)P(A)$$

If A and B both have positive probability, we can exchange the roles of A and B in calculating conditional probabilities:

$$P(B \mid A) = \frac{P(A \mid B)P(B)}{P(A)} \tag{2.21}$$

This formula makes it clear that the conditional probability $P(A \mid B)$ is *not* symmetrical in A and B unless $P(B) = P(A)$.

Example 2.13 Random Prime Numbers

Pick two distinct numbers at random from the set of integers 1, 2, ..., 10. If the first number picked is prime, what is the probability that the second is not prime?

Here we can take as the sample space S all ordered pairs (m, n), where m, n are integers between 1 and 10 and $m \neq n$. (Thus S has $10 \cdot 9 = 90$ elements.) We interpret the phrase "at random" to mean that each point of S has the same probability ($= \frac{1}{90}$). Let

$$\begin{aligned} A &= \text{"The first number picked is prime"} \\ B &= \text{"The second number picked is not prime"} \end{aligned}$$

Since there are four primes (namely 2, 3, 5, and 7) between 1 and 10, the elements in A are determined by four choices for m, followed by nine choices for n. Thus, $|A| = 4 \cdot 9 = 36$. Similarly, the elements of B are determined by six choices for n, followed by nine choices for m, so $|B| = 6 \cdot 9 = 54$. Finally, $A \cap B$ has $4 \cdot 6 = 24$ elements. Thus, by formula (2.19),

$$P(B \mid A) = 24/36 = 2/3$$

We could have arrived at this answer in a more intuitive way by arguing that in picking the second number, we have nine equally likely choices, of which six are not prime (assuming that the first number picked was prime). A similar informal argument gives the value $P(A \mid B) = 4/9$, which agrees with the value $|A \cap B|/|B| = 24/54$ obtained from the more systematic sample-space method. (See Example 2.14, however, where such informal reasoning fails.) Notice that $P(A \mid B) \neq P(B \mid A)$. ■

If we fix any event C with $P(C) > 0$, then we can consider the reduced probability model in which C occurs with probability 1. In this model we calculate all probabilities in terms of the probability function P_C, where

$$P_C(A) = P(A \mid C)$$

Let us check that the Normalization and Additivity axioms for a probability function are satisfied by P_C. It is clear that $P_C(S) = 1$. If A and B are disjoint, then so are $A \cap C$ and $B \cap C$. Since

$$(A \cup B) \cap C = (A \cap C) \cup (B \cap C)$$

by the distributive law, we have

$$P((A \cup B) \cap C) = P(A \cap C) + P(B \cap C)$$

Dividing by $P(C)$, we obtain the Additivity Axiom:

$$P(A \cup B \mid C) = P(A \mid C) + P(B \mid C) \qquad (A \text{ and } B \text{ disjoint}) \qquad (2.22)$$

The validity of the Continuity Axiom for P_C follows directly from the definition; each set A_i is intersected with C. The crucial point in all of these calculations with conditional probabilities is that the denominator $P(C)$ remains fixed and simply plays the role of normalizing the reduced probabilities, so that $P_C(S) = 1$.

Rule of Total Causes

The well-known strategy of divide and conquer has its counterpart in probability theory. Suppose that we have a family $H_1, H_2, \ldots, H_n$ of mutually exclusive events that collectively includes all possible outcomes:

$$S = \bigcup_{i=1}^{n} H_i \tag{2.23}$$

If E is any event, it follows that we can calculate $P(E)$ in stages by using the formula

$$P(E) = \sum_{i=1}^{n} P(E \mid H_i) P(H_i) \tag{2.24}$$

To prove formula (2.24), we note from formula (2.20) that the terms on the right side are just $P(E \cap H_i)$. But by formula (2.23) the event E is the union of the mutually exclusive events $E \cap H_i$. Thus, by the Additivity Axiom we obtain formula (2.24). (By using the Continuity Axiom we can extend formula (2.24) to the case $n = \infty$.)

Bayes' Formula

Suppose that the events H_i in formula (2.23) constitute a complete set of mutually exclusive *hypotheses*; then every outcome of the experiment (every point of the sample space) satisfies precisely one of these hypotheses. The term $P(E \mid H_i)$ in formula (2.24) is the probability that event E will occur, given that hypothesis H_i is true.

Suppose that we now reverse the roles of E and H_i. Informally speaking, we assume that the effect E is known, and we try to determine the probability that the cause H_i is true. This is easily done: we take $B = H_i$ and $A = E$ in formula (2.21), and use the rule of total causes (formula (2.24)) to calculate the denominator in formula (2.21). This yields *Bayes' Formula:*

$$P(H_i \mid E) = \frac{P(E \mid H_i) P(H_i)}{P(E \mid H_1) P(H_1) + \cdots + P(E \mid H_n) P(H_n)}$$

Example 2.14 A Card Trick

Suppose that we have a set of three cards. Card 1 is red on both sides, card 2 is white on both sides, and card 3 has a red side and a white side. The experiment consists of picking one card at random, and then looking at a random side of this card. Suppose that the side of the card observed is red. What is the probability that the other side is white?

An intuitive answer would be $\frac{1}{2}$, since the card picked must be 1 or 3, and each choice is equally likely. This argument is fallacious, however, as it ignores the random selection of a side. We can calculate the correct probability using Bayes' formula. Let H_i be the event "Card i is picked". Picking a card at random implies that $P(H_i) = 1/3$. Let E be the event "Side of the card observed is red". Then clearly

$$P(E \mid H_1) = 1, \quad P(E \mid H_2) = 0, \quad P(E \mid H_3) = 0.5$$

since each side is equally likely to be observed. Thus, by the rule of total causes,

$$P(E) = \left(1 \cdot \frac{1}{3}\right) + \left(0 \cdot \frac{1}{3}\right) + \left(\frac{1}{2} \cdot \frac{1}{3}\right) = \frac{1}{2}$$

(This is intuitively obvious, since there are three red sides out of six sides, and we are picking a side at random.) Substituting these values in formula (2.24), we obtain

$$P(H_3 \mid E) = \frac{P(E \mid H_3)P(H_3)}{P(E)} = \frac{\frac{1}{2} \cdot \frac{1}{3}}{\frac{1}{2}} = \frac{1}{3}$$

Having determined the correct answer, 1/3, we can make sense of it intuitively as follows: there are three red faces on the three cards, but two of these appear on the same card; therefore, in two of every three appearances of a red face, the red/red card is involved, while the red/white card is involved in only one of every three such appearances. ■

Example 2.15 Reliability of Multiple-Choice Exams

Consider a multiple-choice exam that has m choices of answers for each question. Assume that the proportion of correct answers that the student knows is p; if the student doesn't know the correct answer to a question, then he marks an answer at random. Suppose the answer marked to a particular question is correct. What is the probability that the student was guessing?

In this example we have the mutually exclusive hypotheses

$$\begin{array}{lcll} H_1 & = & \text{“Student knows the correct answer”} & (P(H_1) = p) \\ H_2 & = & \text{“Student is guessing”} & (P(H_2) = 1 - p) \end{array}$$

and the observed event E = "Correct answer marked." The conditional probabilities of E, given these hypotheses, are

$$P(E \mid H_1) = 1, \quad P(E \mid H_2) = \frac{1}{m}$$

Hence, by the rule of total causes, $P(E) = 1 \cdot p + [(1/m) \cdot (1 - p)]$; thus by Bayes' formula,

$$P(H_2 \mid E) = \frac{\frac{1}{m} \cdot (1 - p)}{1 \cdot p + \frac{1}{m} \cdot (1 - p)} = \frac{1 - p}{mp + 1 - p}$$

Notice that, for fixed $p > 0$, this conditional probability (which measures the unreliability of the test) can be made small by making m large. The person creating the test doesn't know the value of p, however, and instead must pick a fixed value of m and use the score on the test to estimate p. For example, if $p = \frac{1}{2}$, then the probability is $1/(m + 1)$ that the student found the correct answer by guessing. Thus, for $m = 5$ (a popular choice), if we observe a correct answer on the exam of a student whose true score should be 50%, then the chances are 1 in 6 that the student was guessing. On the other hand, if $p = \frac{3}{4}$, then the probability is $1/(3m + 1)$ that the student found the correct answer by guessing. Thus, for $m = 5$ again, if we observe a correct answer on the exam of a student whose true score should be 75%, the chances are only 1 in 16 that the student was guessing. In this sense, the exam of the better student gives a more reliable measurement of true ability. ■

2.4 Stochastic Independence

Suppose that we have set up a stochastic model of an experiment, with sample space S and probability function $P(\cdot)$, and suppose that A and B are two events in the model with $0 < P(A) < 1$ and $0 < P(B) < 1$ (so A and B are neither completely certain nor completely unlikely). One way to detect a possible cause–and–effect relation between A and B is to compare the probability $P(A)$ that A occurs in a random trial of the experiment with the conditional probability $P(A \mid B)$ that A occurs in a biased trial in which B is known to occur.

Example 2.16

If $B \subset A$ ("B implies A"), then $P(A \cap B) = P(B)$. Thus by the definition of conditional probability we have $P(A \mid B) = P(B)/P(B) = 1$ —no matter what the *a priori* probability of A might be. ■

Example 2.17

If $A \cap B = \emptyset$ ("A and B mutually exclusive"), then $P(A \cap B) = 0$, so $P(A \mid B) = 0$ in this case. As in Example 2.16, this is just what the intuitive definition of conditional probability would suggest. ■

Example 2.18

We calculated in Example 2.14 that if A = "Side of the card observed is red" and B = "red-white card picked," then $P(A) = \frac{1}{2} = P(A \mid B)$. We also calculated that $P(B) = \frac{1}{3} = P(B \mid A)$. Thus in this game the information that B is true doesn't affect the probability that A is true, and vice-versa. (From a gambling point of view, knowing that one of the events has occurred is of no value in betting on the other event). ■

Consider the situation of Example 2.18, where $P(A \mid B) = P(A)$. By formula (2.20), this condition is the same as

$$P(A \cap B) = P(A)P(B). \tag{2.25}$$

We shall say that the events A and B are mutually **stochastically independent** if equation (2.25) holds. From this definition, we see that the property of stochastic independence is symmetric in the two events A and B, as we verified by calculation in Example 3. Also equation (2.25) is equivalent to

$$P(B \mid A) = P(B) \tag{2.26}$$

just as in Example 2.18, provided $P(A) > 0$.

The notion of stochastic independence is fundamental in probability theory. It should not be confused with the property that events are **mutually exclusive**. Indeed, if A and B are mutually exclusive, then $P(A \cap B) = 0$. If A and B are also independent, then by equation (2.25) either A or B (or both) have probability 0 of occurring. In general, the property

"A and B are mutually exclusive"

allows us to calculate the probability of the union $A \cup B$ by adding the probabilities of A and of B. The property

"A and B are stochastically independent"

allows us to calculate the probability of the intersection $A \cap B$ by multiplying the probabilities of A and of B.

Assumptions about the stochastic independence of events play a major role in setting up stochastic models, as we shall see in later chapters. It is important to realize, however, that once the probability function $P(\cdot)$ is defined in a model, then it is a purely *mathematical* question as to whether two events A and B in the model are stochastically independent—that is, whether equation (2.25) holds. Example 2.19 illustrates this.

Example 2.19 Random Primes Revisited

Suppose we pick an integer X at random between 1 and 20 (each integer has probability $\frac{1}{20}$ of being picked). Let A = "X is prime" and B = "X is between 11 and 20." Then $P(A) = \frac{8}{20}$, since there are eight primes in the range 1–20, and $P(B) = \frac{1}{2}$. But there are exactly four primes between 11 and 20, so $P(A \cap B) = \frac{4}{20} = P(A)P(B)$. Thus A and B are mutually independent.

Suppose we change the experiment slightly, by picking the integer X at random between 1 and 21. Now $P(A) = \frac{8}{21}$, $P(B) = \frac{10}{21}$, and $P(A \cap B) = \frac{4}{21} = 0.1905$. But $P(A) \cdot P(B) = 0.1814$, so in this new model A and B are not independent. ■

In general, if events A and B are independent, then so are A^c and B. To verify this, we start by using A to decompose B as the disjoint union

$$B = (B \cap A) \cup (B \cap A^c)$$

Now use the additivity of probability and the given independence of A and B to calculate the probability of each side of this equation:

$$P(B) = P(B)P(A) + P(B \cap A^c)$$

Solving for $P(B \cap A^c)$, we find that

$$P(B \cap A^c) = (1 - P(A))P(B) = P(A^c)P(B)$$

This proves that A^c and B are independent. This calculation also shows that if A and B are independent and we know $P(A)$ and $P(B)$, then we can calculate the probability of all the events that are logical combinations of A and B.

Consider now the case of n events $E_1, E_2, \ldots, E_n$. We shall define this family of events to be **mutually stochastically independent** if for every choice of k events $E_{i_1}, E_{i_2}, \ldots, E_{i_k}$ from this family ($1 \leq k \leq n$), the probability that all of the k events occur is the product of the separate probabilities:

$$P(E_{i_1} \cap \cdots \cap E_{i_k}) = P(E_{i_1}) \cdots P(E_{i_k}) \tag{2.27}$$

For example, in the case of $n = 3$ events A, B, and C, this independence condition requires that the following equations hold:

$$P(A \cap B) = P(A)P(B) \tag{2.28}$$
$$P(A \cap C) = P(A)P(C), \qquad P(B \cap C) = P(B)P(C)$$

and also that

$$P(A \cap B \cap C) = P(A)P(B)P(C) \tag{2.29}$$

Equations (2.28) say that the events A, B, and C are **pairwise independent**. When equations (2.28) hold, then (2.29) says that the occurrence of any pair of these events is independent of the third event. For example, the event $A \cap B$ is independent of the event C.

Example 2.20

In Example 2.19, where the integer X is picked at random between 1 and 20, let

$$C = \text{“Either } X \in \{4, 6, 8, 10\} \text{ or } X \in \{9, 11, 13, 15, 17, 19\}\text{”}$$

and let the events A and B be as in Example 2.19. Then $P(C) = \frac{1}{2}$. By enumerating A and B, we see that

$$|A \cap C| = 4, \quad |B \cap C| = 5 \quad |A \cap B \cap C| = 4$$

Thus, $P(A \cap C) = \frac{4}{20} = P(A)P(C)$ and $P(B \cap C) = \frac{5}{20} = P(B)P(C)$, so the events A, B, and C are pairwise independent. But

$$P(A \cap B \cap C) = 4/20 \neq P(A)P(B)P(C)$$

so the three events A, B, C are not mutually independent. ■

When A, B, and C are mutually independent events, then so are A^c, B, and C. This follows by the same argument given above for A^c and B. Repeating this argument, we find that A^c, B^c, and C^c are mutually independent. Thus we can calculate the probability of any logical combination of A, B, and C in terms of $P(A)$, $P(B)$, and $P(C)$. For example,

$$\begin{aligned} P(A \cup B \cup C) &= 1 - P(A^c \cap B^c \cap C^c) \\ &= 1 - P(A^c)P(B^c)P(C^c) \qquad \text{(by independence)} \\ &= 1 - [(1 - P(A)) \cdot (1 - P(B)) \cdot (1 - P(C))] \end{aligned}$$

We could also do this calculation by the inclusion–exclusion formula of Section 2.2 and equations (2.28) and (2.29). (Verify, by expanding the product above, that the two calculations yield the same result.)

2.5 Repeated Independent Trials

Now that we have a precise formulation of the concept of stochastic independence, we can set up a stochastic model for N independent trials of an experiment under identical conditions. Observe first that we can describe the results of a sequence of N trials by an (ordered) sequence $\omega_1, \omega_2, \ldots, \omega_N$, where ω_i is the outcome of the ith trial. Thus, if S is the sample space for a single trial of the experiment, we can take the set S^N of all N-tuples of elements of S as the sample space for N trials. If we think of the elements of S as being the letters of an alphabet, then S^N consists of all words of length N formed from this alphabet.

Example 2.21 Bernoulli Trials

For an experiment such as a coin toss that has only two possible outcomes, we may take $S = \{0, 1\}$. Thus, S^N consists of words such as $10011\ldots1$ (N binary digits). If 1 corresponds to heads, the event "In four tosses, the first two tosses were heads" consists of the set of all words of the form $11xy$, where x and y can be either 0 or 1. ■

How can we calculate the probability of an event associated with repeated independent trials? Consider first an event $A \subset S$ that can be observed in a single trial of the experiment. Set

$$A^{(k)} = \text{"Event A occurred on the } k\text{th trial"}$$

(i.e. $A^{(k)}$ consists of all words whose kth letter is in the set A, and whose other letters are completely arbitrary). Since we are assuming that each trial of the experiment is performed under identical conditions, we have

$$P(A^{(k)}) = P(A) \quad \text{for } 1 \le k \le N \tag{2.30}$$

Furthermore, if $E_1, E_2, \ldots, E_N$ are events such that the occurrence of E_j depends only on the outcome of trial j, then in a repeated trial model these events are mutually independent. Hence, the probability that all of these events will occur is given by the formula

$$P(E_1 \cap \cdots \cap E_N) = \prod_{j=1}^{N} P(E_j) \tag{2.31}$$

where the probabilities on the right side of formula (2.31) are calculated by application of the single-trial model. ■

Example 2.22 Binomial Probabilities

Suppose that we play a game in which our probability of winning one round is p (with $0 < p < 1$). If we play the game N times, what is the probability that we will win exactly k times?

Solution Let 0 correspond to losing and 1 to winning in one round. Thus, the sample space S^N for N rounds consists of all binary words of length N. According to formula (2.31), a word ω in S^N is assigned probability $p^k q^{N-k}$, where k is the number of times that 1 occurs in ω, and $q = 1 - p$. The number of binary words of length N having exactly k 1's equals the number of choices of k positions out of N in which to put the 1's. Thus, the event

$$E_k = \text{"}k \text{ wins in } N \text{ tries"}$$

has probability

$$P(E_k) = \binom{N}{k} p^k q^{N-k}$$

The events $E_0, E_1, \ldots, E_N$ are mutually exclusive and fill up S^N. The probability normalization $P(S^N) = 1$ implies the binomial identity

$$\sum_{k=0}^{N} \binom{N}{k} p^k q^{N-k} = (p+q)^N = 1$$

■

Example 2.23 Problem of the Points

In Example 2.22, what is the probability that we will win k times before our opponent wins n times?

Solution In the statement of the problem, no limit is mentioned on the number of games to be played. After $k + n - 1$ games have been played, however, either we will have won at least k times, or our opponent will have won at least n times. Thus, the question can be decided by playing $N = k+n-1$ games. The probability is $P(E_{k,n})$, where

$$E_{k,n} = \text{“At least } k \text{ wins in } k+n-1 \text{ games”}$$

We may calculate $P(E_{k,n})$ by using the binomial probabilities for $N = k + n - 1$ trials:

$$P(E_{k,n}) = \sum_{i=k}^{N} \binom{N}{i} p^i q^{N-i} \tag{2.32}$$

For example, the probability of obtaining at least one win before n losses are sustained is calculated by setting $k = 1$ in this formula. The sum in this case is the binomial expansion of $(p+q)^n$ with the term corresponding to $k = 0$ omitted. Hence

$$P(E_{1,n}) = 1 - q^n$$

since $p + q = 1$. (We could have calculated this case directly, since the complement to $E_{1,n}$ consists of a run of n straight losses.)

In general, the probabilities in formula (2.32) must be evaluated numerically. The numerical values are not intuitively obvious, however, and they show that probabilities change under scaling transformations. For example, suppose $p = 0.6$ (you are the favored player). Then numerical calculation shows that in a seven-game series, your probability of winning at least four games before your opponent wins four games is about 0.71. However, in a seventy-game series, your probability of winning at least forty games before your opponent wins forty games increases to 0.964. ■

Example 2.24 Geometric Probabilities

Suppose the experiment consists of picking a real number x at random between 0 and 1. The sample space S for a single trial is the unit interval $(0, 1)$; the sample space for two trials consists of the unit square S^2; and in general, the sample space for N trials consists of the N-dimensional cube S^N. If $E \subset S$ is an interval, the probability that $x \in E$ is the length of E. Thus by formula (2.31) we see that if A and B are intervals in $(0, 1)$, then the probability that $(x, y) \in A \times B$ is just the area of the rectangle $A \times B$. In general, if we pick N numbers $x_1, \ldots, x_N$ independently and randomly in $(0, 1)$, then the probability that a sample point $\omega = (x_1, \ldots, x_N)$ lies in some subset E of the N-cube S^N equals the N-dimensional volume of E. ■

Example 2.25 Meeting for Lunch

We can apply the construction in Example 2.24 to obtain an analytical solution to Example 2.3. The pair of random arrival times X and Y of the two friends corresponds to an outcome of two independent trials of this experiment. The probability that the friends have lunch together is thus given by the area of the region

$$(x, y) \mid 0 < x < 1, 0 < y < 1, |x - y| < 0.25$$

By elementary geometry (see Figure 2.5), this area is $\frac{7}{16} = 0.4375$. ■

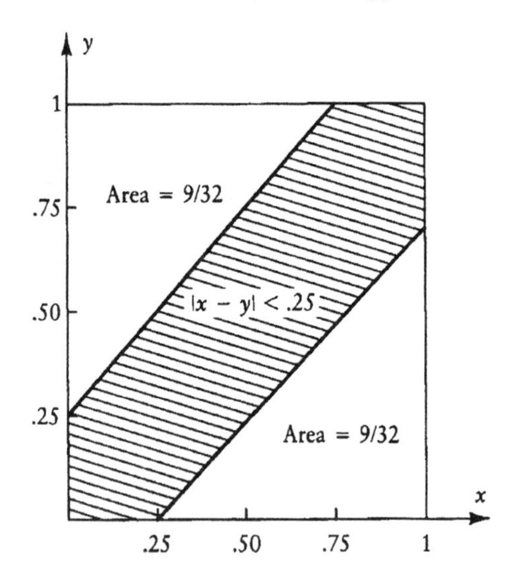

Figure 2.5: Meeting for Lunch

Example 2.26 Infinite Number of Trials

Suppose that E and F are mutually exclusive events in the sample space for a single trial of an experiment, and that $P(E) > 0$. We perform the experiment repeatedly. What is the probability that, in the sequence of trials, E occurs before F does?

Solution In this problem we have to allow an unlimited number of trials, so the sample space is S^∞—the set of all words of infinite length formed out of outcomes for the single trial. The observable events in this model include all events determined by the outcomes of a finite number of trials, and all countable unions and intersections of such events. (This description will suffice for our purposes; a more exact description is beyond the scope of this book.) We shall use two methods to solve this problem.

Method 1 Let E_n denote the event "E occurs on trial n, but neither E nor F occurs on earlier trials." In the notation used at the beginning of this section, we can write

$$E_n = A^{(1)} \cap A^{(2)} \cap \cdots \cap A^{(n-1)} \cap E^{(n)}$$

where $A = E^c \cap F^c = (E \cup F)^c$. Thus, just as in the problem of the points, the occurrence of E_n is determined by the outcome of a finite number of trials. By formulas (2.30) and (2.31), we can calculate

$$P(E_n) = P(A)^{n-1} P(E)$$

Since E and F are disjoint, $P(A) = 1 - P(E \cup F) = 1 - P(E) - P(F)$. Hence

$$P(E_n) = [1 - P(E) - P(F)]^{n-1} P(E) \tag{2.33}$$

In an unlimited sequence of trials E occurs before F if and only if the event E_n occurs for some n. Thus, if $H \subset S^\infty$ is the event "E occurs before F does," then

$$H = \bigcup_{n=1}^{\infty} E_n$$

This shows that H is an observable event. Since the collection of events $\{E_n\}$ is mutually disjoint, it follows that

$$P(H) = \sum_{n=1}^{\infty} P(E_n) \tag{2.34}$$

(Here we have used the Continuity Axiom to allow the infinite range of summation.) Substitute formula (2.33) into formula (2.34); the resulting infinite series is a geometric series with initial term $P(E)$ and ratio $r = 1 - P(E) - P(F) < 1$. (Remember that $P(E) > 0$.) The sum of this series gives the value

$$P(E \text{ occurs before } F) = \frac{P(E)}{P(E) + P(F)}$$

Method 2 Let H be the event "E occurs before F does." We can take advantage of the infinite number of trials allowed in this model by conditioning our argument on the outcome of the first trial. Let $G = (E \cup F)^c$, so that

$$S = E \cup F \cup G \qquad \text{(disjoint)}$$

Obviously

$$P(H \mid E \text{ occurs on first trial}) = 1 \qquad P(H \mid F \text{ occurs on first trial}) = 0$$

What about the conditional probability of H if G occurs on the first trial? If Ed is betting that event E will occur first, and Frank is betting that event F will occur first, then the first trial makes no difference to Ed and Frank when the outcome is neither E nor F. From their point of view, the game might as well have started with the second trial when this happens. Thus,

$$P(H \mid G) = P(H)$$

(This *starting-over argument* is a very useful tool that we will use many times.) Now we can proceed to calculate $P(H)$ using the rule of total causes:

$$\begin{aligned} P(H) &= P(H \mid E^{(1)})P(E) + P(H \mid F^{(1)})P(F) + P(H \mid G^{(1)})P(G) \\ &= P(E) + 0 + P(H)(1 - P(E) - P(F)) \end{aligned}$$

(Recall the notation: $E^{(1)}$ = "E occurs on trial 1".) Solving this equation for $P(H)$, we find the same solution as by method 1, with much less effort. ■

Example 2.27 Gambler's Ruin

Ann and Bob are betting on the outcomes of an unlimited sequence of coin tosses. Ann always bets heads (which appears with probability p on each toss, where $0 < p < 1$), and Bob always bets tails (which appears with the complementary probability $q = 1 - p$.) Ann starts with a stake of a, and Bob with a stake of b. Each time a result of heads occurs, Bob gives Ann \$1, and each time a result of tails occurs, Ann gives Bob \$1. The game stops when one of the players runs out of money.

a. What is the probability that Ann goes broke?

b. What is the probability that Bob goes broke?

c. What is the probability that the game stops?

Solution The neatest way to analyze this problem is to use a starting-over argument based on the outcome of the first trial. Let $N = a + b$ be the total fortune of the two players, which remains constant throughout the game. The probability that a player will go broke depends on the player's initial stake. To make this explicit, we define the events

$$\begin{aligned} A_a &= \text{"Ann starts with } a \text{ and eventually goes broke"} \\ B_b &= \text{"Bob starts with } b \text{ and eventually goes broke"} \end{aligned}$$

and we set

$$f_a = P(A_a) \qquad g_b = P(B_b)$$

We already know some extreme cases of these probabilities:

$$f_N = 0, \qquad g_N = 0 \tag{2.35}$$

$$f_0 = 1, \qquad g_0 = 1 \tag{2.36}$$

These occur when one of the players has all of the money.

Suppose now that $1 \leq a \leq b \leq N-1$. Let E ="Head appears on the first trial." By applying the same starting-over argument that we used in Example 2.25, we see that $P(A_a \mid E) = P(A_{a+1})$ and $P(A_a \mid E^c) = P(A_{a-1})$. Using the rule of total causes, we thus obtain the recursion

$$f_a = p \cdot f_{a+1} + q \cdot f_{a-1} \tag{2.37}$$

Interchanging the two players, we similarly obtain the recursion

$$g_b = q \cdot g_{b+1} + p \cdot g_{b-1} \tag{2.38}$$

We now have the analytical problem of solving for f_a from formula (2.37), taking into account the boundary conditions expressed in equation (2.35). Since $p + q = 1$, we may write the term f_a in formula (2.37) as $p \cdot f_a + q \cdot f_a$. Regrouping terms in this formula according to the coefficients p and q, we obtain

$$q(f_{a-1} - f_a) = p(f_a - f_{a+1})$$

We divide this equation by q, and write $r = p/q$. This gives us the equivalent recursion

$$f_{a-1} - f_a = r(f_a - f_{a+1}) \tag{2.39}$$

We can now successively calculate all the probabilities f_a in terms of f_{N-1}. Start with $a = N-1$ in equation (2.39) and use equation (2.35):

$$f_{N-2} - f_{N-1} = r f_{N-1}$$

Hence, $f_{N-2} = (1+r) f_{N-1}$. Next, we take $a = N-2$ and use this last formula in equation (2.39):

$$f_{N-3} - f_{N-2} = r(f_{N-2} - f_{N-1}) = r^2 f_{N-1}$$

Hence $f_{N-3} = f_{N-2} + r^2 f_{N-1} = (1 + r + r^2) f_{N-1}$. Continuing in this way, we find in general that

$$f_{N-k} = (1 + r + r^2 + \cdots + r^{k-1}) \cdot f_{N-1} \tag{2.40}$$

for $1 \leq k \leq N$. The probability f_{N-1} is still unknown in equation (2.40); we can find it by setting $k = N$ and using the condition $f_0 = 1$:

$$1 = (1 + r + r^2 + \cdots + r^{N-1}) \cdot f_{N-1} \tag{2.41}$$

Recall the formula for the sum of a finite geometric series:

$$1 + r + r^2 + \cdots + r^{k-1} = \frac{1 - r^k}{1 - r}$$

when $r \neq 1$. Use this in equation (2.41) to obtain

$$f_{N-1} = \frac{1 - r}{1 - r^N}$$

when $r \neq 1$. (If $r = 1$ in equation (2.41), then obviously $f_{N-1} = 1/N$.) Using these formulas in formula (2.40) with $k = N - a$, we finally arrive at an explicit formula for the probability that Ann will go broke, starting with an initial stake of a, when the total stake $a + b = N$:

$$f_a = \frac{1 - r^{N-a}}{1 - r^N} \qquad (r = p/q \neq 1) \tag{2.42}$$

When $p = q$, then $f_a = (N - a)/N$.

To calculate the ruin probabilities for Bob, we interchange p and q in the calculations just made. But this is the same as replacing r with $1/r$ in formula (2.42). Thus, we find that

$$g_b = \frac{1 - r^{b-N}}{1 - r^{-N}} \qquad (r = p/q \neq 1)$$

Multiplying the numerator and denominator by $-r^N$ in this formula, we can write it in the equivalent form

$$g_b = \frac{r^b - r^N}{1 - r^N}$$

When $p = q$, then $g_b = (N - b)/N$. In particular, since $a + b = N$, we find that

$$f_a + g_b = \frac{1 - r^b + r^b - r^N}{1 - r^N} = 1$$

Thus *the probability is one that either Ann or Bob will be ruined.* Hence, the game is certain to terminate after a finite number of trials.

Another way of viewing the gambler's ruin problem is as follows. Suppose that Bob has at least b that he is willing to lose. Ann starts with a and sets as her goal to win b from Bob. The probability that she will attain this goal is

$$g_b = \frac{r^b - r^{a+b}}{1 - r^{a+b}}$$

The probability that she will go broke is $1 - g_b$. As in the problem of the points, the numerical values of these probabilities are somewhat unexpected. For example, if Ann's win probability $p = 0.51$ and she starts with \$10, with the goal of winning \$10, then her chance of success is 0.599. If she starts with \$40, with the goal of winning \$40, then her chance of success rises to 0.832 ("the favored rich get richer"). If her win probability increases slightly to $p = 0.52$, then her chance of success, with an initial stake of \$40 and a goal of winning \$40, is 0.961. ∎

Exercises

1. Calculate the probabilities of each of the following poker hands (a set of five cards drawn from deck of fifty-two cards). Each card in the deck has a *suit* and a *denomination*; there are four suits and 13 denominations. Assume that each set of five cards is equally likely.

a. One pair (two cards of denomination a, other cards of denominations b, c, and d with a, b, c, and d distinct).

b. Two pairs (two cards of denomination a, two cards of denomination b, and the remaining card of denomination c, with a, b, and c distinct).

c. Three of a kind (three cards of denomination a, the other cards of denominations b and c with a, b, and c distinct).

d. Four of a kind (four cards of denomination a, the other card of denomination b, with a and b distinct).

(HINT: Enumerate the possible hands by first choosing denominations and then suits.)

2. Calculate the probability that a poker hand has

   a. Exactly two aces

   b. Two aces and three kings

3. A set of four cards is chosen at random from a deck of fifty-two playing cards.

   a. What is the probability that two of the cards are of one suit and two are of another suit?

   b. What is the probability that exactly two of the four cards are hearts?

4. An urn contains n balls numbered 1 to n. An experiment consists of drawing balls at random from the urn (one at a time) n times. Each time a ball is withdrawn its number is recorded on an ordered list, and the ball is placed back in the urn.

   a. Give an explicit description of an appropriate sample space S for this experiment.

   b. What is the probability that each ball will be drawn exactly once?

   c. Use Stirling's formula, $n! \approx (n/e)^n \sqrt{2\pi n}$, to obtain an approximate formula for the probability in part b. when n is large.

5. Suppose that you randomly place N playing pieces on an $N \times N$ checkerboard.

   a. What is the probability that each row of the checkerboard will have only one piece?

   b. Use Stirling's formula, as in the previous exercise, to estimate the probability in part a. Compare your estimate with the exact value when $N = 8$.

6. An urn contains five red balls and fifteen blue balls.

   a. Suppose that you pick a set of four balls at random from the urn. What is the probability that exactly two of the four balls are red?

   b. Suppose that you reach into the urn four times, each time picking a ball at random, observing its color, and then replacing it in the urn. What is the probability of your picking a red ball twice?

   (The experiments in parts (a) and (b) are called sampling *without replacement* and *sampling with replacement*.)

7. An urn contains four red balls, five blue balls, and six white balls. A set of three balls is withdrawn at random from the urn. Calculate the probability that all three balls in the set have the same color.

8. An urn has four red balls and two green balls. Balls are picked at random from the urn, one at a time. If the ball picked is red, it is put back into the urn. If the ball picked is green, it is discarded.

   a. Suppose that two balls have been picked from the urn. What is the probability that no green balls remain in the urn?

   b. Suppose that three balls have been picked from the urn. What is the probability that no green balls remain in the urn?

9. An urn contains eight balls: two red, two yellow, two green, and two blue. A set of four balls is picked at random from the urn.

   a. Let E_c be the event that two balls of a particular color c ($c = R, Y, G$, or B) are in the set of four balls picked. Calculate $P(E_R)$ and $P(E_R \cap E_Y)$.

   b. Calculate the probability that at least two of the balls in the set of four have the same color.

10. Let A, B, and C be subsets of a sample space S. Assume that $P(A) = 0.10$, $P(B) = 0.20$, and $P(C) = 0.30$. Calculate $P(A \cup B \cup C)$ in each of the following cases (each case is a different calculation):

    a. Assume that A, B, and C are mutually exclusive.

    b. Assume that A, B, and C are independent.

    c. Assume that $P(A \cap B) = 0.04$, $P(A \cap C) = 0.05$, $P(B \cap C) = 0.08$, and $P(A \cap B \cap C) = 0.01$.

11. Suppose that a person is picked at random from a population in which 90% of the people are right-handed, 60% of the people are blue-eyed, and 40% of the people have blond hair. In this population, however, 70% of the people with blond hair have blue eyes. Define the events

$$\begin{aligned} A &= \text{"Person picked is right-handed"} \\ B &= \text{"Person picked is blue-eyed"} \\ C &= \text{"Person picked has blond hair"} \end{aligned}$$

    Assume that event A is independent of B, C, and $B \cap C$.

    a. Calculate the probability that the person picked has blue eyes *and* blond hair.

    b. Calculate the probability that the person picked is right-handed *and* has blue eyes *and* has blond hair.

    c. Calculate the probability that the person picked is right-handed *or* has blue eyes *or* has blond hair.

12. A building has seven floors. Suppose four people enter the elevator on the first floor. What is the probability that each person gets off on a different floor? (Assume each person acts independently and is equally likely to choose floors 2-7.)

13. Urn I contains four white and five red balls. Urn II contains four white and three red balls. The experiment consists of two steps: first, draw a ball at random from urn I and place it in urn II; second, draw a ball at random from urn II and place it in urn I. Define the following events:

$$E = \text{“Urn I is unchanged at the end of the experiment (four white, five red)”}$$
$$F = \text{“The ball drawn from urn I was white”}$$

a. Calculate $P(F \mid E)$.

b. Are the events E and F independent?

14. Prove the chain rule for conditional probabilities: if the events F and $F \cap G$ have positive probability, then for any event E,

$$P(E \cap F \cap G) = P(E \mid F \cap G) \cdot P(F \mid G) \cdot P(G).$$

15. In the mixed-up mail problem (Example 2.9), calculate the probability of no match for $N = 2, 4$, and 6 letters. Estimate how large must N be so this probability doesn't depend on N (to three decimal places, say).

16. Consider the mixed-up mail problem (Example 2.9) with n letters and n envelopes.

a. What is the probability that a specific set of k letters is put in the correct envelopes?

(HINT: Enumerate the possible arrangements for the remaining $n - k$ letters.)

b. What is the probability that a specific set of k letters is put in the correct envelopes and each of the remaining $n - k$ letters is put in a wrong envelope?

(HINT: Use part (a) and the chain rule for conditional probabilities.)

c. What is the probability that exactly k letters are put in the correct envelopes, and each of the remaining $n - k$ letters is put in a wrong envelope?

(HINT: Use part (b) and the Rule of Total Causes.)

17. A bridge hand consists of thirteen cards chosen from a playing deck of fifty-two. What is the probability that a bridge hand is void in at least one suit?

(HINT: Let E_i be the event that the hand is void in suit i, for $i = 1, 2, 3, 4$. Apply the exclusion-inclusion formula to the events E_i.)

18. a. If E and F are events in any probability model, prove that

$$P(E \cap F) \geq P(E) + P(F) - 1$$

(this is called *Bonferroni's inequality*).

b. Suppose that 55% of the students in a class are male, and 60% of the students have brown hair. A student is picked at random from the class. Prove that the probability is at least 0.15 that the student is a brown-haired male.

19. Urn A contains three red and three white balls. Urn B contains two red and four white balls. If one ball is randomly picked from each urn, what is the probability that the two balls will be the same color?

20. A committee of four is picked at random from a group of four men and four women. What is the probability that the committee has two men and 2 women?

21. Two cards are randomly selected from a playing deck of fifty-two. What is the probability that one card is an ace and the other one is either ten, jack, king or queen?

22. There are three dice in a box. One is a fair die. The second is a weighted die that always shows 6 when tossed. The third is a weighted die that only shows 1 or 6 when tossed, with each equally likely. Suppose that one of the three dice is selected at random and tossed. If this die shows 6, what is the probability that it is the fair die?

23. Three boxes that are identical in appearance contain the following coins:

    Box A contains two quarters.

    Box B contains one quarter and two dimes.

    Box C contains one quarter and 1 dime.

    A box is selected at random, and a coin is picked at random from this box. Suppose that the coin picked is a quarter. What is the probability that the box picked contains at least one dime?

24. There are two coins: a fair coin and a two-headed coin. One of the coins is selected at random.

    a. The selected coin is flipped once and shows heads. What is the probability that it is the fair coin?

    b. The same coin is flipped a second time and again shows heads. Now what is the probability that it is the fair coin?

    c. Suppose the same coin is flipped n times and shows heads every time. How large would n have to be so that you can be at least 99% certain that the coin is two-headed?

25. A person draws a pair of cards from a deck of fifty-two cards.

    a. The person says, "one of my cards is an ace." What is the probability that the other card in the pair is an ace?

    b. Now the person says, "one of my cards is the ace of spades." Given this additional information, calculate the probability that the other card in the pair is an ace.

26. A bus carrying n passengers makes three stops; we assume that each passenger on board is equally likely to get off at any stop, and that the passengers act independently of each other.

    a. What is the probability that k passengers get off at the first stop?

b. Suppose that four people ride the bus every Monday through Saturday, but only two ride on Sunday. On a certain day (picked at random), we observe that no one gets off at the first stop. What is the conditional probability that the day is Sunday?

27. Consider the problem of diagnosing a rare but fatal disease, which is present in only 1 out of 100,000 people. Suppose that a test for the disease is 90% accurate in detecting its presence when a person actually has the disease. Furthermore, the test is 99% accurate in yielding a negative result when a person does not have the disease.

    a. What proportion of the population will test positively for this disease?

    b. Suppose that you are diagnosed as having the disease by this test. What is the probability that you actually have the disease?

28. In the situation described in the previous exercise, suppose the test is $(100-\delta)\%$ accurate in yielding a negative result when a person does not have the disease. How small does δ have to be in order for the probability to be greater than 1/2 that a person who tests positive for this disease actually has the disease?

29. Let H and E be events having positive probability. Suppose $P(E \mid H) > P(E \mid H^c)$. Show that $P(H \mid E) > P(H)$ in this case. (The *Bayesian interpretation* of this inequality is the following: Suppose H is a hypothesis whose truth is unknown, and E is an event that is more likely to occur when H is true than when H is false. Then we can consider H more likely to be true if we observe that E occurred.)

30. In Example 2.15, let $f(p)$ be the conditional probability that a student marked a correct answer by guessing, where p is the proportion of questions that the student can answer correctly without guessing. Show that $f(p)$ is a strictly monotone decreasing function, for $0 \leq p \leq 1$.

31. A set of three distinct numbers is picked at random from the integers 1 through 5. Define the events

$$\begin{aligned} E &= \{\text{The smallest number in the set is } 1\} \\ F &= \{\text{The largest number in the set is } 5\} \end{aligned}$$

    Are events E and F independent?

32. In Example 2.19, suppose that we pick an integer X at random between 1 and 30. Are the events "X is prime" and "$16 \leq X \leq 30$" independent?

33. Al and Bob play the following game: in each round of the game, Al flips two fair coins and Bob flips three fair coins; if one of the players gets more heads than the other, he wins and the game ends; if both players get the same number of heads, there is a tie and another round of the game is played. Calculate the probabilities of the following events:

    a. Al wins the game in the first round of play.

    b. The first round of the game ends in a tie.

c. Al wins the game in the third round of play.

d. Al wins the game.

34. Suppose that we roll a fair six-sided die. Let $E = \{$ either 1 or 2 shows$\}$, and let $F = \{$ 6 shows$\}$.

a. What is the probability that, in repeated independent rolls, one occurrence of E is observed before one occurrence of F?

b. What is the probability that, in repeated independent rolls, one occurrence of E is observed before two occurrences of F? (Hint: Condition your argument on the outcome of the first trial.)

35. Team A is playing a series of games against team B. The first team to win three games wins the series. On the basis of the teams' past games, you estimate that team A has a 0.6 chance of winning any given game against team B.

a. What is the probability that team A will win the series?

b. Suppose that team A and team B have each won one game. Now what is the probability that team A will win the series?

c. Suppose team A and team B have each won two games. What is the probability that team A will win the series?

36. In successive rolls of a pair a fair dice, what is the probability of getting two sevens before getting six even numbers? (The number that you get on a roll is the sum of the points showing on the two dice.)

37. In the problem of the points (Example 2.23), let $e_{k,n}$ denote the probability of k wins before n losses.

a. Show that, for $k > 0$ and $n > 0$, we have the recurrence

$$e_{k,n} = p \cdot e_{k-1,n} + q \cdot e_{k,n-1}$$

(HINT: Use conditional probability based on the outcome of the first trial).

b. Use the equation in part (a) together with the boundary conditions $e_{0,n} = 1$ and $e_{k,0} = 0$ to calculate $e_{2,3}$. Show that your result agrees with formula (2.32).

38. Alice rolls a die six times and is betting that 1 shows at least once. Barbara rolls a die twelve times and is betting that 1 shows at least twice. Which player is more likely to win her bet?

39. In the gambler's ruin (Example 2.27), suppose that Ann and Bob have total stakes N and agree before starting to play at most k games. Let $f_{a,k}$ be the probability that Ann will go broke when she starts with a and Bob starts with $(N - a)$. Notice that $f_{0,k} = 1$ and $f_{a,k} = 0$ when $a > k$.

a. Find an equation for $f_{a,k}$ in terms of $f_{a+1,k-1}$ and $f_{a-1,k-1}$.

(HINT: Use conditional probability based on the outcome of the first trial.)

b. Use the recursion in part (a) and the boundary conditions to calculate $f_{2,3}$. (Assume that $N \geq 4$.)

c. Give a direct argument for the answer in part (b).

40. You always bet \$1 on red at roulette (win probability 18/38 each round). You get back \$2 if red shows and lose your bet otherwise.

    a. Your initial fortune is \$10 and you decide to continue betting until your fortune increases to \$15 or you go broke. What is the chance that you will win your goal?

    b. Suppose that you achieve your goal in part (a) and decide to continue playing until your fortune increases to \$20. Now what is the chance that you will win your new goal before going broke?

    c. Your initial fortune is \$10 and you decide to continue betting until your fortune increases to \$20 or you go broke. What is the chance that you will win your goal? Explain the relation between this probability and parts (a) and (b).

41. *(Simulation)* In Example 2.1, use column 3 of the random number table (Table 4 at the back of the book) to simulate fifty rolls of a fair die. Compare your results with the simulations coming from columns 1 and 2.

42. *(Simulation)* In the waiting game (Example 2.2), suppose that you roll the die until 1 appears three times. Write a computer simulation of this game, and plot the relative frequencies of the number of rolls required when you simulate 10, 100, and 1000 repetitions of the game.

43. *(Simulation)* In the birthday problem (Example 2.4), write a computer simulation program that determines the relative frequency that three or more people in a room of twenty-five people have the same birthday. Run the simulation for 10, 50, 100, 500, and 1000 rooms, to obtain estimates for the theoretical probability of this event.

44. *(Simulation)* In the mixed-up mail problem (Examples 2.5 and 2.9), write a computer simulation program that determines the relative frequency of exactly k matches between addresses and envelopes, for $0 \leq k \leq N$. Plot the results of the simulation for 10, 100, and 1000 trials.

45. *(Simulation)* Write a simulation program that estimates the area enclosed by the ellipse $x^2 + \kappa^2 y^2 = 1$, where κ is a parameter in the range $0 < \kappa \leq 1$ (see Examples 2.10 and 2.12). Compare the simulation estimates with the exact value π/κ for $\kappa = 0.5, 0.1, 0.05$.

Chapter 3

Distributions and Expectations of Random Variables

Mathematical models of such varied situations as an industrial production system, the checkout lines at a supermarket, or a computer network are based on **observable data**. Examples of such data are the number of defective items produced per day, the number of customers in the line at some moment, or the length of time a computer terminal is in use during the day.

The common element shared by all the models that we shall study is the occurrence of **random events** which influence the working of the system described by the model. The number of defective items manufactured depends on unpredictable failures in the production process; the length of the checkout line fluctuates because the customers arrive randomly and may have few or many items to be tallied. Computer terminal users may be running a short program or a long statistical analysis. Thus the observable quantities of the systems will be random variables.

In this chapter we develop the basic analytical techniques (distribution functions, densities, expectations) used in the study of random variables. We also find how to use computer-generated **pseudo-random numbers** to simulate the sample values of random variables. The sample-space formulation of probability theory from Chapter 2 provides a conceptual framework for these analytical computations. The sample spaces themselves will largely remain in the background.

3.1 Distribution Function of a Random Variable

Recall that an ordinary numerical function f is defined by a rule that calculates a number $f(t)$, for any given value t of the independent variable. In the case of a random variable X associated with a probability model, the point of view is somewhat different. Although X is a function defined on the sample space S for the model, nature (or a random number generator) chooses the outcome ω of the random experiment; we only observe the values $X(\omega)$. By using the probability model, however, we can answer such questions as "What is the probability that the observed value of X is in the range $1 < X < 2$?" We shall write $\{1 < X < 2\}$ to denote this event. In terms of the sample space S, this event corresponds to the set of outcomes

$$\{\omega \in S : 1 < X(\omega) < 2\}$$

The abbreviated notation is very convenient since we shall generally be describing events in terms of random variables—without explicitly mentioning any sample space.

Definition 1. *The* **cumulative distribution function** *of a random variable X is the function*

$$F_X(a) = P\{X \le a\}$$

defined for all real numbers a.

By the normalization of probabilities, the values of the function F_X are between 0 and 1. Using F_X, we can calculate

$$P\{a < X \le b\} = F_X(b) - F_X(a) \tag{3.1}$$

for any pair of numbers $a < b$. To see this, observe that

$$\{X \le b\} = \{X \le a\} \cap \{a < X \le b\} \qquad \text{(disjoint union)}$$

so that, by the Addition Axiom for probabilities,

$$F_X(b) = F_X(a) + P\{a < X \le b\}$$

Rearranging terms, we obtain formula (3.1).

To answer other probability questions about X using F_X, we next observe that F_X is monotone increasing and right-continuous:

$$\lim_{h \to 0+} F_X(b+h) = F_X(b) \tag{3.2}$$

The monotonicity property follows by the same argument that we used to verify formula (3.1). As for the continuity property, let $b_1 > b_2 > \cdots > b_n > \cdots$ be any decreasing sequence of numbers with limit b. The sets $\{X \le b_1\} \supset \{X \le b_2\} \supset \cdots$ are likewise decreasing. Since $x \le b$ if and only if $x \le b_n$ for all n, we have

$$\{X \le b\} = \bigcap_{n=1}^{\infty} \{X \le b_n\}$$

Hence the Continuity Axiom for probabilities implies that

$$P\{X \le b\} = \lim_{n \to \infty} P\{X \le b_n\}$$

This proves (3.2). The same argument can be used to show that

$$P\{X < b\} = \lim_{h \to 0+} F_X(b-h) \tag{3.3}$$

and that F_X satisfies the **normalization condition**

$$\lim_{b \to \infty} F_X(b) = 1, \qquad \lim_{b \to -\infty} F_X(b) = 0 \tag{3.4}$$

Thus all probabilistic questions about X can be answered, in principle, if we know the function F_X. If X and Y are random variables such that $F_X(b) = F_Y(b)$ for all real numbers b, then we say that X and Y are **identically distributed**. This does *not* mean that X and Y are the same random variable.

3.2 Discrete Random Variables

We shall call a random variable X **discrete** if the set of possible values of X is countable (this includes the case that X takes on only a finite number of different values). For a discrete random variable X we define its **probability mass function** as

$$p_X(x) = P\{ X = x \}$$

In this formula x can be an arbitrary real number; however, $p_X(x) = 0$ if the random variable X never takes on the value x.

Example 3.1

Suppose X takes on the values $1, 2, 3, 4, 5$ with respective probabilities 0.1, 0.2, 0.3, 0.25, 0.15. The graph of the probability mass function p_X is shown in Figure 3.1.

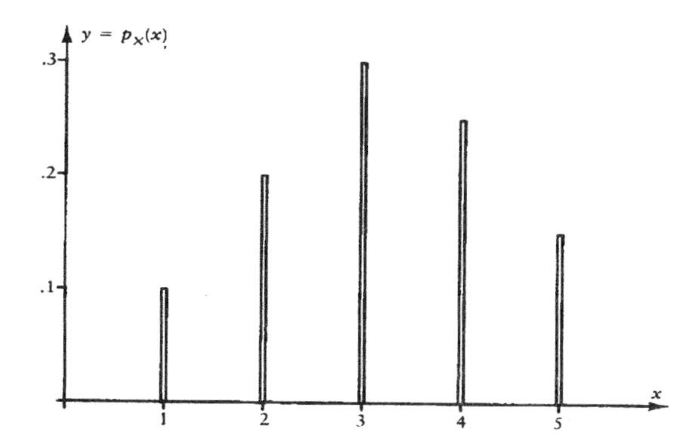

Figure 3.1: Probability Mass Function

The sum of the heights of the stalks in Figure 3.1 is equal to 1. By the Normalization Axiom for probabilities, this property holds for any discrete random variable X. If the values of X are $\{x_1, x_2, \ldots\}$, then

$$\sum_{i=1}^{\infty} p_X(x_i) = 1 \tag{3.5}$$

The cumulative distribution function of X can be obtained from the probability mass function by summation:

$$F_X(a) = \sum_{x \le a} p_X(x) \tag{3.6}$$

The cumulative distribution function F_X for the random variable X as above yields the staircase-shaped graph shown in Figure 3.2, with horizontal steps occurring at each of the values that X can take on. ■

It is always possible to *define* a discrete random variable X by the procedure of Example 3.1. Start with any function $p(x) \ge 0$ (perhaps determined by empirical

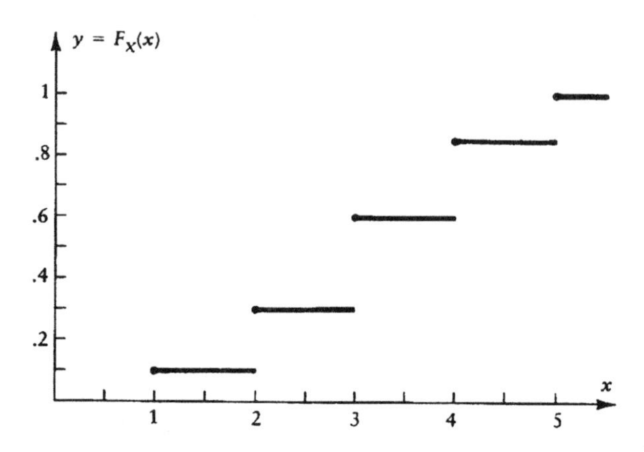

Figure 3.2: Cumulative Distribution Function

frequency data) that is 0 except on a countable set $\{x_1, x_2, ...\}$ and that satisfies formula (3.5); then calculate the probabilities for X in terms of p by summation, as in formula (3.6). In practice, however, most discrete random variables that we will use occur naturally in a particular stochastic model, and the formula for the probability mass function has to be derived using information from the model. Some frequently-occuring random variables are described in Examples 3.2 through 3.6.

Example 3.2 Bernoulli Random Variable

Let X have values 0 and 1, and take $p_X(1) = p$ and $p_X(0) = q$. Here, p is a number between 0 and 1, and $q = 1 - p$. The event $\{X = 1\}$ is called a *success* and occurs with probability p. The event $\{X = 0\}$ is called a *failure* and occurs with probability q. ■

Example 3.3 Binomial Random Variable

Let X be the number of successes in n independent Bernoulli trials, with parameter p. The random variable X takes values $0, 1, \dots, n$ with probabilities

$$p_X(k) = \binom{n}{k} p^k q^{n-k} \tag{3.7}$$

(see Example 2.22), where $q = 1 - p$. In this case, the normalization condition expressed in formula (3.5) holds by the binomial expansion of $(p+q)^n$, since $p+q = 1$.

The probability mass function of X for the values $p = \frac{1}{2}$ and $n = 4$ is shown in Figure 3.3. ■

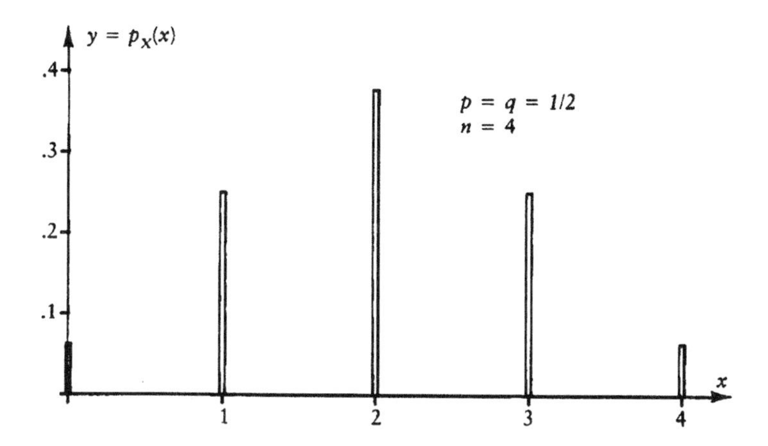

Figure 3.3: Binomial Probabilities

Example 3.4 Geometric Random Variable

Let the random variable N be the number of trials to obtain the first success in a sequence of Bernoulli trials. Thus N takes on the values $1, 2, \ldots$. Since the event $\{N = k\}$ means that failure occurred in trials $1, \ldots, k-1$ and success occurred in trial k, it follows by the independence of successive trials that

$$p_N(k) = q^{k-1}p \tag{3.8}$$

Let us check the normalization condition (formula (3.5)). This condition doesn't follow immediately from the verbal definition of N, since it is conceivable that an unending succession of failures might occur in the model. By the formula for the sum of a geometric series, we have

$$\sum_{k=1}^{\infty} q^{k-1}p = \frac{p}{1-q} = 1$$

Since the event $\{N < \infty\}$ is the union of the mutually exclusive events $\{N = k\}$, for $k = 1, 2, \ldots$, this equation can also be written as

$$P\{N < \infty\} = 1$$

The graph of p_N is shown in Figure 3.4 for $p = \frac{1}{2}$. Of course, the graph continues indefinitely to the right, but the spikes get small very rapidly. ■

If we count the number of failures before the first success we obtain a random variable X that is called the **modified geometric random variable**. The possible values of X are $0, 1, 2, \ldots$. The probability mass function is

$$p_X(k) = q^k p$$

Obviously $X + 1$ is a geometric random variable, since the count goes up by one if we include the first success.

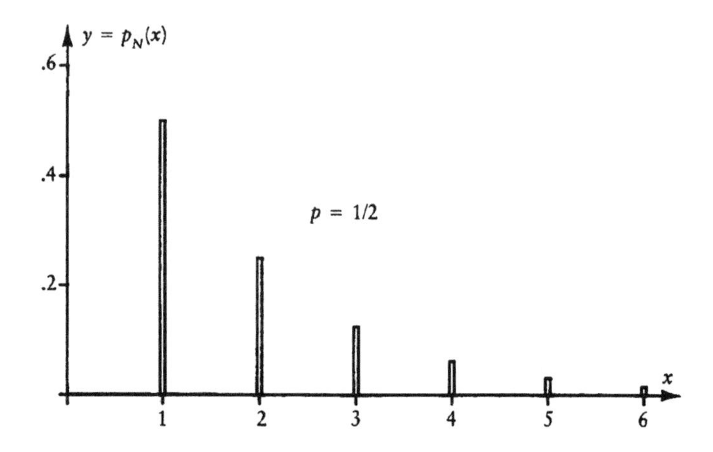

Figure 3.4: Geometric Mass Function

Example 3.5 Negative Binomial Random Variable

The geometric random variable is part of the following family of random variables. We fix a positive integer r, and we let S be the number of Bernoulli trials needed to obtain r successes. The possible values for S are $r, r+1, r+2, \ldots$, and the probability mass function is

$$p_S(k) = \binom{k-1}{r-1} q^{k-r} p^r \tag{3.9}$$

To understand how this formula is derived, observe that the event

$$\{S = k\} = \{r - 1 \text{ successes in trials } 1, \ldots, k - 1\} \cup \{\text{Success in trial } k\}$$

Hence by independence of successive trials,

$$P\{S = k\} = P\{r - 1 \text{ successes in } k - 1 \text{ trials}\} \cdots P\{\text{ Success on } k\text{th trial}\}$$

In Example 3.3 we calculated the probability on the right side of this formula, and this yields equation (3.9). The graph of p_X when $p = \frac{1}{2}$ and $r = 3$ is shown in Figure 3.5. The graph continues indefinitely to the right, with smaller and smaller spikes. Notice the hump that appears, reaching its highest point at $x = 4$ and $x = 5$, is not present in the geometric mass function. The name **negative binomial** is traditional—but not very suggestive—for this family of random variables. A better name would be the **discrete waiting–time random variable**, as we shall see in Chapter 6. ■

Example 3.6 Poisson Random Variable

We define a random variable X that takes on values $0, 1, 2, \ldots$ by using the probability mass function

$$p_X(k) = \frac{\lambda^k}{k!} e^{-\lambda} \tag{3.10}$$

From the power series for the exponential function, we see that p_X satisfies equation (3.5), and hence X is a properly normalized random variable. The primary motivation for studying this family of random variables will become evident in Chapter 7. ■

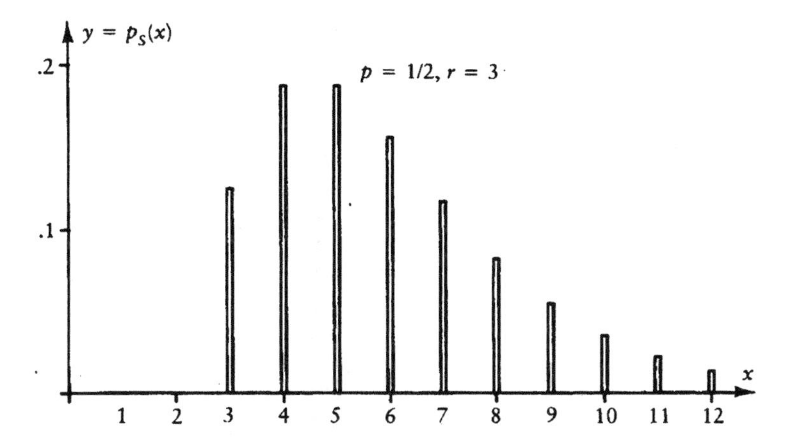

Figure 3.5: Negative Binomial Mass Function

3.3 Continuous Random Variables

For discrete random variables, we obtain the distribution function by summing values of the mass function. Replacing summation by integration, we obtain the notion of a **continuous random variable.** We shall call a random variable X *continuous* if there is a nonnegative function $f(x)$ such that the cumulative distribution function F_X can be calculated by integrating the function f:

$$F_X(a) = \int_{-\infty}^{a} f(x)\,dx \tag{3.11}$$

The function $f(x)$ is called the **probability density** for the random variable X and is often written as $f_X(x)$. (In this notation, the subscript X is just a label; the variable of integration is x.)

By the Normalization Axiom for probabilities, the function f must be nonnegative everywhere and must satisfy

$$\int_{-\infty}^{\infty} f_X(x)\,dx = 1 \tag{3.12}$$

since the integral represents $P\{-\infty < X < \infty\}$. To calculate the probability that an observed value of X lies in the interval (a, b), we simply integrate the density function over this interval:

$$P\{a < X < b\} = \int_{a}^{b} f_X(x)\,dx \tag{3.13}$$

The density function f_X does not have to be continuous—although in all the cases we shall encounter, it will be at least piecewise continuous. The cumulative distribution function F_X, however, is automatically continuous. This has the consequence that

$$P\{X = x\} = 0$$

for every fixed value of x, so the events

$$\{a \le X < b\}, \quad \{a < X \le b\}, \quad \text{and} \quad \{a \le X \le b\}$$

all have the same probability, given by integral (3.13).

Assuming that the density function is piecewise continuous, we can obtain it from the cumulative distribution function by differentiation:

$$f_X(b) = \frac{d}{db} F_X(b) \tag{3.14}$$

(This is a consequence of equation (3.11) and the Fundamental Theorem of Calculus, which asserts that equation (3.14) holds at every point b at which f is continuous.)

To define a continuous random variable X, we can use any function $f(x) \ge 0$ that is integrable and satisfies the normalization condition (3.12); then we calculate the probabilities for X in terms of f by integration, as in (3.13). Following are some examples of continuous random variables that we will be encountering repeatedly. Later we shall see how these random variables arise naturally in stochastic models.

Example 3.7 Uniform Random Variable

Let $f(x) = 1/(\beta - \alpha)$ if x is in the interval $[\alpha, \beta]$, and 0 otherwise. Clearly $f(x)$ satisfies the normalization condition (3.12). If the random variable X has the uniform density $f(x)$, then for any interval $I \subset [\alpha, \beta]$,

$$P\{X \in I\} = \frac{\text{Length}(I)}{\beta - \alpha}$$

When $\alpha = 0$ and $\beta = 1$, sample values of X can be generated using the random number key on a calculator or the RND function in a programming language, as we have already illustrated in Chapter 2. ■

Example 3.8 Exponential Random Variable

Let $f(x) = \lambda e^{-\lambda x}$ for $x \ge 0$, and 0 otherwise. It has the familiar graph shown in Figure 3.6 (for the value $\lambda = 1$). Notice the similarity between this graph and the graph of the discrete geometric mass function shown in Figure 3.4. We integrate f to obtain the cumulative distribution function

$$F(a) = \int_0^a \lambda e^{-\lambda x}\, dx = 1 - e^{-\lambda a}$$

for $a \ge 0$, and $F(a) = 0$ if $a < 0$. In particular, we see that $\lim_{a\to\infty} F(a) = 1$, so that the normalization condition stated in equation (3.12) is satisfied. ■

Example 3.9 Gamma Random Variable

Generalizing Example 3.8, we take

$$f(x) = c(\lambda x)^{r-1} \lambda e^{-\lambda x} \tag{3.15}$$

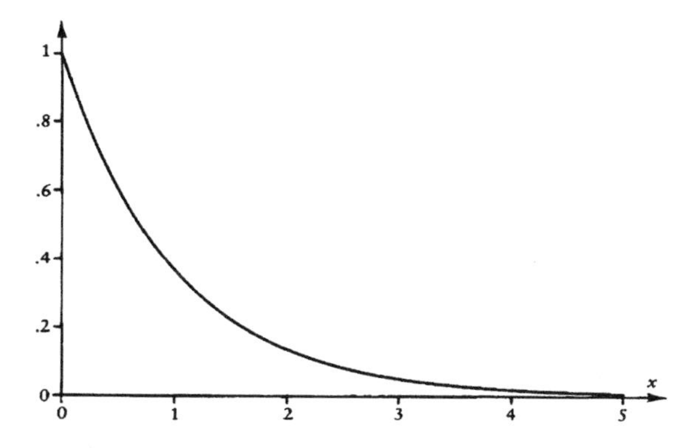

Figure 3.6: Exponential Density

for $x \geq 0$, and $f(x) = 0$ for $x < 0$. Here c is a normalizing constant that we must choose to satisfy equation (3.12). The case $r = 1$ is the exponential random variable. (We shall see in Chapter 4 how n observations of an exponential random variable naturally lead to a gamma random variable with $r = n$.)

To determine the normalizing constant c, we make a change of variable $y = \lambda x$ in integral (3.12). This yields the formula

$$\int_0^\infty f(x)\,dx = c \int_0^\infty y^{r-1} e^{-y}\,dy$$

We define the **gamma function** as

$$\Gamma(r) = \int_0^\infty y^{r-1} e^{-y}\,dy \tag{3.16}$$

By definition the normalizing constant c in formula (3.15) is then $1/\Gamma(r)$. To evaluate this constant numerically, we integrate by parts in (3.16) and get $\Gamma(r+1) = r\Gamma(r)$.

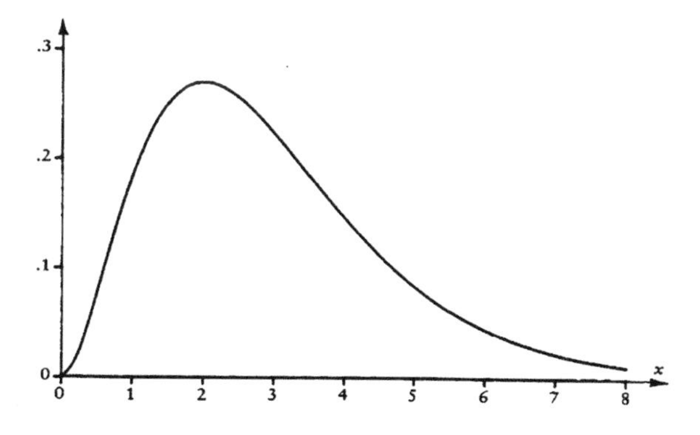

Figure 3.7: Gamma Density

Since $\Gamma(1) = 1$, we conclude that for integer values $r = n$, we have

$$\Gamma(n) = (n-1)! \tag{3.17}$$

(For nonintegral values of r, we can calculate $\Gamma(r)$ by numerical integration in the interval $0 < r < 1$ and use the recursion to obtain $\Gamma(r+1)$, $\Gamma(r+2)$, and so on.) The graph of this density function is given in Fig. 3.7 (with $\lambda = 1$ and $r = 3$). Notice its similarity to the graph of the discrete mass function for the negative binomial random variable in Figure 3.5. ■

Example 3.10 Normal Random Variable

The function

$$\phi(z) = c \exp\left(-\frac{z^2}{2}\right)$$

(where $c > 0$ is the normalizing constant) is nonnegative and symmetrical around 0. It has the bell-shaped graph shown in Figure 3.8.

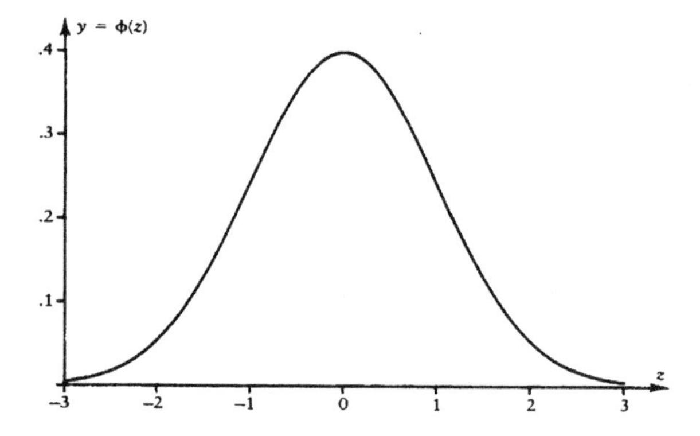

Figure 3.8: Standard Normal Density

Notice the similarity between this graph and that of the binomial mass function in Figure 3.3. We can determine the value for c (which turns out to be approximately 0.39894), so that the area under this graph is 1, by the following trick: Write

$$\left(\int_{-\infty}^{\infty} \phi(z)\,dz\right)^2 = c^2 \int_{-\infty}^{\infty}\int_{-\infty}^{\infty} e^{-(x^2+y^2)/2}\,dxdy$$

Here we have replaced the integration variable z with x in one factor and with y in the other. The product of the two single integrals becomes a double integral that is easy to evaluate if we change to polar coordinates (r, θ), where $r^2 = x^2 + y^2$. In this change of variables the element of area $dxdy$ is replaced by $r\,dr d\theta$:

$$\begin{aligned}\int_{-\infty}^{\infty}\int_{-\infty}^{\infty} e^{-(x^2+y^2)/2}\,dxdy &= \int_0^{2\pi}\int_0^{\infty} e^{-r^2/2}\,rdrd\theta \\ &= -2\pi \cdot e^{-r^2/2}\Big|_{r=0}^{r=\infty} = 2\pi\end{aligned}$$

Thus the value of the normalizing constant is

$$c = (2\pi)^{-1/2}$$

The random variable Z with probability density $\phi(z)$ is called the **standard normal random variable**. The cumulative distribution function of Z is denoted by Φ:

$$\Phi(a) = (2\pi)^{-1/2} \int_{-\infty}^{a} \exp\left(-\frac{z^2}{2}\right) dz$$

Values of $\Phi(a)$ for $0 \leq a \leq 4.99$ are given in Table 3 at the end of the book. Because of the symmetry of the density $\phi(z)$, we can calculate the probability that Z falls in an interval symmetric around 0 in terms of F as follows:

$$\begin{aligned} P\{-a \leq Z \leq a\} &= P\{Z \leq a\} - P\{Z < -a\} \\ &= P\{Z \leq a\} - P\{Z > a\} \\ &= \Phi(a) - \big(1 - \Phi(a)\big) \\ &= 2\Phi(a) - 1 \end{aligned}$$

for any value of $a \geq 0$.

More generally, let $-\infty < \mu < \infty$ and $\sigma > 0$, and define

$$f(x) = \frac{1}{\sigma(2\pi)^{1/2}} \exp -\frac{1}{2}\left(\frac{x-\mu}{\sigma}\right)^2$$

Then $f(x)$ is the probability density obtained from $\phi(z)$ by the scaling change of variable $x = \sigma z + \mu$. (The factor $1/\sigma$ that multiplies the exponential comes from the substitution rule in integrals: $dz = dx/\sigma$, under this change of variable.) The random variable X, with probability density $f(x)$, is called a **normal random variable with parameters** μ, σ. ■

3.4 Expectation of a Random Variable

In Chapter 2, we examined many examples of events associated with stochastic models. For each event A—such as a complete mismatch of letters and envelopes in the mixed-up mail model—the key problem was to calculate the number $P(A)$. This number gave us a quantitative measure of the significance of the event A. We also obtained estimates for this number by simulating many trials of the model and calculating the relative frequency with which A occurred in the simulation.

Now we make a similar calculation for a random variable X in a stochastic model. We want a fixed *number* μ that represents the average value of X. This number should have the property that, if we simulate n trials of the model and obtain observed values $x_1, \ldots, x_n$ for X, then the **sample average**

$$\frac{x_1 + x_2 + \cdots + x_n}{n}$$

should be approximately μ when n is large. This means that the number μ is the fair stake associated with X: if we pay \$$\mu$ and win \$$X$ on each trial, our net cost in the long run will be zero. If there is such a number μ, we will call it the **average value** or **expectation** of X, and we write $\mu = \mu_X = E[X]$. To find a general formula for μ, we consider first the following special case.

Example 3.11

Let A be an event in a probability model, and define the **indicator random variable** for A by

$$I_A = \begin{cases} 1 & \text{if } A \text{ occurs} \\ 0 & \text{if } A \text{ doesn't occur} \end{cases}$$

Then I_A is a Bernoulli random variable with parameter $p = P(A)$. If we set $X = I_A$ and simulate n trials of the model, then the sample average

$$\frac{x_1 + x_2 + \cdots + x_n}{n} = \frac{\text{Number of trials where } A \text{ occurs}}{\text{Total number of trials}}$$

is the relative frequency of A. From the simulation examples in Chapter 2, we know that this relative frequency is approximately $P(A)$ when n is large (this will be proved mathematically in Section 4.6). Therefore we should *define* the expectation of an indicator random variable to be

$$E[I_A] = P(A) \tag{3.18}$$

A certain abuse of language occurs here: the number $P(A)$ is not one of the values taken on by the random variable I_A, unless the event A has probability 0 or 1. ■

Example 3.11 gives the first step in the general definition of $E[X]$. Suppose next that, in a stochastic model, X is a discrete random variable that takes on the distinct values $c_1, c_2, \ldots, c_d$, with probabilities $p_1, p_2, \ldots, p_d$. Write A_i for the event $\{X = c_i\}$, so that $P(A_i) = p_i$. Consider a simulation of n trials of the model, and let $x_1, \ldots, x_n$ be the observed values of X. We can calculate the relative frequency f_i with which A_i occurs in the simulation by means of the formula

$$f_i = \frac{\text{Number of } X \text{ values that equal } c_i}{n} \tag{3.19}$$

Hence the sample average of the observed values of X can be written in terms of these relative frequencies as

$$\frac{x_1 + x_2 + \cdots + x_n}{n} = c_1 f_1 + c_2 f_2 + \cdots + c_d f_d \tag{3.20}$$

(To see this, group the repeated x values together on the left side of equation (3.20); the number of times that c_i appears is nf_i, by formula (3.19).) From Example 3.11, we know that $f_i \approx p_i$ when the number of trials is large. Hence, to obtain the fundamental property of the expected value we must *define*

$$E[X] = \sum_{i=1}^{d} c_i p_i \tag{3.21}$$

Notice that, when X is an indicator random variable for an event A_1, then $c_1 = 1$, $p_1 = P(A_1)$, and definition (3.21) reduces to definition (3.18). In terms of the probability mass function p_X of X, we can write definition (3.21) as

$$E[X] = \sum_c c p_X(c) \tag{3.22}$$

For a general discrete random variable X we use formula (3.22) as the definition of $E[X]$. Of course, only the countable set of numbers c such that $p_X(c) > 0$ actually occur in this sum, and $E[X]$ is only defined if this series is absolutely convergent. (We shall prove in Section 4.6 that this definition of $E[X]$ satisfies the fundamental large sample average property—the Law of Large Numbers.)

The examples of random variables introduced in the previous sections all involve a certain number of parameters; when the parameters are given specific numerical values, then the random variable is completely specified. Certain natural parameters are associated with most random variables. The basic natural parameter is the **expectation** $E[X]$, which we will also call the **mean** of X. The problem then arises of calculating $E[X]$ in terms of the given parameters.

Example 3.12

Let X be a geometric random variable with parameter p. We can calculate $E[X]$ as follows:

$$\begin{aligned} E[X] &= p\sum_{n=1}^{\infty} nq^{n-1} = p\frac{d}{dq}\left(\sum_{n=0}^{\infty} q^n\right) = p\frac{d}{dq}\left(\frac{1}{1-q}\right) \\ &= \frac{p}{(1-q)^2} = \frac{1}{p} \end{aligned}$$

Here, we have used the formula for the sum of a geometric series. (The trick of writing the series for $E[X]$ as the derivative of a known power series will be turned into a systematic technique—the moment-generating function—in Chapter 4.) This answer is reasonable from the probabilistic definition of X. As an example, for a fair die, $p = \frac{1}{6}$, so on average we can expect that $1/p = 6$ tosses will be required to roll a 1 for the first time. ■

Suppose X is a continuous random variable, with probability density function $f(x)$. By analogy with formula (3.22), we define the expectation of X by the integral

$$E[X] = \int_{-\infty}^{\infty} x f(x)\, dx \tag{3.23}$$

provided this integral is absolutely convergent. (We shall prove in Section 4.6 that this definition of $E[X]$ satisfies the Law of Large Numbers and therefore $E[X]$ can be estimated from large sample averages.) A mechanical analogy is helpful in understanding this definition: if $f(x)$ is viewed as the **mass density** of a slender rod lying on the x-axis, the point $\mu = \mu_X$ on the x-axis is the **balance point** (center of mass) of the rod. This follows because the turning moment of the rod around the point μ is zero:

$$\begin{aligned} \int_{-\infty}^{\infty} (x-\mu) f(x)\, dx &= \int_{-\infty}^{\infty} x f(x)\, dx - \mu \int_{-\infty}^{\infty} f(x)\, dx \\ &= \mu - \mu = 0 \end{aligned}$$

In an evaluation of $E[X]$ for a particular random variable, the appropriate limits of integration depend on the range of x-values for which the density f_X is not zero.

Example 3.13

Suppose X is an exponential random variable with parameter λ. The effective limits of integration are from 0 to ∞, and we can calculate $E[X]$ using integration by parts:

$$E[X] = \int_0^\infty \lambda x e^{-\lambda x}\, dx = -\, x e^{-\lambda x}\Big|_{x=0}^{x=\infty} + \int_0^\infty e^{-\lambda x}\, dx = 1/\lambda$$

■

Example 3.14

Suppose that X is a random variable that takes on only nonnegative values. We can calculate $E[X]$ in this situation by integrating the tail of the probability distribution of X:

$$E[X] = \int_0^\infty P\{X > a\}\, da \tag{3.24}$$

To see that formula (3.24) is valid, suppose that X is a continuous random variable with probability density function $f(x)$. Then

$$\int_0^\infty P\{X > a\}\, da = \int_0^\infty \left(\int_a^\infty f(x)\, dx \right) da$$

If we interchange the order of integration in this double integral, we get

$$\int_0^\infty \left(\int_0^x da \right) f(x)\, dx = \int_0^\infty x f(x)\, dx = E[X]$$

as claimed. A similar calculation works (replacing integrals by sums) when X is discrete.

As an example for formula (3.24), let X be an exponential random variable with parameter λ. Then from Example 3.8,

$$P\{X > a\} = e^{-\lambda a}$$

Hence by formula (3.24)

$$E[X] = \int_0^\infty e^{-\lambda a}\, da = -\frac{1}{\lambda} e^{-\lambda a}\Big|_{a=0}^{a=\infty} = 1/\lambda$$

which agrees with the calculation in Example 3.13.

3.5 Functions of a Random Variable

An observed value of a random variable X is a number x, so it makes sense to calculate other numbers from it, such as $2x + 5$, x^2, e^x, or (more generally) $g(x)$, where g is an ordinary function of a real variable. The random distribution of the numbers x determined by the probability mass function or density of X then gives rise to a random distribution of the numbers $g(x)$, and we write $g(X)$ to denote the corresponding random variable. The probability distribution of $g(X)$ can in principle be calculated from the probability distribution of X, since the event $\{g(X) \leq a\}$ is the same as the event $\{X \in I_a\}$, where $I_a = \{x : g(x) \leq a\}$. We illustrate this in some examples.

Example 3.15 Scaling Transformations

Suppose $g(x) = \alpha x + \beta$ is a linear function, where α and β are constants and $\alpha > 0$. The cumulative probability distribution of the random variable $Y = g(X)$ is easily calculated. For any constant c,

$$\begin{aligned} P\{Y \leq c\} &= P\{\alpha X + \beta \leq c\} = P\left\{X \leq \frac{c-\beta}{\alpha}\right\} \\ &= F_X\left(\frac{c-\beta}{\alpha}\right) \end{aligned}$$

For example, if X is uniformly distributed on (0, 1), then Y is uniformly distributed on $(\beta, \alpha + \beta)$. This follows from preceding formula, since the values of Y are in the interval $(\beta, \alpha + \beta)$ and since

$$F_Y(y) = \frac{y-\beta}{\alpha} \quad \text{for } \beta \leq y \leq \alpha + \beta$$

■

Example 3.16 Chi-squared Random Variable

Let Z be a standard normal random variable and let $Y = Z^2$. Then Y is called a **chi-squared random variable** with *one degree of freedom*. If $b > 0$ then the event $\{Y \leq b\}$ is the same as $\{-b^{1/2} \leq Z \leq b^{1/2}\}$. Thus,

$$P\{Y \leq b\} = P\{-b^{1/2} \leq Z \leq b^{1/2}\}$$

By Example 3.10, we can express this probability in terms of the cumulative distribution function $\Phi(a)$ for Z, by

$$P\{Y \leq b\} = 2\Phi(b^{1/2}) - 1$$

The probabilities for Y can then be calculated from this equation using Table 3 (at the back of the book) for the standard normal random variable. For example,

$$P\{Y \leq 4\} = 2\Phi(2) - 1 = 2(0.9772) - 1 = 0.9544$$

We can obtain the probability density function for Y by differentiating this equation for $F_Y(b)$ with respect to b, and then using the chain rule:

$$\begin{aligned} f_Y(b) &= 2\frac{d}{db}\Phi(b^{1/2}) = 2\phi(b^{1/2}) \cdot \frac{d}{db}b^{1/2} \\ &= \phi(b^{1/2}) \cdot b^{-1/2} \end{aligned}$$

where $\phi(x)$ is the density for Z. Substituting the formula for ϕ from Example 3.10, we find that

$$f_Y(b) = cb^{-1/2}e^{-b/2}$$

where the constant $c = (2\pi)^{-1/2}$. We can recognize this density function as a *gamma* density, with parameters $\lambda = \frac{1}{2}$ and $r = \frac{1}{2}$. ■

Example 3.17 Cauchy Random Variable

Let Θ be a random angle, uniformly distributed on the interval $(-\pi/2, \pi/2)$. Set $X = \tan\Theta$. Then X takes all real values. Just as in Example 3.16, we can determine the distribution function of X by a direct probability calculation:

$$\begin{aligned} P\{X \le b\} &= P\{\tan\Theta \le b\} = P\{-\pi/2 < \Theta \le \arctan b\} \\ &= \frac{1}{\pi}\left[\arctan(b) - \frac{\pi}{2}\right] \end{aligned}$$

(Here the factor $1/\pi$ is the normalizing constant for the uniform density of Θ.) If we now differentiate this equation relative to the variable b, we obtain the probability density of X:

$$f_X(b) = \frac{1}{\pi(1+b^2)}$$

This is called the **Cauchy density**.

The graph of the Cauchy density is bell-shaped and symmetrical around 0, just as the normal density (Fig. 3.8) is. However, $1/(1+x^2)$ decreases to zero very slowly as $x \to \infty$, while the decrease of the normal density is extremely rapid. This means that sample values of X tend to be very dispersed, and consequently X does not have an average value in a meaningful sense. To see this, remember that, for any random variable X, the number $E[X]$ is only defined when the sum or integral involved converges *absolutely*—that is, when the non-negative random variable $|X|$ has finite expectation. For the Cauchy random variable, we have

$$E[|X|] = \lim_{b\to\infty}\int_{-b}^{b} |x| f_X(x)dx$$

Using the symmetry of the density function f_X around 0, we can evaluate this integral as

$$\frac{1}{\pi}\int_0^b \frac{2x\,dx}{1+x^2} = \frac{1}{\pi}\log(1+b^2)$$

Letting $b \to \infty$, we find that $E[|X|] = \infty$. ■

For more complicated functions $g(x)$, the explicit calculation of the distribution function of $g(X)$ can be quite unwieldy. To calculate $E[g(X)]$, however, we can use the following formula (which has been dubbed the **Law of the Unconscious Statistician**):

$$E[g(X)] = \begin{cases} \sum_x g(x) p_X(x) & (X \text{ discrete}) \\ \int_{-\infty}^{\infty} g(x) f_X(x)\,dx & (\text{X continuous}) \end{cases}$$

Proof of the Law of the Unconscious Statistician When X is a discrete random variable, the proof is easy. We view X as a function $X(\omega)$ on a discrete sample space S. Then $g(X)$ is the function whose value at the sample point ω is $g(X(\omega))$. The definition of expectation in Section 3.4 can be rewritten in terms of S as

$$E[g(X)] = \sum_{\omega \in S} g(X(\omega)) P\{\omega\} \tag{3.25}$$

Next, we group the terms in the sum according to the values of X. If these values are $x_1, x_2, \ldots$, we define

$$A_i = \{\omega \in S \, : \, X(\omega) = x_i\}$$

Since the function $g(X(\omega))$ has the same value $g(x_i)$ for all sample points $\omega \in A_i$, we can write formula (3.25) as a double sum:

$$E[g(X)] = \sum_i \Big(\sum_{\omega \in A_i} P\{\omega\} \Big) g(x_i) = \sum_i g(x_i) P(A_i)$$

But $P(A_i) = p_X(x_i)$, so this last formula is just what we want.

Now consider a continuous random variable X. Suppose first that the function $g(x)$ is nonnegative for all x. Then by Example 3.14,

$$E[g(X)] = \int_0^{\infty} P\{g(X) > y\}\,dy = \int_0^{\infty} \Big(\int_{I_y} f_X(x)\,dx \Big) dy \tag{3.26}$$

where I_y is the set of x values for which $g(x) > y$. Let R be the region in the xy plane between the graph of $y = g(x)$ and the x axis. Then the iterated integral in formula (3.26) is just the double integral of $f_X(x)$ over R, set up by first scanning the region R for fixed y and then integrating on y (see Fig. 3.9, where the horizontal scanning lines have been indicated).

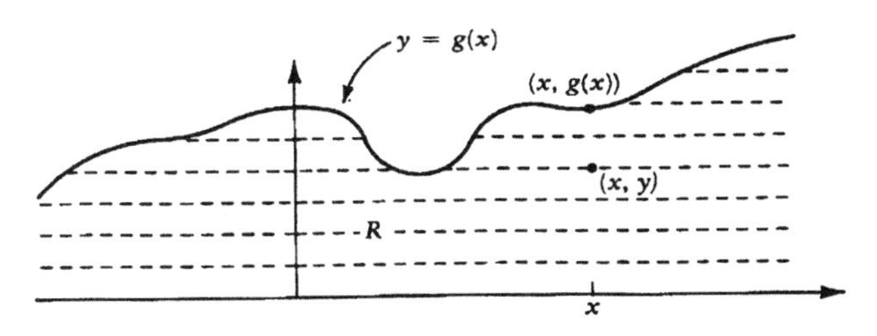

Figure 3.9: Region $R = \{(x, y) \, : \, y < g(x)\}$

If we interchange the order of integration in formula (3.26) the new y limits are between 0 and $g(x)$, and the x limits are from $-\infty$ to ∞:

$$E[g(X)] = \int_{-\infty}^{\infty} \left(\int_{0}^{g(x)} dy \right) f_X(x)\, dx = \int_{-\infty}^{\infty} g(x) f_X(x)\, dx$$

This gives the desired formula when $g \geq 0$. In general, any function g can be split as $g(x) = h(x) - k(x)$, where $h(x)$ and $k(x)$ are non-negative functions. By the linearity of the expectaton of a random variable, we obtain the formula in general.

Example 3.18 Moments of a Random Variable

As a first application of the Law of the Unconscious Statistician, we take a random variable X and form its powers X^n, where n is a nonnegative integer. The expected value of the random variable X^n is called the nth **moment** of X:

$$E[X^n] = \begin{cases} \displaystyle\sum_x x^n p_X(x) & \text{(X discrete)} \\ \displaystyle\int_{-\infty}^{\infty} x^n f_X(x)\, dx & \text{(X continuous)} \end{cases} \tag{3.27}$$

(In formula (3.27) we assume that the sums or integrals are absolutely convergent.) We will discuss moments using moment-generating functions in Chapter 4.

Example 3.19 Estimating a Random Variable

Suppose X is a random variable. *What constant c is the best approximation to X?* To make this question well-posed, we use the **mean-square error**

$$e(c) = E[(X - c)^2]$$

to measure the goodness of the approximation. We calculate $e(c)$ by Law of the Unconscious Statistician:

$$e(c) = \sum_x (x - c)^2 p_X(x) \tag{3.28}$$

when X is discrete (or the analogous integral formula when X is continuous). From formula (3.28), it is clear that $e(c) \geq 0$; expanding the square inside the expectation, we obtain the formula

$$e(c) = E[X^2] - 2cE[X] + c^2 \tag{3.29}$$

Thus $e'(c) = -2E[X] + 2c$, and $e''(c) = 2$. It follows that the value

$$c = E[X]$$

minimizes $e(c)$. Thus, the mean μ_X is the best deterministic estimate of X in the mean-square sense. (Here we are assuming that X has a finite second moment.) ■

The other natural parameter of a random variable X (besides its mean) is its **variance**. The variance of X, which measures the dispersion of values away from the mean μ_X, is defined to be

$$\text{Var}(X) = E[(X - \mu_X)^2] \tag{3.30}$$

(assuming that this expectation exists). We can calculate Var(X) in terms of the first and second moment of X by using formula (3.29) with $c = \mu_X$:

$$\text{Var}(X) = E[X^2] - E[X]^2$$

If Var$(X) = 0$, then, in formula (3.28), with $c = \mu_X$, every term equals zero. Hence, $p_X(x) = 0$ for every value of $x \neq \mu_X$. A similar argument applies for a continuous random variable (if we ignore events of probability zero), and we conclude that *a random variable with zero variance is constant* (with probability 1).

From definition (3.30) is is clear that Var$(X) \geq 0$. It is customary to define the **standard deviation** of X to be

$$\sigma_X = [\text{Var}(X)]^{1/2}$$

which has the same units of measurement as X. In light of Example 3.19, we can think of σ_X as the *random size* of X (non-random quantities thus having random size zero).

After developing some more probabilistic tools in the next chapter, we shall calculate the natural parameters μ_X and σ_X in terms of the given parameters for the families of random variables introduced in Section 3.2 and 3.3.

3.6 Simulation of a Random Variable

We have already seen in Chapter 2 how to use tables of random numbers or a random number generator to obtain simulated relative frequencies for many elementary stochastic models. In this section we determine how to modify the output of a random number generator to obtain simulated values of a random variable.

Example 3.20 Discrete Random Variable

Suppose we want to simulate a discrete random variable X that takes on the values 1, 2, 3, 4, 5 with respective probabilities 0.1, 0.2, 0.3, 0.25, 0.15 (see Example 3.2). We can obtain such a random variable by taking a suitable function $g(U)$ of a uniform $(0, 1)$ random variable U. Define

$$g(u) = \begin{cases} 1 & \text{for } 0 < u \leq 0.1 \\ 2 & \text{for } 0.1 < u \leq 0.3 \\ 3 & \text{for } 0.3 < u \leq 0.6 \\ 4 & \text{for } 0.6 < u \leq 0.85 \\ 5 & \text{for } 0.85 < u < 1 \end{cases}$$

It is easy to verify that the random variable $X = g(U)$ has the desired probability distribution. For example,

$$P\{y = 4\} = P\{0.6 < U \leq 0.85\} = 0.25$$

by the uniform distribution of the values of U. We can generate a sequence of sample values of X by starting at any randomly chosen point in Table 4 (at the back of the book) and moving down the table to determine consecutive u values (take each five-digit entry and place a decimal point to the left). Then $x = g(u)$ is the corresponding sample X value. For example, if we pick the top of column 4 as the starting point, then the first sample value is $x_1 = g(0.02011) = 1$. The next four sample X values determined by going down this column are $5, 5, 4, 2$. ■

The technique used in Example 3.20 obviously applies to any random variable X that takes on n values: divide the interval $(0, 1)$ into n segments corresponding to the values of X, with the length of each segment equal to the probability that X will take on the corresponding value. For a discrete random variable that takes on infinitely many values, a similar method applies (with some additional work), as in the following example:

Example 3.21 Modified Geometric Random Variable

Suppose we try to find a function $g(u)$ so that $X = g(U)$ will have a modified geometric distribution with parameter p, $0 < p < 1$, whenever U is uniform on $(0, 1)$. We start by calculating the tail of the distribution function of X (see Example 3.4)

$$\begin{aligned} P\{X \geq n\} &= \sum_{k=n}^{\infty} q^k p = \frac{pq^n}{1-q} \\ &= q^n = e^{n \log q} \end{aligned}$$

using the formula for the sum of a geometric series (here $q = 1 - p$). Consider the inverse of the function $u = e^{y \log q}$ which is $y = \log u / \log q$. Since we want a function taking integral values, this suggests setting

$$g(u) = \mathrm{INT}\left(\frac{\log u}{\log q}\right)$$

where $\mathrm{INT}(z)$ denotes the integer part of the real number z. The graph of $x = g(u)$ is shown in Figure 3.10, where we have taken the independent variable u on the *vertical* axis, and where $q = e^{-1/2}$.

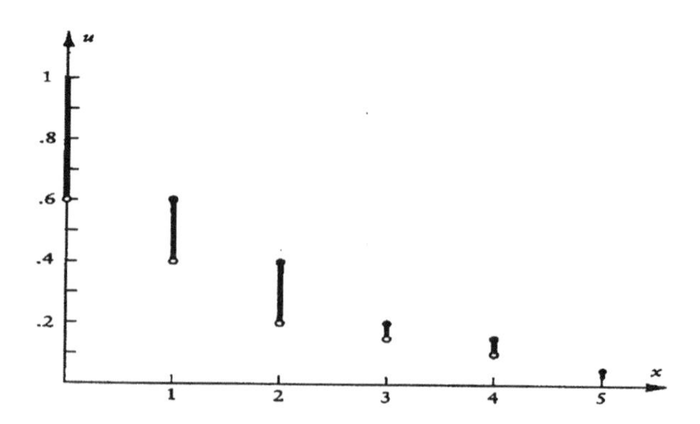

Figure 3.10: Simulation of Modified Geometric Random Variable

We now verify that, if U is a uniform random variable on $(0, 1)$, then the random variable $X = g(U)$ will have a modified geometric distibution. We start by observing that

$$X = n \quad \text{if and only if} \quad n \leq \frac{\log U}{\log q} < n + 1$$

(see Figure 3.10). Since $\log q < 0$, this inequality on $\log U$ is equivalent to

$$\log(q^n) \geq \log U > \log(q^{n+1})$$

Exponentiating the terms in this inequality, we find that

$$X = n \quad \text{if and only if} \quad q^n \geq U > q^{n+1}$$

But the probability that U satisfies these inequalities is

$$q^n - q^{n+1} = q^n(1 - q) = q^n p$$

This is exactly the probability mass function of a modified geometric random variable

To demonstrate this simulation technique, suppose that we take $q = e^{-1/2}$ as in Figure 3.10, so that $g(u) = \text{INT}(-2\log(u))$. If we start at row 6 and move down column 4 of Table 4 to generate values of u, the successive sample X values are $g(0.42751) = 1$, $g(0.69994) = 0$, $g(0.07972) = 5$, $g(0.10281) = 4$, and so on (see Fig. 3.10). By the definition of a modified geometric random variable, these X values simulate the lengths of losing streaks in successive rounds of a game in which our win probability each round is $p = 1 - e^{-1/2} = 0.39347$. ■

To generate sample values of an exponential random variable using a random number generator, we use the following continuous version of Example 3.21.

Lemma 3.1. *If U is a uniformly distributed random variable on the interval $(0, 1)$, and if $\mu > 0$, then the random variable*

$$X = -\mu \log U$$

is exponentially distributed, with mean μ.

Proof Set $\lambda = 1/\mu$. Then $U = e^{-\lambda X}$. Thus, for any number t, the inequality $X > t$ is equivalent to $U < e^{-\lambda t}$. Hence, the function $u = e^{-\lambda x}$ sets up a one-to-one correspondence between the interval (t, ∞) of the x-axis and the interval $(0, e^{-\lambda t})$ of the u-axis (see the bold segments of the two axes in Figure 3.11). Thus the events $\{X > t\}$ and $\{U < e^{-\lambda t}\}$ are the same.

It follows that

$$P\{X > t\} = P\{U < e^{-\lambda t}\} = e^{-\lambda t}$$

if $t \geq 0$. (This probability equals 1 if $t < 0$, since $e^{-\lambda t} > 1$ in that case.) Hence, the cumulative distribution function of X describes an exponential random variable with parameter λ, as claimed. ■

To simulate other continuous random variables we can often use a method similar to that of Lemma 3.1, called the **probability integral transform**. Let $F(x)$ be the cumulative distribution function of a random variable X. Given a value u of a uniform $(0, 1)$ random variable U, we solve the equation

$$u = F(x) \tag{3.31}$$

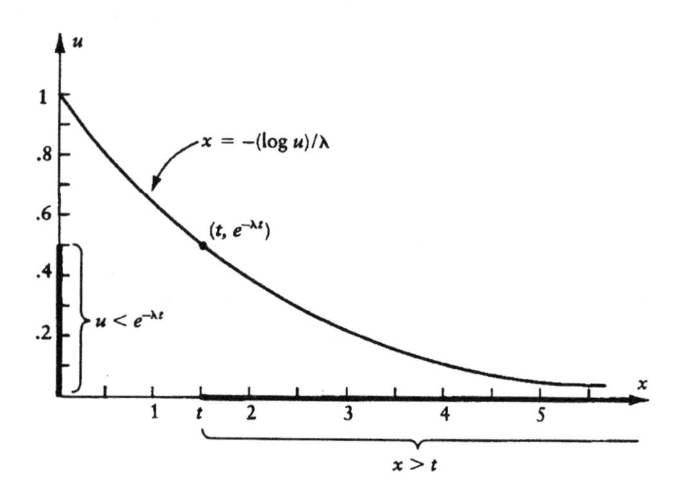

Figure 3.11: Simulation of Exponential Random Variable ($\lambda = 2$)

for x. (We assume that $F(x)$ is strictly monotone increasing between its limiting values 0 and 1, so that equation (3.31) has a unique solution for every u between 0 and 1.) This value of x furnishes the sample value of X.

Example 3.22 Normal Random Variable

Suppose X is a standard normal random variable. The cumulative distribution function $F(x)$ (usually denoted by $\Phi(x)$) has graph shown in Figure 3.12. In this figure, we have labeled the vertical axis u. Given a particular u value, we determine the point x so that (x, u) is on the graph of F.

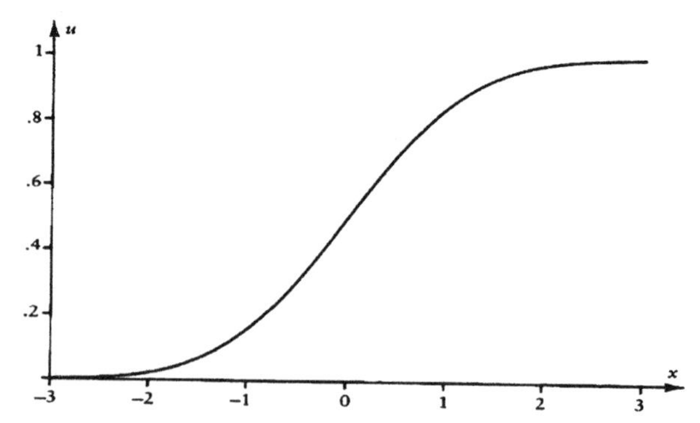

Figure 3.12: Distribution Function of Standard Normal Random Variable

For example, the first entry of column 5 in Table 4 gives $u = 0.81647$. Using linear interpolation in Table 3, we find the corresponding X value to be $x = .9002$ (to four decimal places). As an alternate to this procedure, we could use a table of random values of a standard normal random variable. This example shows how such a table can be constructed from Tables 3 and 4. (In Example 4.27 we will obtain a simpler way of generating sample values of a normal random variable.) ■

We now prove the validity of the probability integral transform method by the same argument we used in Lemma 3.1. Write $x = g(u)$ for the solution to equation (3.31). Our simulation method then produces sample values of the random variable $g(U)$. We must show that $g(U)$ has the same probability distribution as X. Fix b between 0 and 1, and let $a = g(b)$. Then

$$P\{g(U) \leq a\} = P\{U \leq b\} = b$$

since U is uniform on $(0, 1)$ (see Fig. 3.13). But the relation between a and b can also

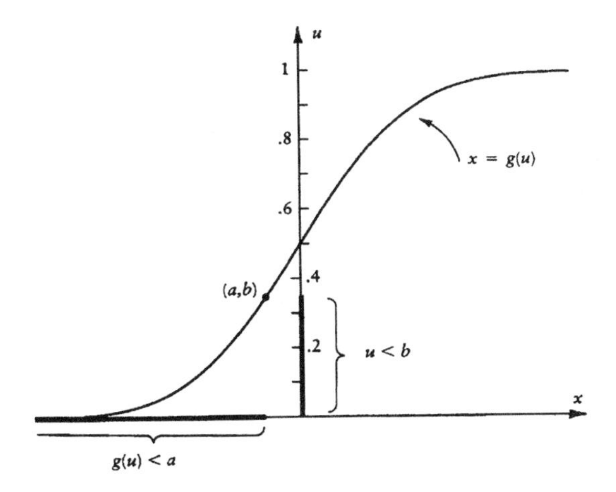

Figure 3.13: Probability Integral Transform

be written as $b = F(a)$. Hence,

$$P\{g(U) \leq a\} = F(a)$$

As b ranges over $(0, 1)$, a ranges over all the possible values of $g(U)$. This shows that the random variable $g(U)$ has cumulative distribution function F.

Exercises

1. Suppose the discrete random variable X takes on the values 0, 1, 3, and 6 with respective probabilities 0.3, 0.2, 0.4, and 0.1. Calculate μ_X, σ_X^2, and plot the cumulative distribution function $F_X(x)$ for $-1 \leq x \leq 7$.

2. If a fair coin is successively flipped, find the probabilities of the following events:

   a. Two heads occur in five flips.

   b. Two heads occur in five flips, *and* the second appearance of heads occurs on the fifth flip.

3. An urn contains ten balls, numbered 1 to 10. A set of three balls is withdrawn at random from the urn. Let X be the smallest of the numbers of the three balls chosen. Determine the range and the probability mass function of the random variable X.

4. Urn I contains one white ball and two red balls. Urn II contains two white balls and one red ball. Do the following experiment: first, draw a ball at random from urn I and place it in urn II; then draw a ball at random from urn II and place it in urn I. Let X be the number of white balls in urn I after the second drawing. Calculate the probability mass function of X.

5. You roll a fair die once or twice, according to the following rules: If the die shows 6 on the first roll, you win \$10 and get to roll again (otherwise, the game is over); if the die shows a 6 on the second roll, you win an additional \$30. Let X be the total amount you win.

   a. Calculate the probability mass function of X.

   b. What is a fair price to pay for playing this game?

6. Let X be a binomial random variable with parameters n, p. Plot the probability mass function $p_X(k) = P\{X = k\}$ for the values $n = 4$, $p = 0.5$ and $n = 4$, $p = 0.7$. Show that, for general values of n, p with $np - q$ not an integer, $p_X(k)$ first increases monotonically and then decreases monotonically as a function of k, and show that $p_X(k)$ has its maximum value when k is the smallest integer greater than $np - q$.

   (HINT: Consider $p_X(k+1)/p_X(k)$.)

7. Suppose that X has a binomial distribution, with parameters $n = 6$ and $p = 0.3$. Use the result of Exercise 6 to calculate the *most likely value* for X—that is, the value of k for which $P\{X = k\}$ is a maximum. Compare this value to the average value $E[X]$.

8. Suppose that a coin has probability p of coming up heads on one flip. The coin is flipped six times under identical conditions. Calculate the probabilities of the following events:

   a. Exactly three flips show heads.

   b. The first, third, and sixth flips show heads.

   c. The third head occurs on the fifth flip.

9. Let X be a random variable with probability density

$$f(x) = \begin{cases} cx(1-x) & \text{for } 0 \le x \le 1 \\ 0 & \text{otherwise} \end{cases}$$

   a. What is the value of the constant c in this formula?

   b. Calculate μ_X and σ_X.

   c. Calculate the cumulative distribution function $F_X(x)$, and plot it in the range $-1 \le x \le 2$.

10. If the random variable X is uniformly distributed on $(0, 1)$, calculate $E[X^n]$ and $\mathrm{Var}(X^n)$ by using the Law of the Unconscious Statistician.

11. Suppose that a battery used in a communications satellite has an exponentially distributed random lifetime X (measured in years) with parameter $\lambda = \frac{1}{3}$.

    a. What is the probability that a new battery will fail in its first year of service?

    b. What is the probability that a new battery will fail in its second year of service (that is, its lifetime will be more than 1 year but less than 2 years)?

    c. If a particular battery of this type is still functioning after 1 year of use, what is the probability that it will fail sometime during the second year of use?

12. Let X be a continuous random variable with probability density $f(x)$. Let c be a positive constant. Show that the random variable $Y = cX$ has probability density $(1/c)f(y/c)$. (HINT: $P\{Y \leq a\} = P\{X \leq a/c\}$. Write this probability as an integral and make the change of variable $x = y/c$.)

13. Suppose that X is a gamma random variable with parameters λ and n. Let c be a positive constant. Use the result of Exercise 12 to show that the random variable $Y = cX$ is a gamma random variable with parameters λ/c and n.

14. Let X be a continuous random variable with probability density function

$$f(x) = \begin{cases} 2xe^{-x^2} & \text{for } x > 0 \\ 0 & \text{for } x \leq 0 \end{cases}$$

    Set $Y = X^2$.

    a. Calculate $P\{Y \leq 4\}$.

    b. Find the probability density function of Y.

15. Let U be a random variable that is uniformly distributed on $(0, 1)$. Define a random variable X in accordance with the following rule:

$$X = \begin{cases} 0 & \text{if } 0 < U \leq 0.3 \\ 1 & \text{if } 0.3 < U \leq 0.5 \\ 3 & \text{if } 0.5 < U \leq 0.9 \\ 6 & \text{if } 0.9 < U < 1.0 \end{cases}$$

    Show that X has the same probability distribution as the random variable in Exercise 1.

16. Let X be a geometric random variable with parameter p. Take a large integer n, define $Y = X/n$, and set $\lambda = np$. Show that Y is approximately an exponential random variable with parameter λ, in the sense that

$$P\{Y \leq a\} \approx 1 - e^{-\lambda a}$$

    for all $a > 0$. This approximation becomes more and more exact as $n \to \infty$, $p \to 0$ with $np = \lambda$ fixed.

    (HINT: Use the formula for the sum of a finite geometric series and the relation $e^x = \lim_{n\to\infty}(1 + x/n)^n$.)

17. Let X be a negative binomial random variable. Verify that the formula given in the text for the probability mass function is a bona fide distribution:

$$\sum_{n=0}^{\infty} P\{X = n\} = 1$$

(HINT: This sum is the probability that $X < \infty$. Use the waiting-time interpretation of X and the property that $P\{N < \infty\} = 1$ when N is a geometric random variable.)

18. (Simulation) Write a program to generate sample values of a negative binomial random variable. Compare the relative frequencies generated by 100 and 1000 samples (for parameters $p = 0.5$ and $r = 3$) with the probability mass function given in Figure 3.5

(HINT: See Exercise 29 of Chapter 2.)

19. A basket of 50 apples contains 2 rotten apples. You buy a bag of 10 apples picked at random from the basket. Let X be the number of rotten apples in your bag.

a. Calculate the probability mass function of X.

b. What is the expected number of rotten apples in your bag?

c. What is the *most likely* number of rotten apples in your bag? (This means the value of X that has the largest probability of occurring.)

20. Team A and Team B are playing a series of games; the first team to win three games wins the series. Suppose that in each game Team A has a $5/9$ win probability, independent of the outcomes of the other games. Let X be the number of games played in the series. Calculate the probability mass function of X.

21. Suppose that the number of fires that occur each day in a certain city is a Poisson random variable with parameter $\lambda = 3$.

a. What is the probability of at most two fires in one day?

b. If one fire has already occurred on a particular day, what is the probability that there will be at least one more fire this day?

22. A roulette wheel has 38 numbers—0 through 36 and 00. Suppose that you always bet that the outcome will be the number 7. Each bet costs you \$1, and you stop playing when you win for the first time. What is a fair payoff for this game?

23. Suppose X is a binomial random variable, with parameters n, p, and suppose Y is a Poisson random variable with parameter $\lambda = np$. Then in Chapter 8, Proposition 8.1, it is shown that $P\{X = k\}$ can be approximated by $P\{Y = k\}$ when n is large, p is small, and k is of moderate size.

a. Check this approximation property when $n = 20$, $p = 0.1$, and $k = 2$ by calculating $P\{X = 2\}$ and $P\{Y = 2\}$.

b. Suppose you buy 1000 lottery tickets, and each ticket has a 1/500 chance of winning. Use the Poisson approximation to estimate the probability that you hold at least two winning tickets.

24. You flip a fair coin until a head appears for the first time. If this happens on the nth flip, you win $\$2^n$. Let X be the amount that you win.

    a. What are your chances of winning at least a million dollars?

    b. Is there a *fair price* to pay for playing this game?

25. Let X be a geometric random variable with parameter $p = 0.1$.

    a. Find a general formula for $P\{X > k\}$ for $k = 1, 2, 3, \ldots$.

    b. Plot the cumulative distribution function of X in the range $0 \leq x \leq 10$.

    c. Let Z be an exponential random variable with parameter $\lambda = 1$, and let $Y = 10Z$. Plot the cumulative distribution function of Y in the range $0 \leq y \leq 10$, and compare with the graph in b.

26. a. Suppose X is a discrete random variable that takes the values $k = 0, 1, 2, \ldots$. Prove that

$$E[X] = \sum_{k \geq 0} P\{X > k\}$$

    (HINT: Adapt the proof in Example 3.14, replacing the integrals by sums.)

    b. Apply the result of part (a) when X is a geometric random variable.

27. Suppose buses arrive at a bus stop at 8:00, 8:15, 8:30, ... If the random arrival time of a passenger at the stop is uniformly distributed between 8:10 and 8:30, calculate the following:

    a. The probability that the passenger waits less than 5 minutes for a bus.

    b. The probability that the passenger waits more than 10 minutes for a bus.

28. Suppose that the yearly rainfall in a certain city is a normal random variable, with parameters $\mu = 30$ inches and $\sigma = 10$ inches.

    a. What is the probability that the rainfall in a year will exceed 40 inches?

    b. If more than 30 inches of rain have already fallen by September 1 in a particular year, then what is the probability that the total rainfall in this year will exceed 40 inches?

29. Let X be a normal random variable with parameters $\mu = 2$ and $\sigma = 3$. Using Table 3 at the back of the book, calculate the following probabilities:

    a. $P\{X < 4\}$

    b. $P\{X < 1\}$

    c. $P\{0 < X < 5\}$

    d. $P\{1 < X < 4\}$

    (HINT: Express the events in terms of the standard normal random variable $Z = (X - 2)/3$.)

30. Let X be a normal random variable with parameters μ and σ. Verify that μ is the expectation of X and σ is the standard deviation of X.

    (HINT: Use the change of variable $x = \sigma z + \mu$ in the integrals.)

31. Suppose that X is an exponential random variable with parameter λ. Find a value m such that $P\{X > m\} = 0.5$ (m is called the **median** of X). Compare m with the expected value μ_X.

32. Suppose that X is an Gamma random variable with parameters $r = 2$ and $\lambda = 1$.

    a. Calculate $E[X]$.

    b. Find the value of x at which the probability density $f(x)$ of X attains its maximum (this value is called the **mode** of X).

33. Let $\Gamma(r)$ be the gamma function (see Example 3.9).

    a. Use the result of Example 3.16 to prove that $\Gamma(\frac{1}{2}) = \sqrt{\pi}$.

    (HINT: Compare the formula for the density of Y in Example 3.16 with the formula for the density of the gamma random variable.)

    b. Use the result in part (a) to calculate $\Gamma(\frac{3}{2})$, and then find a general formula for $\Gamma(n + \frac{1}{2})$ for any positive integer n.

    c. A random variable X having a gamma distribution with paramaters $\lambda = 1/2$ and $r = n/2$, with n an integer, is called a **chi-squared random variable** with n degrees of freedom. Use the result in part (b) to write down a formula for the probability density function of X.

34. The lifetime (time until failure) of a machine is a continuous random variable $X \geq 0$. Suppose that $P\{X > a\} = \exp(-a^\beta)$ for all numbers $a > 0$, where $\beta > 0$ is a fixed parameter (assume that X is measured in years).

    a. Suppose the machine is one year old and still functioning. What is the probability that it will function for at least two more years?

    b. Calculate the probability density function of X.

    c. Suppose $\beta = 2$. Calculate $E[X]$ in two ways: first by using the probability density, and then by using formula (3.24).

    (HINT: See Example 3.10).

35. Let X be the number of times that heads show in two flips of a fair coin. Set $Y = X^2$.

    a. Find the probability mass function of Y.

    b. Calculate $E[Y]$ and $E[Y^2]$ using the result of part (a).

    c. Calculate $E[Y]$ and $E[Y^2]$ using the Law of the Unconscious Statistician.

36. Let X be a continuous random variable that is uniformly distributed on $(0, 1)$. Let $Y = e^X$.

    a. Find the probability density function for Y.

    b. Calculate the moments $E[Y^n]$ using the probability density function for Y from part (a).

    c. Use the Law of the Unconscious Statistician to calculate the moments of Y.

37. Let X be a continuous random variable that is uniformly distributed on $(0, 1)$. If we estimate X by a constant c, then the mean-square error in our estimate is $e(c) = E[(X - c)^2]$.

    a. Plot the graph of $e(c)$ for $0 \leq c \leq 1$.

    b. What value of c minimizes $e(c)$, and what is the minimum?

38. Suppose you toss a fair coin twice. Let X be the number of heads that show.

    a. Find a function $g(u)$ so that when U is a uniform $(0, 1)$ random variable, then $g(U)$ has the same probability distribution as X.

    b. Use the result of part (a) and the first five entries in column 2 of Table 4 to generate five sample values of X.

39. Suppose you toss a fair coin repeatedly. Let X be the number of tosses you make to obtain the first head.

    a. Find a function $g(u)$ so that when U is a uniform $(0, 1)$ random variable, then $g(U)$ has the same probability distribution as X.

    b. Use the result in part (a) and the first five entries in column 3 of Table 4 to generate five sample values of X.

40. The lifetime of a certain type of eletronic device is an exponential random variable X with mean $\mu = 200$ hours.

    a. Find a function $g(u)$ so that when U is a uniform $(0, 1)$ random variable, then $g(U)$ has the same probability distribution as X.

    b. Use the result in part (a) and the first five entries in column 10 of Table 4 to generate five sample values of X.

41. Suppose that the yearly rainfall in a certain city is a normal random variable, with parameters $\mu = 30$ inches and $\sigma = 10$ inches. Use the method of Example 3.22 and the first five entries in column 1 of Table 4 to simulate five years of rainfall measurements.

42. Obtain an algorithm for simulating the continuous random variable having density function $f(x) = (1/2)\exp(-|x|)$, for $-\infty < x < \infty$.

    (HINT: Use the probability integral transform method.)

43. Obtain an algorithm for simulating the continuous random variable for which $P\{X > a\} = \exp(-a^\beta)$ for all $a > 0$.

    (HINT: Use the probability integral transform method.)

Chapter 4

Joint Distributions of Random Variables

A single stochastic model usually involves many random variables. For calculating probabilities associated with several random variables, we introduce the **joint distribution** of a collection of random variables. The notion of **stochastic independence** of events has a counterpart for random variables, and the observations associated with repeated independent trials lead to a study of sums of independent random variables. We prove the Law of Large Numbers, which relates relative frequencies (described in Chapter 2) to mathematical expectations. We also obtain the Central Limit Theorem, which shows the basic role played by the normal random variable in large-sample models.

4.1 Joint Mass Functions and Densities

Suppose that we have set up a stochastic model for some experiment with random outcomes, as described in Chapters 1 and 2. According to the general formulation of mathematical probability, this requires defining a sample space S and a probability function P on the subsets of S. We saw in Example 3.11 that every event $A \subset S$ corresponds to a unique indicator random variable I_A. Thus we can study the stochastic model directly in terms of random variables if we wish, without explicitly using the sample space (see the examples in Section 1.2). The sample space still plays an important role conceptually, however, as the common domain for all the random variables in a model.

We use the symbol $\mathbb{R}^n$ to denote the set of all n-tuples of real numbers. Let $X_1, \ldots, X_n$ be random variables associated with some stochastic model. We shall say that these random variables have a **joint discrete distribution** if there is a non-negative function $p(x_1, x_2, \ldots, x_n)$ of n real variables which is zero except at a countable set of points in $\mathbb{R}^n$, such that

$$P\{X_1 = x_1; X_2 = x_2; \ldots; X_n = x_n\} = p(x_1, x_2, \ldots, x_n)$$

for all points $(x_1, x_2, \ldots, x_n) \in \mathbb{R}^n$. We shall call $p(x_1, x_2, \ldots, x_n)$ the **joint proba-**

bility mass function (p.m.f.) for the set of random variables. Obviously the function p satisfies the constraint

$$\sum_{x \in \mathbb{R}^n} p(x) = 1 \tag{4.1}$$

Example 4.1 Probability Tables

In the case of two discrete random variables X and Y that only take on a finite number of values, we can present their joint p.m.f. in tabular form. Suppose X and Y each have values 0 and 1, with the joint distribution given by Table 4.1. In the table, the rows correspond to the values of X, and the columns correspond to the values of Y.

Table 4.1 Joint Distribution

$X \setminus Y$	0	1	$p_X(x)$
0	0.2	0.3	0.5
1	0.4	0.1	0.5
$p_Y(y)$	0.6	0.4	

For a fixed value x of X, the sum of the elements of the corresponding row equals $p_X(x)$. Similarly, the column sums give $p_Y(y)$. These sums are displayed as an extra column (for X) and an extra row (for Y) on the margin of the table. For this reason, the individual probability mass functions of X and Y are often called the **marginal distributions**. We see that X is a Bernoulli random variable with parameter 0.5, while Y is a Bernoulli random variable with parameter 0.4. Notice that the sum of all the entries in the table (excluding the marginal entries) is 1, as required by formula (4.1). ■

Knowing the joint distribution of a family of discrete random variables, we can obtain the distribution of any one of them, say X_1, by summing over all values of the remaining random variables $X_2, \ldots, X_n$, as in Example 4.1. Thus

$$p_1(x) = \sum_{x_2, \ldots, x_n} p(x, x_2, \ldots, x_n) \tag{4.2}$$

is the probability mass function of X_1.

We say that the collection $X_1, X_2, \ldots, X_n$ of random variables has a **joint continuous distribution** if there exists a nonnegative integrable function $f(x_1, x_2, \ldots, x_n)$ of n real variables that satisfies

$$P\{X_1 \le b_1; \ldots; X_n \le b_n\} = \int_{-\infty}^{b_1} \cdots \int_{-\infty}^{b_n} f(x_1, \ldots, x_n)\, dx_1 \cdots dx_n \tag{4.3}$$

for all choices of upper limits $b_1, \ldots, b_n$. We call f the **joint probability density** of the random variables $X_1, \ldots, X_n$. It satisfies the normalizing condition

$$\int_{-\infty}^{\infty} \cdots \int_{-\infty}^{\infty} f(x_1, \ldots, x_n)\, dx_1 \cdots dx_n = 1$$

If we write $F(b_1, \ldots, b_n)$ for the probability in formula (4.3), then, just as in the one-variable case, we can obtain the joint density f from the joint distribution function F

by differentiation:

$$f(x_1, \ldots, x_n) = \left(\frac{\partial^n}{\partial x_1 \cdots \partial x_n} \right) F(x_1, \ldots, x_n) \tag{4.4}$$

(see formula (3.14)).

Knowing the joint distribution of $X_1, \ldots, X_n$, we can obtain the distribution of any one of the random variables, say X_1, by integrating over all values of the remaining random variable $X_2, \ldots, X_n$, just as in the discrete case. Thus,

$$f_1(x) = \int_{-\infty}^{\infty} \cdots \int_{-\infty}^{\infty} f(x, x_2, \ldots, x_n)\, dx_2 \cdots dx_n \tag{4.5}$$

is the probability density function of X_1.

Example 4.2 Random Point in a Circle

Consider the stochastic model for picking a point at random in a circle of radius 1 in $\mathbb{R}^2$ (see Example 2.10). Since the circle has total area π, the random coordinates X and Y have a joint continuous distribution with density

$$f(x, y) = \begin{cases} \frac{1}{\pi} & \text{if } r < 1 \\ 0 & \text{otherwise} \end{cases}$$

where $r^2 = x^2 + y^2$. By formula (4.5), the probability density of X is

$$f_1(x) = \int_{-\sqrt{1-x^2}}^{\sqrt{1-x^2}} \frac{dy}{\pi} = \frac{2}{\pi}\left(1 - x^2\right)^{1/2}$$

when $-1 < x < 1$, and zero otherwise. ■

Let X and Y be random variables defined in a stochastic model. Assume that the expectations $E[X]$ and $E[Y]$ both exist. Then the random variable $X + Y$ is also defined in the model. A fundamental property of expectations in probability theory is that

$$E[X + Y] = E[X] + E[Y] \tag{4.6}$$

Reasoning heuristically by the Law of Large Numbers (see Section 3.4), we can derive formula (4.6) as follows. Simulate the model n times to get n pairs of sample values (x_1, y_1), (x_2, y_2), ..., (x_n, y_n) of (X, Y). Set $z_i = x_i + y_i$. Then the n values $z_1, \ldots, z_n$ are sample values for the random variable $Z = X + Y$. By the Law of Large Numbers, the sample mean

$$\frac{z_1 + z_2 + \cdots + z_n}{n}$$

tends to $E[Z]$ as n gets large. But this sample mean is obvously the sum

$$\frac{x_1 + x_2 + \cdots + x_n}{n} + \frac{y_1 + y_2 + \cdots + y_n}{n}$$

of the sample means for X and Y . These tend to $E[X]$ and $E[Y]$, respectively, as n gets large. So formula (4.6) follows.

When X and Y are discrete random variables, we can prove (4.6) directly using the sample space S for the model. By Equation (3.25),

$$\begin{aligned} E[X+Y] &= \sum_{\omega \in S} \left(X(\omega)+Y(\omega)\right) P\{\omega\} \\ &= \sum_{\omega \in S} X(\omega) P\{\omega\} + \sum_{\omega \in S} Y(\omega) P\{\omega\} \\ &= E[X]+E[Y] \end{aligned}$$

An analytical proof of formula (3.25) in general requires a theory of integrals beyond the level of this book.

We can always consider the n random variables $X_1, \ldots, X_n$ as the coordinates of a **random point X** in $\mathbb{R}^n$. For a discrete distribution, the joint probability mass function $p(x)$ is given by $P\{\mathbf{X}=x\}$, where now $x \in \mathbb{R}^n$. If A is a subset of R^n, the probability that $\mathbf{X}$ lies in A can be calculated by summing the values of $p(x)$ over all points $x \in A$ at which $p(x) \neq 0$ (this is a countable set). For a joint continuous distribution, integral (4.3) gives the probability that $\mathbf{X} \in B$, where B is the n-dimensional region described by the inequalities $x_i \leq b_i$, for $i = 1, 2, \ldots, n$.

If $g(x)$ is a real-valued function defined on $\mathbb{R}^n$, we can plug in the random point

$$\mathbf{X} = (X_1, \ldots, X_n)$$

to get a real-valued random variable $g(\mathbf{X})$. The expectation of this random variable can then be calculated by means of the Law of the Unconscious Statistician, as in Section 3.5 (the sum or integral now being taken over $\mathbb{R}^n$, using the joint probability mass function or density). Following is an example of this technique.

Example 4.3 Random Distance to a Point

Take X and Y as in Example 4.2, and let R be the random distance between the point (X, Y) and the origin. Then, by the Law of the Unconscious Statistician, the average value of R is

$$E[R] = \int_{-\infty}^{\infty}\int_{-\infty}^{\infty} \left(x^2+y^2\right)^{1/2} f(x,y)\, dx dy$$

where $f(x, y)$ is the joint density from Example 4.2. If we write this integral in polar coordinates (r, θ), using the element of area $r\, dr d\theta$, we get

$$E[R] = \frac{1}{\pi}\int_0^{2\pi}\int_0^1 r^2\, dr d\theta = \int_0^1 2r^2\, dr = \frac{2}{3}$$

Note that the average value of R is *not* half the radius of the circle. However, if we draw a random circle centered at 0 having random radius R, then the expected area of this circle is

$$E[\pi R^2] = \int_0^{2\pi}\int_0^1 r^3\, dr d\theta = \int_0^1 2\pi r^3\, dr = \frac{\pi}{2}$$

Thus a circle with randomly chosen radius has has one-half the area of the unit circle. ■

Consider a random variable $Z = X + Y$. We have calculated the basic parameter $E[Z]$ in terms of $E[X]$ and $E[Y]$. What about $\text{Var}(Z)$? Write $\mu = E[X]$ and $\nu = E[Y]$. Then

$$\begin{aligned} \text{Var}(Z) &= E\big[(X + Y - \mu - \nu)^2\big] \\ &= E\big[(X - \mu)^2 + 2(X - \mu)(Y - \nu) + (Y - \nu)^2\big] \\ &= E\big[(X - \mu)^2\big] + 2E\big[(X - \mu)(Y - \nu)\big] + E\big[(Y - \nu)^2\big] \end{aligned}$$

where we have used formula (4.6) to write the expectation of a sum of random variables as the sum of the expectations. In this last formula, we recognize the first and last terms as the variances of X and Y, respectively. The middle term is something new. We define the *covariance* of the pair X, Y to be

$$\text{Cov}(X, Y) = E[(X - \mu)(Y - \nu)] \tag{4.7}$$

where $\mu = E[X]$ and $\nu = E[Y]$. Then the formula for the variance of Z can be written as

$$\text{Var}(X + Y) = \text{Var}(X) + 2\,\text{Cov}(X, Y) + \text{Var}(Y)$$

If we expand the product on the right side of formula (4.7), we get another formula for covariance:

$$\text{Cov}(X, Y) = E[XY] - E[X]E[Y] \tag{4.8}$$

Notice that, if $X = Y$, then $\text{Cov}(X, X) = \text{Var}(X)$, and formula (4.8) becomes the familiar formula for $\text{Var}(X)$.

The covariance satisfies the **Cauchy-Schwarz Inequality**

$$|\,\text{Cov}(X, Y)| \le \sigma_X \sigma_Y \tag{4.9}$$

To prove this inequality, we may assume that $\mu_X = \mu_Y = 0$ by subtracting the means from X and Y. We may also assume that $\sigma_X = \sigma_Y = 1$ by dividing X and Y by their standard deviations. Then we have $E[X^2] = 1$, $E[Y^2] = 1$, and $\text{Cov}(X, Y) = E[XY]$. But

$$0 \le E[(X \pm Y)^2] = E[X^2] \pm 2E[XY] + E[Y^2] = 2(1 \pm E[XY])$$

Hence $-1 \le E[XY] \le 1$, which is equivalent to inequality (4.9).

Example 4.4 Covariance of Indicator Random Variables

Let A and B be two events in a stochastic model, and let $X = I_A$ and $Y = I_B$ be the corresponding indicator random variables. Then $XY = I_{A \cap B}$, so we have

$$E[X] = P(A), \quad E[Y] = P(B), \quad E[XY] = P(A \cap B).$$

Hence by formula (4.8),

$$\text{Cov}(X, Y) = P(A \cap B) - P(A)P(B)$$

Assume that $0 < P(B) < 1$ (B is neither impossible nor certain). Since $P(A \cap B) = P(A \mid B) \cdot P(B)$, we can write

$$\operatorname{Cov}(X, Y) = \big(P(A \mid B) - P(A)\big) P(B)$$

Thus the sign of the covariance has a simple interpretation in this case:

$$\operatorname{Cov}(X, Y) > 0 \quad \text{if and only if} \quad P(A \mid B) > P(A)$$

In words: "When B occurs, then A is more likely to occur." Also note that $\operatorname{Cov}(X, Y) = 0$ if and only if the events A and B are independent. ■

Similar formulas apply to the sum of any number of random variables. Given random variables $X_1, \ldots, X_n$, define the random variable $S = X_1 + X_2 + \cdots + X_n$. Then by (4.6), S has expectation

$$\mu_S = \mu_1 + \mu_2 + \cdots + \mu_n$$

where $\mu_i = E[X_i]$. To calculate the variance of S, first note that

$$(S - \mu_S)^2 = \sum_{i=1}^{n} (X_i - \mu_i)^2 + 2 \sum_{i=1}^{n-1} \left\{ \sum_{j=i+1}^{n} (X_i - \mu_i)(X_j - \mu_j) \right\}$$

Taking the expected value of both sides, we get

$$\operatorname{Var}(S) = \sum_{i=1}^{n} \operatorname{Var}(X_i) + 2 \sum_{i=1}^{n-1} \left\{ \sum_{j=i+1}^{n} \operatorname{Cov}(X_i, X_j) \right\} \tag{4.10}$$

just as in the case $n = 2$.

Example 4.5 Variance of Random Matchings

Recall the mixed-up mail problem (Examples 2.5 and 2.9). Define

$$X_i = \begin{cases} 1 & \text{if the } i\text{th envelope contains the correct letter} \\ 0 & \text{otherwise} \end{cases}$$

for $i = 1, \ldots, n$, where n is the number of envelopes. Each X_i is an indicator random variable, and

$$S = X_1 + \cdots + X_n$$

is the total number of letters that match their envelopes. Since the probability is $1/n$ that a particular letter and envelope match, we have

$$E[X_i] = \frac{1}{n}$$

Hence, by formula (4.6), $E[S] = n \cdot (1/n) = 1$; therefore, on average, only one letter will be in its correct envelope, no matter how large n is.

To calculate Var(S), we first observe that, when $i \neq j$, then

$$\begin{aligned} E[X_i X_j] &= P\{X_i = 1 \quad \text{and} \quad X_j = 1\} \\ &= P\{X_i = 1 | X_j = 1\} \cdot P\{X_j = 1\} \\ &= \frac{1}{n-1} \cdot \frac{1}{n} \end{aligned}$$

(if the jth letter is in its envelope, then there are $n - 1$ equally likely choices for the ith letter). Hence, by formula (4.8) we find that

$$\text{Cov}(X_i, X_j) = \frac{1}{n(n-1)} - \left(\frac{1}{n}\right)^2 = \frac{1}{n^2(n-1)}$$

for all values of $i \neq j$. Since X_i is an indicator random variable, we have

$$E\left[X_i^2\right] = P\{X_i^2 = 1\} = P\{X_i = 1\} = \frac{1}{n}$$

Thus,

$$\text{Var}(X_i) = \frac{1}{n} - \left(\frac{1}{n}\right)^2 = \frac{n-1}{n^2}$$

for all values of i.

Now observe that on the right side of formula (4.10), the summation of the covariances has $\binom{n}{2}$ terms (all choices of i, j with $1 \leq i < j \leq n$). Hence

$$\begin{aligned} \text{Var}(S) &= n \cdot \text{Var}(X_1) + 2\binom{n}{2} \cdot \text{Cov}(X_1, X_2) \\ &= \frac{n(n-1}{n^2} + \frac{2n(n-1)}{2n^2(n-1)} = 1 \end{aligned}$$

Thus Var(S) is also independent of n. ∎

4.2 Independent Random Variables

The major difficulty in studying several random variables simultaneously is to analyze the interactions among them. In picking a point at random inside a circle (Section 4.1, Example 2), the value of X influences the value of Y, since these two random variables are constrained by the condition $X^2 + Y^2 \leq 1$. At the opposite extreme, the situation where a collection of random variables do not interact can be described as follows:

Definition 2. *The random variables $X_1, X_2, ..., X_n$ are* **mutually independent** *if the events*

$$\{X_1 \leq b_1; \ldots; X_n \leq b_n\}$$

are stochastically independent for every choice of $b_1, \ldots, b_n$. Thus the joint probability distribution function

$$F(b_1, \ldots, b_n) = P\{X_1 \leq b_1; \ldots; X_n \leq b_n\}$$

is the product

$$P\{X_1 \le b_1\} \cdots P\{X_n \le b_n\}$$

of the separate distribution functions.

Note that independence is a property of a *set* of random variables; it is possible for the three pairs of random variables $\{X, Y\}$, $\{Y, Z\}$, and $\{X, Z\}$ to be mutually independent, even though the triple $\{X, Y, Z\}$ is not mutually independent. (Take the indicator functions of the three events in Example 2.20.)

Suppose that the random variables $X_1, X_2, \ldots, X_n$ have a joint discrete distribution. Let $p_1(x_1)$, $p_2(x_2)$, ..., $p_n(x_n)$ be the individual probability mass functions of $X_1, X_2, \ldots, X_n$. The definition of independence of this set of random variables is equivalent to the factorization property

$$p(x_1, x_2, \ldots, x_n) = p_1(x_1)p_2(x_2) \cdots p_n(x_n)$$

of the joint probability mass function.

Example 4.6 Independent Bernoulli Random Variables

Let X and Y be mutually independent Bernoulli random variables with parameters 0.5 and 0.4 respectively. The joint distribution of X and Y is given in Table 4.2.

Table 4.2 Independent Bernoulli Random Variables

$X \setminus Y$	0	1	$p_X(x)$
0	0.3	0.2	0.5
1	0.3	0.2	0.5
$p_Y(y)$	0.6	0.4	

Notice that, because of the independence of X and Y, each entry in the table is the product of the marginal entries from the corresponding row and column. Comparing this table to Table 4.1, which has the same marginal entries, we conclude that the random variables in Example 4.1 are *not* independent. ■

When the random variables $X_1, \ldots, X_n$ have a joint continous distribution, then it follows from formula (4.3) that the property of independence is the same as the factorization of the joint probability density:

$$f(x_1, x_2, \ldots, x_n) = f_1(x_1) f_2(x_2) \cdots f_n(x_n)$$

Here, $f_i(x_i)$ is the marginal probability density function of X_i.

Example 4.7 Random Point in a Circle

In Example 4.2, the random variables X and Y are *not* independent. To see this, observe (for example) that

$$P\{X > 0.8 \quad \text{and} \quad Y > 0.8\} = 0$$

since the region $\{x > 0.8 \quad \text{and} \quad y > 0.8\}$ does not intersect the unit circle. On the other hand, it is obvious—by geometry—that

$$P\{X > 0.8\} > 0 \quad \text{and} \quad P\{Y > 0.8\} > 0$$

Thus, the joint probability is not the product of the separate probabilities; and so the definition of independence is not satisfied.

If we use the natural geometric symmetry of this example and consider the polar coordinates (R, Θ) of the random point (where Θ is measured between 0 and 2π), instead of rectangular coordinates, then we do get an independent pair of random variables. Indeed, if $0 \leq r \leq 1$ and $0 \leq \theta \leq 2\pi$, then, by geometry, we have

$$P\{R \leq r \text{ and } \Theta \leq \theta\} = \frac{\text{Area(Sector of radius } r\text{, Angle } \theta)}{\text{Area(Unit Circle)}} = \frac{r^2\theta}{2\pi}$$

Taking the partial derivative $\partial^2/\partial r \partial\theta$, and using formula (4.3), we obtain the joint density

$$f(r, \theta) = \begin{cases} \frac{r}{\pi} & \text{for } 0 \leq r \leq 1 \text{ and } 0 \leq \theta \leq 2\pi \\ 0 & \text{otherwise} \end{cases}$$

By integrating $f(r, \theta)$ between $\theta = 0$ and $\theta = 2\pi$, we obtain the marginal density of R:

$$f_R(r) = 2r \quad \text{for } 0 \leq r \leq 1$$

Similarly, we find that the marginal density of Θ is

$$f_\Theta(\theta) = \frac{1}{2\pi} \quad \text{for } 0 \leq \theta \leq 2\pi$$

(integrate $f(r, \theta)$ between $r = 0$ and $r = 1$). Since the product of these two marginal densities gives the joint density, we conclude that R and Θ are independent. ∎

Example 4.8 Sum of Independent Poisson Random Variables

Let X and Y be mutually independent Poisson random variables with parameters λ and μ, respectively. What is the probability mass function for the random variable $Z = X + Y$?

Solution We first observe that the possible values for Z are $n = 0, 1, 2, \ldots$, and that the event $\{Z = n\}$ is the union of the events

$$E_k = \{X = k \quad \text{and} \quad Y = n - k\}$$

for $k = 0, 1, \ldots, n$. Since X and Y are independent, we have

$$P(E_k) = P\{X = k\} \cdot P\{Y = n - k\}$$

The events E_k are mutually exclusive, so

$$P\{Z = n\} = \sum_{k=0}^{n} P\{X = k\} \cdot P\{Y = n - k\} \tag{4.11}$$

If we now substitute the Poisson probabilities (see Example 3.6) in formula (4.11), we get

$$\begin{aligned} P\{Z=n\} &= e^{-(\lambda+\mu)} \sum_{k=0}^{n} \frac{\lambda^k}{k!} \cdot \frac{\mu^{n-k}}{(n-k)!} \\ &= \frac{(\lambda+\mu)^n}{n!} e^{-(\lambda+\mu)} \end{aligned}$$

by the binomial formula. Thus Z is also a Poisson random variable, with parameter $\lambda + \mu$. ■

Example 4.9 Comparison of Independent Random Variables

Suppose X and Y are mutually independent continuous random variables. Calculate $P\{X < Y\}$.

Solution Denote the density functions of X and Y by $f(x)$ and $g(y)$, respectively. By independence, the joint probability density of X and Y is thus $f(x)g(y)$. The probability to be calculated is the integral of this function over the set $\{x < y\}$ in the xy-plane:

$$\begin{aligned} P\{X < Y\} &= \int_{-\infty}^{\infty} \left\{ \int_{-\infty}^{y} f(x)\,dx \right\} g(y)\,dy \\ &= \int_{-\infty}^{\infty} F(y)g(y)\,dy \end{aligned}$$

where F is the distribution function of X.

For example, if X and Y are uniformly distributed on $(0, 1)$, then $F(y) = y$ and $g(y) = 1$ for $0 < y < 1$. Hence,

$$P\{X < Y\} = \int_0^1 y \cdot 1\,dy = \frac{1}{2}$$

The answer in this case is intuitively obvious because of the symmetry of the problem in X and Y. ■

When X and Y are mutually independent continuous random variables, then we can calculate the distribution function of the random variable $Z = X + Y$ by setting up a double integral as in Example 4.9. Thus,

$$\begin{aligned} P\{Z \le z\} &= P\{X + Y \le z\} \\ &= \int_{-\infty}^{\infty} \left\{ \int_{-\infty}^{z-y} f(x)\,dx \right\} g(y)\,dy \\ &= \int_{-\infty}^{\infty} F(z-y)g(y)\,dy \end{aligned}$$

where F is the distribution function of X and g is the density function of Y. If we differentiate the last integral relative to the variable z and recall that $(\partial/\partial z)F(z-y) = f(z-y)$, then we get the density function $h(z)$ of the random variable Z:

$$h(z) = \int_{-\infty}^{\infty} f(z-y)g(y)\,dy \tag{4.12}$$

The integral on the right side of formula (4.12) is called the **convolution** of the functions f and g.

Example 4.10 Moving Averages

Suppose X and Y are independent and that Y is uniformly distributed on $(-1, 1)$. Then the density $g(y)$ of Y is zero outside this interval and has the constant value $\frac{1}{2}$ inside. In this case formula (4.12) becomes

$$h(z) = \frac{1}{2}\int_{-1}^{1} f(z-y)\,dy$$

If we make the change of variable $x = z - y$ in this integral, we can write

$$h(z) = \frac{1}{2}\int_{z-1}^{z+1} f(x)\,dx$$

This version of the convolution formula shows that the density $h(z)$ is a **moving average** of the density $f(x)$, since it is the average of $f(x)$ over the interval of width 2 centered at z. Such a moving average smooths out the fluctuations in the density $f(x)$.

For example, suppose that X is also uniform on $(-1, 1)$. Then the range of $X + Y$ is $(-2, 2)$, so $h(z)$ is zero when $|z| > 2$. Also $f(x)$ is 0 when $|x| > 1$. Thus for z in the range $-2 \le z \le 0$, we have $z - 1 < -1$, so the integral for $h(z)$ is cut off below -1 by $f(x)$:

$$\begin{aligned} h(z) &= \frac{1}{2}\int_{-1}^{z+1} f(x)\,dx = \frac{1}{4}\int_{-1}^{z+1} dx \\ &= \frac{1}{4}(2+z) \quad \text{for } -2 \le z \le 0 \end{aligned}$$

Similarly, for z in the range $0 \le z \le 2$, we have $z + 1 > 1$, so the integral for $h(z)$ is cut off above 1 by $f(x)$:

$$\begin{aligned} h(z) &= \frac{1}{2}\int_{z-1}^{1} f(x)\,dx = \frac{1}{4}\int_{z-1}^{1} dx \\ &= \frac{1}{4}(2-z) \quad \text{for } 0 \le z \le 2 \end{aligned}$$

This density function is tent shaped and symmetrical around 0 (see Fig. 4.1). Observe that the density for $X + Y$ has no jumps, even though the densities for X and Y are rectangular and do have jumps.

If we repeat this moving average process one more time, the change in the shape of the probability density is even more remarkable. Let $f_3(z)$ be the density function for the sum

$$Z = X_1 + X_2 + X_3$$

where the random variables X_i are mutually independent and uniform on $(-1, 1)$. This density is the moving average of the function graphed in Figure 4.1. Hence it is 0 outside the interval $(-3, 3)$ and is symmetrical around 0. We can calculate it by considering the cases $|z| \le 1$ and $1 < |z| \le 3$ separately, as we did for $h(z)$ above.

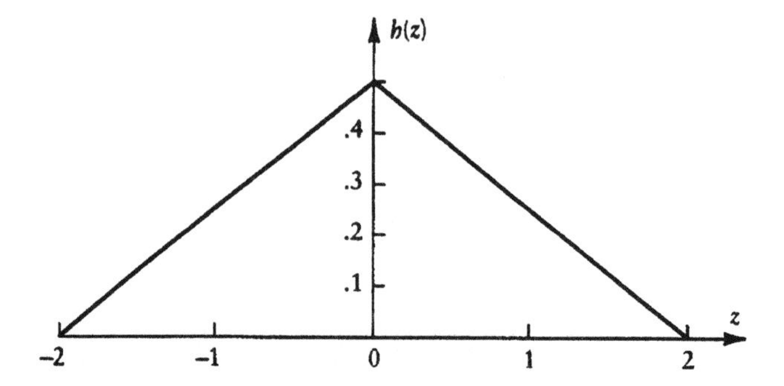

Figure 4.1: Density for $X + Y$ (Independent Uniform Random Variables)

Since $h(z)$ is piecewise linear, it follows that f_3 is piecewise quadratic. The explicit formula is

$$f_3(z) = \begin{cases} \frac{1}{8}(3 - z^2) & \text{when } |z| \leq 1 \\ \frac{1}{16}(3 - |z|)^2 & \text{when } 1 < |z| \leq 3 \end{cases}$$

The graph of this density consists of the smoothly joined parabolic arcs shown in Figure 4.2.

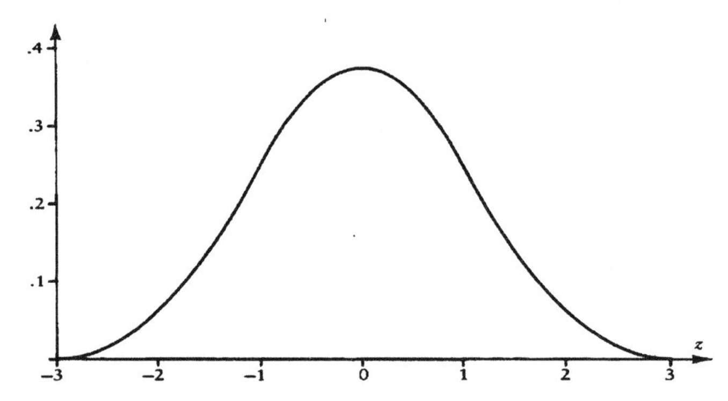

Figure 4.2: Density for $X_1 + X_2 + X_3$ (Independent Uniform Random Variables)

Notice how closely the graph in Figure 4.2 resembles the bell-shaped normal density curve in Figure 3.8. ■

When X and Y are nonnegative random variables, the limits of integration in the convolution formula (4.12) become 0 and z, since $g(y) = 0$ for $y < 0$ and $f(z - y) = 0$ for $y > z$. Thus,

$$h(z) = \int_0^z f(z - y)g(y)\, dy \tag{4.13}$$

in this case. Formula (4.13) is the continuous analogue of formula (4.11) in Example 4.8.

Example 4.11 Sum of Independent Exponential Random Variables

Suppose X and Y are independent exponential random variables with parameter λ. What is the distribution of the random variable $Z = X + Y$?

Solution The probability density function of Z is 0 for $z < 0$ and can be calculated for $z \geq 0$ by formula (4.13):

$$\begin{aligned} h(z) &= \int_0^z \lambda^2 e^{-\lambda(z-y)} \, e^{-\lambda y} \, dy \\ &= \lambda^2 e^{-\lambda z} \int_0^z dy \\ &= \lambda^2 z e^{-\lambda z} \end{aligned}$$

We recognize this as a *gamma density*, with parameters 2, λ. ■

From the Law of the Unconscious Statistician and the factorization of the joint probability density, we get the following useful theoretical result.

Theorem 4.1. *Suppose X and Y are mutually independent random variables. Then, for any pair of functions $g(x)$ and $h(y)$, the random variables $g(X)$ and $h(Y)$ are mutually independent. Furthermore, if $g(X)$ and $h(Y)$ have finite expectations, then $E[g(X)h(Y)] = E[g(X)]E[h(Y)]$.*

In particular, if X and Y are mutually independent and have finite first and second moments, then we may take $g(x) = x - \mu_X$ and $h(y) = y - \mu_Y$ in Theorem 4.1 to obtain

$$\mathrm{Cov}(X, Y) = E[X - \mu_X]E[Y - \mu_Y] = 0$$

Thus, when the random variables $X_1, X_2, \ldots, X_n$ are mutually independent and $S = X_1 + \cdots + X_n$, then formula (4.10) simplifies to

$$\mathrm{Var}(S) = \sum_{i=1}^{n} \mathrm{Var}(X_i) \tag{4.14}$$

Ratios of Random Variables

Let X and Y be continuous random variables. In many situations we want to compare X with Y. One way to do this is to form the ratio $Z = X/Y$, which is always defined since $P\{Y = 0\} = 0$. We will calculate the probability density of Z under the assumption that $Y > 0$.

If the joint density of X and Y is $f_{X,Y}(x, y)$, then we can calculate the cumulative distribution function of Z as follows: Since the event $\{ Z < a \}$ is the same as $\{ X < aY \}$, we have

$$\begin{aligned} F_Z(a) &= P\{ X < aY \} \\ &= \int_0^\infty \left\{ \int_{-\infty}^{ay} f_{X,Y}(x, y) \, dx \right\} dy \end{aligned}$$

In the inner integral we make the change of variable $x = yz$, $dx = y\,dz$ (for fixed y). The limits of integration become $-\infty < z < a$, and so

$$F_Z(a) = \int_0^\infty \left\{ \int_0^a f_{X,Y}(yz, y)\, y\, dz \right\} dy$$

Now interchange the order of integration to get

$$F_Z(a) = \int_0^a \left\{ \int_0^\infty f_{X,Y}(yz, y) y\, dy \right\} dz$$

Since $f_Z(z) = (d/dz)F_Z(z)$, we conclude that

$$f_Z(z) = \int_0^\infty f_{X,Y}(yz, y) y\, dy$$

In the special case that X and Y are mutually independent, we have $f_{X,Y}(x, y) = f_X(x) f_Y(y)$ and this formula becomes

$$f_Z(z) = \int_0^\infty f_X(yz) f_Y(y) y\, dy$$

Example 4.12 Ratio of Gamma Random Variables

Let X and Y be mutually independent gamma random variables with parameters $(\alpha, 1)$ and $(\beta, 1)$, respectively. Their joint density function is

$$f(x, y) = \frac{1}{\Gamma(\alpha)\Gamma(\beta)} x^{\alpha-1} y^{\beta-1} e^{-(x+y)}$$

Hence the ratio $Z = X/Y$ has probability density

$$\begin{aligned} f_Z(z) &= \frac{1}{\Gamma(\alpha)\Gamma(\beta)} \int_0^\infty (yz)^{\alpha-1} y^\beta e^{-(1+z)y}\, dy \\ &= \frac{z^{\alpha-1}}{\Gamma(\alpha)\Gamma(\beta)} \int_0^\infty y^{\alpha+\beta-1} e^{-(1+z)y}\, dy \end{aligned}$$

for $z > 0$ (the density is 0 for $z \leq 0$). Making the change of variable $t = (1 + z)y$ in the integral and using the definition of the Gamma function, we see that

$$f_Z(z) = \frac{\Gamma(\alpha+\beta)}{\Gamma(\alpha)\Gamma(\beta)} z^{\alpha-1} (1+z)^{-(\alpha+\beta)}$$

This is the **Fischer distribution** that occurs in statistical tests of variances of two populations. ■

If $X > 0$ and $Y > 0$, then another way to compare X and Y is to form the *proportion*

$$U = \frac{X}{X+Y}$$

which is a random variable with values between 0 and 1. We can easily obtain the probability density $f_U(u)$ of U from that of $Z = X/Y$. Indeed, if $0 < a < 1$ then

$$\begin{aligned} P\{U < a\} &= P\{X < a(X+Y)\} = P\left\{\frac{X}{Y} < \frac{a}{1-a}\right\} \\ &= F_Z\left(\frac{a}{1-a}\right) \end{aligned}$$

Hence

$$f_U(u) = \frac{d}{du} F_Z\left(\frac{u}{1-u}\right) = f_Z\left(\frac{u}{1-u}\right)\frac{1}{(1-u)^2}$$

Example 4.13 Beta Random Variable

Let X and Y be as in Example 4.12. Then $U = X/(X+Y)$ has density

$$\begin{aligned} f(u) &= \frac{\Gamma(\alpha+\beta)}{\Gamma(\alpha)\Gamma(\beta)}\left(\frac{u}{1-u}\right)^{\alpha-1}\left(1+\frac{u}{1-u}\right)^{-(\alpha+\beta)}(1-u)^{-2} \\ &= \frac{\Gamma(\alpha+\beta)}{\Gamma(\alpha)\Gamma(\beta)} u^{\alpha-1}(1-u)^{\beta-1} \end{aligned}$$

The random variable U is said to have a **Beta distribution**. The family of Beta random variables is very useful for modeling, since it includes the uniform density (when $\alpha = \beta = 1$), densities symmetric around $\frac{1}{2}$ (when $\alpha = \beta$), and densities concentrated near 0 ($\alpha = 1$ and β large) or near 1 (α large and $\beta = 1$). As a byproduct of our calculation, we have evaluated the *Beta integral*

$$\int_0^1 u^{\alpha-1}(1-u)^{\beta-1}\,du = \frac{\Gamma(\alpha)\Gamma(\beta)}{\Gamma(\alpha+\beta)} \quad \text{for } \alpha > 0 \text{ and } \beta > 0,$$

since we know that $\int_0^1 f(u)\,du = 1$. ■

4.3 Moment-generating Functions

Suppose that X is a random variable that has finite moments $\mu_n = E[X^n]$ for all values $n = 1, 2, \ldots$. We have already seen that μ_1 and μ_2 determine the mean and standard deviation of X, which are its two basic scaling parameters. The higher moments yield further information about X. Indeed, for most commonly used random variables, the complete set of moments $\{\mu_n\}_{n\geq 1}$ uniquely determines the distribution of X.

A powerful technique for making calculations with a random variable X consists of using the moments μ_n to form an auxiliary function $\phi_X(t)$ that depends on the (nonrandom) variable t. This function can then be manipulated by the usual methods of calculus and algebra, and the results translated back into properties of the random variable X.

To define this **moment-generating function** $\phi_X(t)$, we start with the function $g(x) = e^{tx}$, where t is a fixed number, and we define $\phi_X(t)$ to be the expectation of the random variable $g(X)$:

$$\phi_X(t) = E[e^{tX}]$$

We can calculate this expectation by the Law of the Unconscious Statistician:

$$\phi_X(t) = \begin{cases} \sum_x e^{tx} p_X(x) & \text{(X discrete)} \\ \int_{-\infty}^{\infty} e^{tx} \phi_X(x)\, dx & \text{(X continuous)} \end{cases} \tag{4.15}$$

(Notice that the summation or integration is with respect to x; t is held fixed.)

An alternate way of calculating $\phi_X(t)$ is to use the Taylor series expansion of the exponential function in formula (4.15) and carry out the summation (integration) term by term. This gives the formula

$$\phi_X(t) = \sum_{n=0}^{\infty} \mu_n \frac{t^n}{n!} \tag{4.16}$$

In order for $\phi_X(t)$ to be defined and to be equal to the series (4.16), the series or integral in formula (4.15) must converge absolutely when $|t| \leq \epsilon$, for some positive ϵ. This is a much stronger condition than the assumption that X has finite moments of all orders, and it is not satisfied by all random variables.

The appearance of the moments μ_n as the coefficients in series (4.16) explains the name *moment-generating function*. In practice, formula (4.15) for $\phi_X(t)$ is usually the easiest to use.

Example 4.14 Binomial Random Variable

Let X be a binomial random variable with parameters n, p. Set $q = 1 - p$ as usual. Then

$$\phi_X(t) = \sum_{k=0}^{n} \binom{n}{k} e^{tk}\, p^k q^{n-k} = \left(pe^t + q\right)^n$$

by the binomial expansion. ■

Example 4.15 Geometric Random Variable

Let X be a geometric random variable with parameter p. Then

$$\begin{aligned} \phi_X(t) &= \sum_{k=1}^{\infty} e^{tk}\, pq^{k-1} = \frac{pe^t}{1 - qe^t} \\ &= \frac{p}{e^{-t} - q} \end{aligned}$$

Here, we summed a geometric series with ratio qe^t, so we must restrict $t < -\log q$ to make this series converge. (Notice that $-\log q > 0$, since we are assuming that $q < 1$.) Thus $\phi_X(t)$ is defined for values of t near 0, which is all that is needed to calculate all moments of X from $\phi_X(t)$. ■

Example 4.16 Poisson Random Variable

Let X be a Poisson random variable, with parameter λ. Then

$$\begin{aligned}\phi_X(t) &= \sum_{k=0}^{\infty} \frac{\lambda^k e^{tk}}{k!} e^{-\lambda} = e^{-\lambda} \exp\left(\lambda e^t\right) \\ &= \exp\left(\lambda(e^t - 1)\right)\end{aligned}$$

In this case, the formula is valid for all values of t. ■

Example 4.17 Gamma Random Variable

Let X be a Gamma random variable with parameters $r > 0, \lambda > 0$. Then

$$\begin{aligned}\phi_X(t) &= \frac{1}{\Gamma(r)} \int_0^{\infty} e^{tx} \left(e^{-\lambda x} (\lambda x)^{r-1}\right) dx \\ &= \frac{\lambda^r}{\Gamma(r)} \int_0^{\infty} e^{-(\lambda - t)x} x^{r-1} \, dx \\ &= \frac{\lambda^r}{(l - t)^r} \quad \text{if } t < l\end{aligned}$$

Here, to evaluate the integral, we made the change of variable $y = (\lambda - t)x$ and used formula (3.16) of Section 3.3. The restriction $t < \lambda$ is needed to make the integral defining $\phi_X(t)$ converge. ■

Properties of Moment-generating Functions

Moment-generating functions have three key properties:

(i) *The nth derivative of $\phi_X(t)$ at $t = 0$ is the nth moment of the random variable X:*

$$E[X^n] = \left(\frac{\partial}{\partial t}\right)^n \phi_X(t)\Big|_{t=0}$$

This follows from formula (4.16).

(ii) *Adding mutually independent random variables X and Y corresponds to multiplying their moment-generating functions:*

$$\phi_{X+Y}(t) = \phi_X(t) \cdot \phi_Y(t) \qquad (X \text{ and } Y \text{ independent})$$

This follows from the multiplicative property $e^{t(X+Y)} = e^{tX} e^{tY}$ of the exponential function, and from the fact that the expectation of random variable $e^{tX} e^{tY}$ is the product of the expectations, by the Theorem 4.1.

(iii) *The probability distibution of X is uniquely determined by $\phi_X(t)$*: If X and Y are random variables, and

$$\phi_X(t) = \phi_Y(t) \quad \text{for all } t \text{ near } 0$$

then X and Y are *identically distributed* (of course they are not necessarily the *same* random variable).

The uniqueness property (iii) is deeper than the others; it follows from the fact that a random variable X that has a moment-generating function is uniquely determined by its complete set of moments. Thus, if we know $\phi_X(t)$, then we can determine all the moments of X by using property (i).

4.4 Families of Random Variables

With the moment-generating function now available as a tool, we return to study the families of random variables introduced in Chapter 3. We relate the parameters occurring in the mass functions and densities of these random variables to the natural parameters $E[X]$ and $\mathrm{Var}(X)$. (The results are summarized in Table 1 and Table 2 at the back of the book.)

Example 4.18 Bernoulli and Binomial

If $X_1, X_2, \ldots, X_n$ are independent Bernoulli random variables, each with parameter p, then X_i has moment-generating function $pe^t + q$, where $q = 1 - p$. Hence by property (ii) of moment-generating functions, the sum $X = X_1 + X_2 + \cdots + X_n$ has moment-generating function $(pe^t + q)^n$. This agrees with the calculation made in Example 4.14, since we know that X has a binomial distribution, with parameters n, p. Furthermore, $E[X] = nE[X_1] = np$, and $\mathrm{Var}(X) = n \cdot \mathrm{Var}(X_1) = npq$ by formula (4.14).

By the uniqueness property of moment-generating functions (or by direct probability interpretation), we see that the sum of two independent binomial random variables with parameters n_1, p and n_2, p is again a binomial random variable with parameters $n_1 + n_2, p$. ■

Example 4.19 Geometric and Negative Binomial

If $X_1, X_2, \ldots, X_n$ are independent geometric random variables, each with parameter p, then X_i has moment-generating function $p/(e^{-t} - q)$. Hence, by Example 4.15 and property (ii) of moment-generating functions, we find that the sum $S = X_1 + X_2 + \cdots + X_n$ has moment-generating function

$$\frac{p^n}{\left(e^{-t} - q\right)^n}$$

By the probabilistic definition of the geometric and negative binomial random variable, however, it is obvious that S has a negative binomial distribution, with parameters n, p. Thus we obtain the formula for the moment-generating function of a negative binomial random variable, without having to evaluate an infinite series.

By property (i) of the moment-generating function,

$$\begin{aligned} E[X_1] &= \frac{d}{dt}\, p\left(e^{-t} - q\right)^{-1}\Big|_{t=0} \\ &= pe^{-t}\left(e^{-t} - q\right)^{-2}\Big|_{t=0} \\ &= \frac{1}{p} \end{aligned}$$

as we already calculated in Example 3.12. Differentiating once more, we get

$$\begin{aligned} E[X_1^2] &= \frac{d}{dt}\left(pe^{-t}(e^{-t}-q)^{-2}\right)\Big|_{t=0} \\ &= \left(2pe^{-2t}(e^{-t}-q)^{-3} - pe^{-t}(e^{-t}-q)^{-2}\right)\Big|_{t=0} \\ &= \frac{2-p}{p^2} \end{aligned}$$

Thus, we find that

$$\operatorname{Var}(X_1) = E[X_1^2] - E[X_1]^2 = \frac{2-p-1}{p^2} = \frac{q}{p^2}$$

From this we calculate $E[S] = nE[X_1] = n/p$ and hence

$$\operatorname{Var}(S) = n\operatorname{Var}(X_1) = \frac{nq}{p^2}$$

by formula (4.14). ■

Example 4.20 Poisson

Suppose X and Y are mutually independent Poisson random variables, with parameters λ and μ. Set $Z = X + Y$. Then Z is Poisson, with parameter $\nu = \lambda + \mu$, by Example 4.16. This also follows by the uniqueness property of moment-generating functions, since the moment-generating function of Z is the product

$$\exp\left(\lambda(e^t - 1)\right) \cdot \exp\left(\mu(e^t - 1)\right) = \exp\left(\nu(e^t - 1)\right)$$

We can calculate the mean and variance of a Poisson random variable by differentiating $\phi_X(t)$. Write $\phi(t)$ for $\phi_X(t)$. Then $\phi'(t) = \lambda e^t\phi(t)$ and

$$\phi''(t) = \lambda e^t\phi(t) + \lambda^2 e^{2t}\phi(t)$$

Hence

$$E[X] = \phi'(0) = \lambda$$

while

$$\begin{aligned} \operatorname{Var}(X) &= E[X^2] - E[X]^2 = \phi''(0) - \phi'(0)^2 \\ &= \lambda + \lambda^2 - \lambda^2 = \lambda \end{aligned}$$

Thus the mean and variance of a Poisson random variable have the same numerical value; so the random variable $Z = X + Y$ has both mean and variance equal to $\lambda + \mu$. ■

Example 4.21 Exponential and Gamma

If $X_1, X_2, \ldots, X_n$ are independent exponential random variables, each with parameter λ, then X_i has generating function $\lambda/(\lambda - t)$. Hence, the sum $X = X_1 + X_2 + \cdots + X_n$ has moment-generating function $\lambda^n/(\lambda - t)^n$. By the uniqueness property of moment-generating functions and by Example 4.15, we conclude that X has a Gamma distribution, with parameters n, λ (see Example 4.11 for a direct calculation, when $n = 2$). This gives an intuitive meaning to the integer parameter n, which we will use often.

By the same argument, if X and Y are mutually independent Gamma random variables, with parameters n_1, λ and n_2, λ, then $X + Y$ is again a Gamma random variable, with parameters $n_1 + n_2, \lambda$. Setting

$$\phi(t) = \frac{\lambda}{\lambda - t}$$

we have

$$\phi'(t) = \frac{\lambda}{(\lambda - t)^2} \quad \text{and} \quad \phi''(t) = \frac{2\lambda}{(\lambda - t)^3}$$

Hence, the mean of an exponential random variable is $\phi'(0) = 1/\lambda$, as calculated earlier, and the variance is $\phi''(0) - \phi'(0)^2 = 1/\lambda^2$. Since X is the sum of n such independent random variables, we have

$$E[X] = \frac{n}{\lambda} \quad \text{and} \quad \text{Var}(X) = \frac{n}{\lambda^2}$$

■

Example 4.22 Normal

Let Z be a standard normal random variable ($\mu = 0$ and $\sigma = 1$). Then

$$\begin{aligned} \phi_Z(t) &= \frac{1}{(2\pi)^{1/2}} \int_{-\infty}^{\infty} e^{tz} \exp\left(-\frac{z^2}{2}\right) dz \\ &= \frac{1}{(2\pi)^{1/2}} \exp\left(\frac{t^2}{2}\right) \int_{-\infty}^{\infty} \exp\left(-\frac{(z-t)^2}{2}\right) dz \end{aligned}$$

where we have combined the exponentials in the integral and completed the square. Making the change of variable $x = z - t$, we find that the second integral evaluates to $(2\pi)^{1/2}$. Hence,

$$\phi_Z(t) = \exp\left(\frac{t^2}{2}\right) = 1 + \frac{t^2}{2} + \frac{1}{2!}\left(\frac{t^2}{2}\right)^2 + \cdots$$

Thus, the first and second derivatives of $\phi_Z(t)$ at $t = 0$ are 0 and 1, respectively, so that

$$E[Z] = 0 \quad \text{and} \quad \text{Var}(Z) = 1$$

If X is a normal random variable with parameters μ, σ, then we can write

$$X = \sigma Z + \mu$$

where Z is a standard normal random variable (see Example 3.10). It follows that

$$\begin{aligned}\phi_X(t) &= E[e^{t\sigma Z+t\mu}] = e^{\mu t} E[e^{t\sigma Z}] \\ &= e^{\mu t}\phi_Z(\sigma t) = \exp\left(\mu t + \frac{(\sigma t)^2}{2}\right)\end{aligned} \tag{4.17}$$

In particular,

$$E[X] = \mu \quad \text{and} \quad \operatorname{Var}(X) = \sigma^2$$

Thus, the two parameters used in Section 3.3 for a normal random variable are its natural parameters (mean and standard deviation).

Suppose X and Y are independent normal random variables. Set $W = X + Y$. Since $\phi_W(t) = \phi_X(t) \cdot \phi_Y(t)$, we see from (4.17) that

$$\phi_W(t) = \exp\left(\mu t + \frac{(st)^2}{2}\right)$$

where $\mu = E[X] + E[Y]$ and $\sigma^2 = \operatorname{Var}(X) + \operatorname{Var}(Y)$. By the uniqueness property of moment-generating functions, we conclude that *the sum of independent normal random variables is normal*. ■

4.5 Markov and Chebyshev Inequalities

Suppose that X is a random variable. Up to this point, we have concentrated on calculating exact numerical values for probabilities such as $P\{0 \le X \le 1\}$. Already for the normal random variable, however, these probabilities cannot be expressed in terms of the elementary functions of calculus, and they can only be calculated approximately. More generally, the probability mass function or density function may not be explicitly known—even for a random variable that arises as a combination of known random variables. Thus, it is important to have methods of *estimating* probabilities, using only the natural parameters $E[X]$ and $\operatorname{Var}(X)$. Fortunately, several simple but powerful such estimates are available. Here is the most basic one.

Markov's Inequality *Suppose X is a nonnegative random variable with finite expectation μ. Then for any number $c > 0$,*

$$P\{X \ge c\} \le \frac{\mu}{c} \tag{4.18}$$

Proof Assume that X is a continuous random variable with density $f(x)$. Then, by definition,

$$\mu = \int_0^\infty x f(x)\,dx$$

Here, the lower limit of integration is 0, since we are assuming that X is nonnegative. Hence, the integrand is nonnegative everywhere in the range of integration. If we delete the interval $(0, c)$ from this range, the integral does not increase. Thus,

$$\mu \ge \int_c^\infty x f(x)\,dx \tag{4.19}$$

But in this contracted range of integration, we have the inequality $xf(x) \geq cf(x)$. Substituting this inequality into integral (4.19) and factoring the constant c out of the integral, we obtain Markov's inequality:

$$\mu \geq c \int_c^\infty f(x)\,dx = cP\{X \geq c\}$$

The same proof works in the discrete case, with sums replacing integrals. ■

Example 4.23 Repeated Coin Tosses

Let X be the number of times heads appear in $10,000$ **flips of a** fair coin. Then X is a binomial random variable, with parameters $n = 10,000$ and $p = \frac{1}{2}$. Hence $\mu = np = 5000$ in this case, and Markov's inequality reads

$$P\{X \geq c\} \leq \frac{5000}{c}$$

If $c < 5000$, then the right side is greater than one, and we gain no information. If $c > 10,000$, then the left side is 0, and again the inequality is vacuous. For $c = 8000$, the inequality yields the estimate $P\{X \geq 8000\} \leq \frac{5}{8}$. ■

It is evident from Example 4.23 that, by itself, Markov's inequality is fairly crude. It serves as the basic tool, however, for obtaining other estimates. In particular, if we know both the mean and the variance of a random variable, we can obtain the following fundamental estimate for deviations of a random variable away from its mean.

Chebyshev's Inequality *Suppose that the random variable X has mean μ and variance σ^2. Then, for any constant $\delta > 0$,*

$$P\{|X - \mu| \geq \delta\} \leq \left(\frac{\sigma}{\delta}\right)^2 \tag{4.20}$$

Proof From the definition of variance, we have

$$\sigma^2 = E[|X - \mu|^2]$$

If we define the random variable $Y = |X - \mu|^2$, then Y is nonnegative with expectation σ^2. From Markov's inequality, with $c = \delta^2$, we obtain the estimate

$$P\{Y \geq \delta^2\} \leq \left(\frac{\sigma}{\delta}\right)^2 \tag{4.21}$$

But the event $\{Y \geq \delta^2\}$ is the same as the event $\{|X - \mu| \geq \delta\}$. Hence, Chebyshev's inequality follows from estimate (4.21). ■

Example 4.24 Repeated Coin Tosses Revisited

Take X as in Example 4.23. Then $\sigma^2 = npq = 2500$, so $\sigma = 50$. With $\delta = 3000$, we get the estimate

$$P\{|X - 5000| \geq 3000\} \leq \left(\frac{50}{3000}\right)^2 = 2.8 \times 10^{-4} \tag{4.22}$$

Let's compare this estimate with the one in Example 4.23. Since the mass function of X is symmetrical around the value 5000, we have

$$P\{X \geq 8000\} = \frac{1}{2} P\{|X - 5000| \geq 3000\} \leq 1.4 \times 10^{-4}$$

by estimate (4.22). The previous estimate (obtained by using only the mean of X) is 4000 times bigger than this estimate, which uses both the mean and the variance of X. ■

An alternative formulation of the Chebyshev inequality is sometimes more informative. Given a random variable X with finite mean μ and variance σ^2, we define the corresponding *standardized* random variable as

$$Z = \frac{X - \mu}{\sigma} \tag{4.23}$$

We can recover X from Z by the linear transformation $X = \sigma Z + \mu$. In terms of sample values, the passage from X to Z amounts to taking μ as the origin of coordinates and taking σ as the unit of measurement. Observe that

$$E[Z] = \frac{1}{\sigma} E[X - \mu] = 0 \quad \text{and} \quad \text{Var}(Z) = \frac{1}{\sigma^2} E[(X - \mu)^2] = 1$$

Thus, Chebyshev's inequality for Z reads

$$P\{|Z| \geq \delta\} \leq \frac{1}{\delta^2} \tag{4.24}$$

4.6 Law of Large Numbers

The mathematical theory of probability was created as a tool for interpreting statistical data (Section 2.1). The repeated independent trials model (Section 2.5) is the basic stochastic model in this setting. If the outcome of each trial is either *success* or *failure*, then we can describe the results of n trials by a set of mutually independent random variables $X_1, \ldots, X_n$. Each X_i is a Bernoulli random variable with parameter p, and $X_i = 1$ if the ith trial is successful. The proportion of successes in n trials is the random variable $(X_1 + \cdots + X_n)/n$. This proportion is a random variable (in fact, a binomial random variable with parameters n, p), but we would expect that by the Law of Averages it should be close (in some sense) to the constant p when the number of trials is large. We have already used this principle in a heuristic way in Section 2.1 for simulation of relative frequencies via random-number generators and in Sections 3.4 and 4.1 to motivate the definition of *expectation* of a random variable. In this section we obtain a precise mathematical formulation of this principle.

Let $X_1, X_2, \ldots$ be a sequence of random variables. We assume that for every value of n, the set of random variables $X_1, \ldots, X_n$ is mutually independent. We shall also assume that each X_i has the same probability distribution (these assumptions about the sequence $\{X_i\}$ are sometimes abbreviated as **i.i.d.r.v.**: *independent, identically distributed random variables*). Such a sequence of random variables typically occurs as a series of repeated independent measurements of some randomly varying quantity.

Define the nth **sample mean** as

$$\overline{X}_{(n)} = \frac{1}{n} \cdot (X_1 + \cdots + X_n) \tag{4.25}$$

We assume that X_i has a finite expectation μ and variance σ^2 (these are the same for every i, by assumption). From definition (4.25) it is obvious that

$$E[\overline{X}_{(n)}] = \mu$$

just as for X_n. However, the averaging effect created by the independent random fluctuations of the individual terms on the right side of (4.25) results in much less fluctuation in the sample mean. Indeed, by formula (4.14), we have

$$\operatorname{Var}(\overline{X}_{(n)}) = \frac{1}{n^2} \operatorname{Var}(X_1 + \cdots + X_n) = \frac{\sigma^2}{n} \tag{4.26}$$

Thus, when n is significantly larger than σ^2, it is very unlikely for $\overline{X}_{(n)}$ to have values far from its mean μ. Applying Chebyshev's inequality to $\overline{X}_{(n)}$, we obtain

$$P\{|\overline{X}_{(n)} - \mu| \geq \delta\} \leq \frac{\sigma^2}{n\delta^2} \tag{4.27}$$

for any number $\delta > 0$. By letting $n \to \infty$ in formula (4.27), we obtain one quantitative formulation of the Law of Averages:

Theorem 4.2 (Law of Large Numbers). *The sample means converge* **in probability** *to the expected value μ as the sample size $n \to \infty$. By definition, this means that for every fixed $\delta > 0$,*

$$\lim_{n \to \infty} P\{|\overline{X}_{(n)} - \mu| \geq \delta\} = 0$$

Example 4.25 Repeated Tosses of a Die

Suppose we roll a fair die n times. The proportion of the rolls in which 1 appears is $(X_1 + \cdots + X_n)/n$, where $X_1, \ldots, X_n$ constitute a sequence of independent Bernoulli random variables with $p = \frac{1}{6}$ (since $X_i = 1$ if the ith roll shows 1). In this case, $\mu = \frac{1}{6}$ and $\sigma^2 = pq = \frac{5}{36}$. Suppose that we take $\delta = 0.01$ in formula (4.27). Then we get the estimate

$$P\{\,|\overline{X}_{(n)} - \frac{1}{6}| \geq 0.01\,\} \leq \frac{50,000}{36n} \tag{4.28}$$

Taking n large enough, we can make the right side of estimate (4.28) as small as we please. (The small factor δ^2 in the denominator on the right side of (4.28) requires us to take n quite large to get small probabilities, however.)

Let S_n be the total number of times that 1 shows in n rolls. Then $S_n = n \cdot \overline{X}_{(n)}$, so that

$$P\{\,|\overline{X}_{(n)} - \frac{1}{6}| \geq 0.01\,\} = P\{\,|S_n - \frac{n}{6}| \geq \frac{n}{100}\,\}$$

Taking $n = 60,000$, for example, we can assert that the event

$$\{S_{60,000} \geq 10,600 \quad \text{or} \quad S_{60,000} \leq 9,400\}$$

has probability less than $\frac{5}{36 \cdot 6} = 0.023$. ■

Example 4.26 Estimating the Mean

Suppose the concentration μ (parts per billion) of a particular toxic substance in a water reservoir must be determined by making a sequence of measurements. Because of the random errors associated with the extremely low concentrations being measured, we have to consider the ith measurement as a sample value of a random variable X_i with mean μ. The characteristics of the measuring apparatus indicate that the standard deviation of X_i is 10% of its mean. Provided no drastic changes occur in the measuring apparatus or the watershed around the reservoir while the measurements are made, it is reasonable to assume that $X_1, X_2, \dots$ is a sequence of i.i.d.r.v. How many measurements must be taken in order to be 90% *sure* that the sample mean $\overline{X}_{(n)}$ is within 1% of the true mean μ?

Solution We shall interpret the phrase 90% *sure* to mean that

$$P\{|\overline{X}_{(n)} - \mu| \geq 0.01\mu\} \leq 0.1$$

To achieve this, we take $\delta = 0.01\mu$ in formula (4.27). We are assuming that $\sigma = \mu/10$, so, with this choice of δ, the right side of formula (4.27) becomes $100/n$ (note that the unknown parameter μ cancels from both numerator and denominator). Thus, we need $n \geq 1000$ to obtain a 90% reliable estimate of μ within a 1% relative error. ■

The version of the Law of Large Numbers in Theorem 4.2 is often called the *weak* Law. There is also a *strong* Law of Large Numbers, which asserts that, in an unlimited sequence of repeated independent trials, the sequence of sample means is certain to converge to μ. Symbolically, this can be stated as

$$P\{\lim_{n\to\infty} |\overline{X}_{(n)} - \mu| > 0\} = 0 \tag{4.29}$$

Although (4.29) looks superficially like the weak Law, it is a different assertion (in particular, notice that the limit of random variables is taken *before* probabilities are calculated, whereas the weak Law is an assertion about limits of numerical probabilities). Its proof requires substantial mathematical developments (including a careful definition of events of *probability zero*) that are beyond the scope of this book. We shall use the strong Law of Large Numbers on several occasions later in the book, however, in the form

Long-term sample averages = Probability averages

The utility of this principle (called in physics the **ergodic hypothesis**) lies in the fact that information about the sample means, obtained by real data or through simulation, conveys information about theoretical expecations, and vice versa.

Although the Law of Large Numbers, alias the Law of Averages, is the fundamental link between mathematical probability and experimental relative frequency data, consider the following instance where it does not apply.

Example 4.27 Independent Cauchy Random Variables

Suppose we take a sequence $U_1, U_2, \dots$ of independent random variables that are uniformly distributed on $(0, 1)$. We make a scale transformation

$$Y_i = \frac{\pi}{2}(2U_i - 1)$$

to obtain random variables that are uniform on $(-\pi/2, \pi/2)$. Now we form the sequence of random variables $X_i = \tan(Y_i)$. These random variables are mutually independent and have the Cauchy distribution (Example 3.17).

Consider the sample means

$$\overline{X}_{(n)} = \frac{X_1 + \cdots + X_n}{n}$$

Since the Cauchy density is symmetrical around 0, the values of X_i are equally likely to be positive or negative. By the Law of Averages, we might expect that, in the long run, the positive and negative values would balance. This might make the sample values of $\overline{X}_{(n)}$ tend to cluster around 0 when n is large.

To test this hypothesis experimentally, we run a computer simulation that generates uniform (0, 1) random numbers $u_1, u_2, \ldots, u_n$, and then calculates

$$x_i = \tan\left(\frac{\pi}{2} \cdot (2u_i - 1)\right)$$

and the sample mean $\overline{x}_{(n)} = (x_1 + \cdots + x_n)/n$. Table 4.3 gives some data from such a simulation.

Table 4.3 Simulation of Sample Means of Cauchy Distribution

n	100	500	1000	5000	10,000
SAMPLE MEAN	−0.6651	−0.2810	−1.1297	2.6421	−1.0959

The data in Table 4.3 certainly don't suggest that the sample mean is converging to 0 as n gets large. This doesn't contradict the Law of Large Numbers, strictly speaking (recall from Chapter 3 that the theoretical expectation $E[X_i]$ doesn't exist as an absolutely convergent integral for a Cauchy random variable). This is a rather legalistic explanation, however, and we might be tempted to run the simulation for larger and larger values of n, just to see if anything new happens.

In this case there is a better way to understand the situation. Introduce the Cauchy density with *parameter* $t > 0$:

$$f_t(x) = \frac{t}{\pi}\left(\frac{1}{t^2 + x^2}\right)$$

(for $t = 1$ we have the ordinary Cauchy density). This family of densities fits into the same pattern as the families we studied in Section 4.4: namely, if X and Y are independent random variables with Cauchy densities having parameters s and t, respectively, then $X + Y$ has Cauchy density with parameter $s + t$. (This can be proved by calculating the convolution integral for the Cauchy densities via partial fractions, or by more advanced methods using the Fourier transform.)

Assuming this property of the Cauchy densities, we see that, if each X_i is a Cauchy random variable (with parameter $t = 1$), then the sum

$$S_n = X_1 + \cdots + X_n$$

has a Cauchy density with parameter $t = n$. Hence the sample mean $\overline{X}_{(n)} = (1/n) \cdot S_n$ has density

$$n \cdot f_n(nx) = \frac{n^2}{\pi}\left(\frac{1}{n^2 + (nx)^2}\right) = \frac{1}{\pi}\left(\frac{1}{1 + x^2}\right)$$

(see Exercise 12 of Chapter 3). But this is the same Cauchy density that we started with! So the sample means $\overline{X}_{(n)}$ are all Cauchy random variables with parameter 1, and show no clustering tendency around 0 whatsoever. This explains the lack of convergence in our simulation data: Table 4.3 represents sample values of a single, unchanging probability distribution, no matter how large the value of n. So a sample value obtained with $n = 10,000$ is no more likely to be near zero than a value obtained with $n = 100$. ■

4.7 Central Limit Theorem

Let $X_1, X_2, \ldots$ be a sequence of i.i.d.r.v., as in Section 4.6, with common mean μ and variance σ^2. We already know from the Law of Large Numbers that when n is large, then the sample mean $\overline{X}_{(n)}$ has a probability distribution that is concentrated around μ. In this section we obtain an approximation to this distribution. This approximation is *universal* in the sense that it only depends on the natural parameters μ and σ of the sequence $\{X_i\}$. It also shows the fundamental role played by the normal random variable.

To describe the approximation, we first perform a scaling transformation to spread out the fluctuations of the sample mean. Recall from Section 4.6 that the nth sample mean has variance σ^2/n which tends toward 0 as $n \to \infty$. The corresponding standarized random variable

$$Z_n = \frac{\sqrt{n}}{\sigma}(\overline{X}_{(n)} - \mu)$$

has mean 0 and variance 1. We can also write

$$Z_n = \frac{1}{\sqrt{n}} \sum_{i=1}^{n} \frac{X_i - \mu}{\sigma} \tag{4.30}$$

(Notice that the factor $\sqrt{n}$ is now in the denominator in (4.30).) Let

$$F_n(b) = P\{Z_n \leq b\}$$

be the cumulative distribution function of Z_n, and let

$$\Phi(b) = \frac{1}{\sqrt{2\pi}} \int_{-\infty}^{b} \exp\left(-\frac{z^2}{2}\right) dz \tag{4.31}$$

be the cumulative distribution function of the standard normal random variable. (Values of Φ are given in Table 3 at the end of the book.)

Theorem 4.3 (Central Limit Theorem). *The random variable Z_n is approximately normal when n is large. For every number b,*

$$\lim_{n\to\infty} F_n(b) = \Phi(b) \tag{4.32}$$

Before outlining the proof of this result, we give some examples to illustrate its use.

Example 4.28 DeMoivre–Laplace Approximation

Suppose we roll a fair die 12 times. Let X be the number of times that 1 appears. Then X is a binomial random variable with parameters $n = 12, p = \frac{1}{6}$. Observe that $X = X_1 + \cdots + X_{12}$, where X_i is the indicator random variable for the event "1 appears on roll i." Thus X is the sum of twelve independent Bernoulli random variables, each having parameter $p = \frac{1}{6}$. The mean of X is $\mu = np = 2$, and the standard deviation of X is $\sigma = \sqrt{npq} = \sqrt{5/3}$. In Figure 4.3, we compare the probability mass function for X with a normal density curve having these same values for μ and σ.

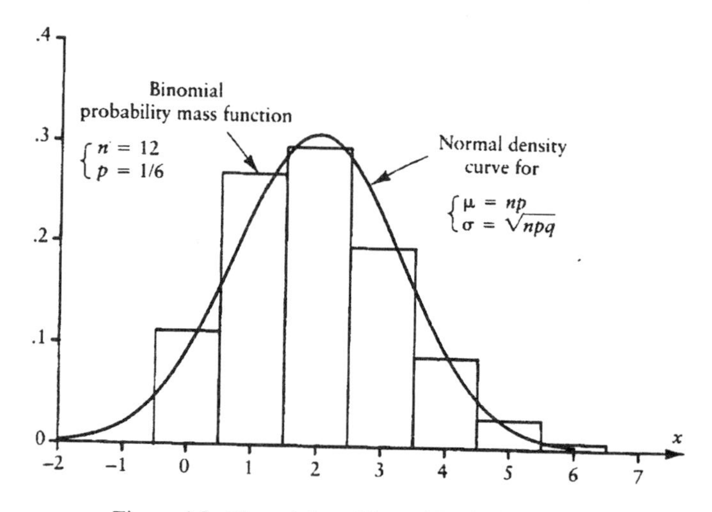

Figure 4.3: Binomial vs. Normal Probabilities

The binomial probability mass function appears in Figure 4.3 as a bar graph, with each bar having width 1 and being centered at the corresponding X-value. (The values of X greater than 6 have such small probabilities that they don't appear in the graph.) Notice how the continuous normal density curve tends to pass through the middle of most bars. Even though the Central Limit Theorem refers to the case $n \to \infty$, we see from Figure 4.3 that, already for $n = 12$, the probability mass function of X is quite close to a normal density having the same mean and standard deviation.

In general, suppose X is a binomial random variable with parameters n, p. Then

$$X = X_1 + \cdots + X_n$$

where $\{X_i\}$ is a sequence of independent Bernoulli random variables with parameter p. We can thus use the Central Limit Theorem to approximate $P\{X \leq k\}$ when n is large.

For better accuracy in passing from the integer-valued X to the real-valued Z, we observe that

$$P\{X \leq k\} = P\{X \leq k + 0.5\}$$

for any integer k. Replacing k by $k + 0.5$ before applying the Central Limit Theorem is called a **continuity correction**, and it corresponds to plotting the probability mass function of X as a bar graph, as in Figure 4.3. Notice that $P\{X \leq k\}$ is the sum of the areas of the rectangles up to $x = k + 0.5$ in Figure 4.3. When we approximate this area by the area under the normal curve up to the same x-value, then the error is the signed sum of the areas of the triangular regions between the normal curve and the bars. These regions are alternately above and below the normal curve, and so the errors tend to cancel.

The standardized random variable corresponding to X is

$$Z_n = \frac{X - np}{\sqrt{npq}}$$

where $q = 1 - p$. (Use equation (4.30) with $\mu = p$ and $\sigma = \sqrt{pq}$.) Thus,

$$\begin{aligned} P\{X \leq k\} &= P\left\{Z_n \leq \frac{k + 0.5 - np}{\sqrt{npq}}\right\} \\ &\approx \Phi\left(\frac{k + 0.5 - np}{\sqrt{npq}}\right) \end{aligned}$$

where the approximation comes from the Central Limit Theorem.

As a numerical example of this approximation, we return to the example of twelve rolls of a fair die. What is the probability that 1 appears at most once? Here $p = \frac{1}{6}$, $n = 12$, and the exact probability is

$$P\{X \leq 1\} = \binom{12}{0}\left(\frac{5}{6}\right)^{12} + \binom{12}{1}\left(\frac{5}{6}\right)^{11}\left(\frac{1}{6}\right) = 0.38$$

The normal approximation in this case is

$$\Phi\left(\frac{1.5 - 2}{\sqrt{1.667}}\right) = \Phi(-0.387) = 1 - 0.65 = 0.35$$

obtained from Table 3 (at the back of the book), using the symmetry relation $\Phi(-b) = 1 - \Phi(b)$. This approximation has an error of 8%.

If we double the number of rolls to $n = 24$ and want the probability that 1 appears at most two times, then the exact probability becomes

$$P\{X \leq 2\} = 0.212$$

while the normal approximation gives

$$\Phi\left(\frac{2.5 - 4}{\sqrt{3.33}}\right) = \Phi(-0.822) = 0.206$$

The error is now about 3%. Taking n ever larger, we would find the error in the approximation getting smaller and smaller. ■

Example 4.29 Simulation of a Normal Random Variable

In Section 3.6, we used the probability integral method to simulate a normal random variable. A simpler method is based on the Central Limit Theorem, as follows.

Let $U_1, U_2, \ldots$ be a sequence of independent random variables, uniformly distributed on $(0, 1)$. We can generate sample values $u_1, u_2, \ldots$ of such a sequence by using a random number table such as Table 4, or by calling the standard random number generator in a computer. Since each X_i has mean $\mu = \frac{1}{2}$ and variance $\sigma^2 = \frac{1}{12}$, the sum $X_1 + X_2 + \cdots + X_n$ has mean $n/2$ and variance $n/12$. The standardized random variable

$$Z_n = \sqrt{\frac{12}{n}}\left(X_1 + X_2 + \cdots + X_n - \frac{n}{2}\right)$$

has an approximately standard normal distribution, by the Central Limit Theorem.

For example, when $n = 3$, the density $f_3(z)$ of Z_3 and the normal density $\phi(z)$ are compared in Figure 4.4. (See Example 4.10 for the calculation of $f_3(z)$.) The densities are fairly close, except near 0.

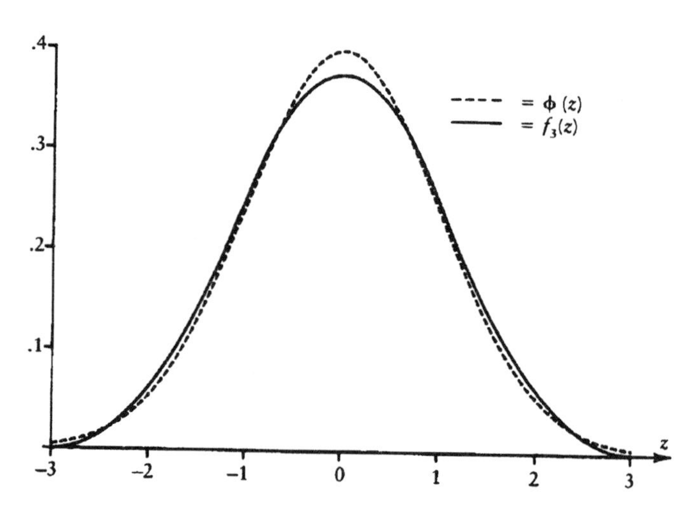

Figure 4.4: Density of Z_3 vs. Normal Density

To obtain a better approximation to the normal random variable, we can take a somewhat larger value of n—say $n = 12$ (with this choice the variance of the sum is 1). To generate sample Z-values, we make twelve calls to the random number generator and get successive sample values $u_1, u_2, \ldots, u_{12}$. We take

$$z = u_1 + u_2 + \cdots + u_{12} - 6$$

as the corresponding sample value of a normal random variable. Of course, the values of z are certain to lie in the interval $(-6, 6)$, and in this respect the simulation range differs from the unlimited range of a true normal random variable. We can calculate, however, that $1 - \Phi(6) < 10^{-9}$, so sample normal values outside $(-6, 6)$ have extremely small probability of occurring (see Exercise 20 at the end of the chapter).

■

Example 4.30 Estimating the Mean

Let's reconsider Example 4.26 with the aid of the Central Limit Theorem. Recall that we wanted to determine how big n should be so that

$$P\{|\overline{X}_{(n)} - \mu| \geq 0.01\mu\} \leq 0.1$$

Since $\sigma = \mu/10$ in this case, the standardized version of this inequality reads

$$P\left\{|Z_n| \geq \sqrt{\frac{n}{100}}\right\} \leq 0.1 \tag{4.33}$$

From the Central Limit Theorem, the probability in (4.33) is approximately $2\big(1 - \Phi(\sqrt{n/100})\big)$. Substituting this, we find the condition

$$\Phi\left(\sqrt{\frac{n}{100}}\right) \geq 0.95 \tag{4.34}$$

that n must satisfy. From Table 3 we see that inequality (4.34) will be true if

$$\sqrt{\frac{n}{100}} \geq 1.65$$

Squaring both sides, we find that $n \geq 273$ suffices. Notice that, by using the estimate coming from the Central Limit Theorem in place of Chebyshev's Inequality, we have reduced the number of required measurements from 1000 to less than 300. ■

Proof of Central Limit Theorem We may assume that $\mu = 0$ and $\sigma = 1$, since we can arrange this by a scaling transformation $X_i \to (X_i - \mu)/\sigma$. We shall assume that the moment-generating function $\phi(t)$ for the random variable X_i exists for t near 0 (the theorem can be proved under much weaker assumptions). Let $\phi_n(t)$ be the moment-generating function of Z_n. We shall show that

$$\lim_{n\to\infty} \phi_n(t) = \exp\left(\frac{t^2}{2}\right) \tag{4.35}$$

for t near zero. We recognize the right side of equation (4.35) as the moment-generating function of a standard normal random variable. The Central Limit Theorem is then a consequence of (4.35) and a general result about moment-generating functions: If the moment-generating functions of the random variables Z_n converge to the moment-generating function of Z for all values of t near 0, then equation (4.32) holds for the distribution functions. (This is the **Continuity Theorem** for moment-generating functions, which we shall not prove. It is a very strong version of the Uniqueness Theorem for moment generating functions.)

Assuming the Continuity Theorem, we only have to prove (4.35). By equation (4.30), with $\mu = 0$ and $\sigma = 1$, we can write

$$Z_n = \frac{X_1 + \cdots + X_n}{\sqrt{n}}$$

Thus

$$\exp(tZ_n) = \exp\left(\frac{tX_1}{\sqrt{n}}\right) \cdots \exp\left(\frac{tX_n}{\sqrt{n}}\right) \tag{4.36}$$

Now we calculate the moment-generating function of Z_n using (4.36) and the independence of $X_1, \ldots, X_n$:

$$\phi_n(t) = E\Big[\exp\Big(\frac{tX_1}{\sqrt{n}}\Big)\cdots\exp\Big(\frac{tX_n}{\sqrt{n}}\Big)\Big] = \big(\phi(t/\sqrt{n})\big)^n \tag{4.37}$$

To determine the behavior of the right side of equation (4.37) when n is large, we calculate the Taylor expansion of $\phi(t)$ around $t = 0$. (By our assumptions, $\phi(t)$ can be expanded in a Taylor series for t near zero; see equation (4.16).) Since the first two moments of X_i are $\mu_1 = 0$ and $\mu_2 = 1$, we have

$$\phi(t) = 1 + \frac{t^2}{2} + R(t) \tag{4.38}$$

where $R(t)$ is the sum of the terms of degree ≥ 3 in the series. Using equation (4.38) in (4.37), we find that

$$\phi_n(t) = \left(1 + \frac{t^2}{2n} + R\Big(\frac{t}{\sqrt{n}}\Big)\right)^n \tag{4.39}$$

Write $R(t) = t^3 r(t)$, where $r(t)$ is a convergent power series in t. (This is possible, since the leading term in the series for $R(t)$ is $t^3\mu_3/3!$) Then

$$1 + \frac{t^2}{2n} + R\Big(\frac{t}{\sqrt{n}}\Big) = 1 + \frac{t^2}{2n}\big(1 + \epsilon(t,n)\big) \tag{4.40}$$

where $\epsilon(t,n) = \big(2t/\sqrt{n}\big)\, r\big(t/\sqrt{n}\big)$. We observe that $\lim_{n\to\infty} \epsilon(t,n) = 0$. Now substitute the right side of equation (4.40) into (4.39) and take logs:

$$\log\phi_n(t) = n\log\Big(1 + \frac{t^2}{2n}\big(1 + \epsilon(t,n)\big)\Big) \tag{4.41}$$

Now recall that $\log(1+u) \approx u$ when u is near 0. Using this approximation in equation (4.41), we find that

$$\lim_{n\to\infty} \log\phi_n(t) = \frac{t^2}{2}\lim_{n\to\infty}\big(1 + \epsilon(t,n)\big) = \frac{t^2}{2}$$

Hence the limit in equation (4.35) is $\exp\big(t^2/2\big)$, as claimed. ∎

Exercises

1. The joint distribution of two discrete random variables X and Y is given by the following table:

$X \setminus Y$	0	1	2
0	0.1	0.1	0.3
1	0	0.2	0
2	0.1	0.2	0

a. Find the (marginal) distributions of X and Y.

b. Find the expectation of X and Y.

c. Find the probability mass function of the random variable $Z = X - Y$.

d. Find the expectation of Z in two ways: by using the probability mass function for Z, and by convolution of the probability mass functions of X and Y.

2. Suppose X and Y are continuous random variables with joint density function

$$f(x, y) = \begin{cases} x^2 y e^{-xy} & \text{for } 1 < x < 2 \text{ and } 0 < y < \infty \\ 0 & \text{otherwise} \end{cases}$$

a. Calculate the (marginal) densities of X and Y.

b. Calculate $E[X]$ and $E[Y]$.

c. Calculate $\mathrm{Cov}(X, Y)$.

3. Let $X_1, X_2, \ldots, X_n$ be independent random variables uniformly distributed on $(0, 1)$. Define the random variable $M = \max\{X_1, X_2, \ldots, X_n\}$.

a. Show that $P\{M \leq x\} = x^n$ for $0 \leq x \leq 1$.

(HINT: Express the event $\{M \leq x\}$ as an intersection of independent events.)

b. Calculate the probability density function for M and use it to calculate $E[M]$.

4. Let X and Y be independent normal random variables with mean 0 and variance 1. Define

$$R = (X^2 + Y^2)^{1/2}$$

a. Calculate $P\{R \leq a\}$ for any number $a > 0$.

(HINT: Use a double integral.)

b. Find the probability density function for the random variable R.

5. An unbalanced coin, having probability $\frac{1}{4}$ of landing heads, is flipped until the third head appears. Let N denote the number of flips required. Calculate $E[N]$ and $\mathrm{Var}(N)$.

6. An urn contains 9 balls numbered 1–9. An experiment consists of picking a ball at random from the urn, recording the number on the ball, and returning the ball to the urn. Define the events

$$\begin{aligned} A &= \{\text{The number on the ball picked is a multiple of 2 }\} \\ B &= \{\text{The number on the ball picked is a multiple of 3 }\} \end{aligned}$$

Let I_A and I_B be the indicator random variables for these events.

a. Calculate the expectations, the variances, and the covariance of the random variables I_A and I_B.

b. Suppose the experiment is performed n times, with each trial independent of the others. Let X be the total number of times that event A occurs, and let Y be the total number of times that event B occurs. Calculate the covariance of X and Y.

7. At a party, the N guests compare their birthdays. Let X be the number of *distinct* birthdays represented. Find $E[X]$ and $\text{Var}(X)$ under the assumption that birthdays are equally likely to occur on each of the 365 days of the year.

   (HINT: Introduce an indicator random variable X_i for the ith day of the year to indicate if that birthday is represented at the party.)

8. Sam and Sarah enter a store at the same moment. Suppose that Sam's (random) shopping time is uniformly distributed on $(10, 20)$, and that Sarah's shopping time is uniformly distributed on $(15, 25)$. What is the probability that Sam finishes shopping before Sarah? (Assume that the shopping times are mutually independent.)

9. Let X be uniformly distributed on $(0, 1)$, and let Y be a continuous random variable with probability density function $g(y) = 2y$ for $0 < y < 1$, and $g(y) = 0$ outside this range. Assume that X and Y are mutually independent.

   a. Calculate $P\{Y \leq X\}$.

   b. Set $Z = X + Y$. Calculate the probability density function of Z.

10. An experiment consists of picking a point at random that lies uniformly inside the triangle with vertices $(0, 0)$, $(0, 1)$, $(4, 0)$ in the xy plane. Let the random variables X and Y be the coordinates of the point picked.

    a. Find the joint density function of X and Y.

    b. Find the marginal densities $f_X(x)$ and $f_Y(y)$.

    c. Are X and Y independent?

11. Suppose the continuous random variable X is uniformly distributed between 0 and 1, that the continuous random variable Y is exponential with parameter $\lambda = 1$, and that X, Y are independent.

    a. Calculate the probability that $X + Y \geq 1$.

    b. Calculate the expected value of the random variable $Y^2 e^X$.

12. A circular target of radius 3 is marked with concentric circles of radii 1 and 2, dividing the target into three zones. A hit in the inner zone is worth ten points, in the middle zone is worth four points, and in the outer zone is worth one point. Calculate the expected number of points obtained by a dart thrower for whom the impact point of the dart has the following random distribution:

    a. uniform over the target.

    b. density $f(x, y) = c(9 - x^2 - y^2)$, where x and y are the coordinates of the impact point in a coordinate system centered on the target and c is the normalizing constant for this density.

13. Derive the formula for the probability density of $X_1 + X_2 + X_3$ in Example 4.10, where the random variables X_i are independent and uniformly distributed on $(-1, 1)$.

14. Suppose that the random variable X takes on each of the values -1, 0, and 1 with probability $\frac{1}{3}$.

    a. Calculate the moment generating function of X.

    b. Calculate $E[X]$ and $E[X^2]$ in two different ways: first by using the probability mass function for X, and then by differentiating the moment-generating function.

15. Suppose a random variable X has moment-generating function

$$f_X(t) = \frac{2e^t + 1}{3 - t}$$

    Calculate $E[X]$ and $\text{Var}(X)$.

16. A factory produces chocolate bars. The weight of an individual bar (in grams) is a random variable with mean 50 and standard deviation 5. Suppose you buy 100 such bars. Let X be the average (sample mean) of the weights of all these bars.

    a. Use Chebyshev's inequality to find a number c such that

$$P\{49 \leq X \leq 51\} \geq c$$

    b. Estimate $P\{49 \leq X \leq 51\}$ by using the Central Limit Theorem. Compare this estimate with your result in part (a).

17. A balanced coin is flipped 100 times. Let X be the number of times that the coin shows heads. Obtain a good numerical estimate for $P\{X > 55\}$.

18. A pair of fair dice is rolled 500 times. Let X be the number of times that the sum of the numbers showing on the dice is 7. Obtain a good numerical estimate for $P\{X \leq 100\}$.

19. Chebyshev's inequality asserts that

$$P\{\,|X| \geq 2\,\} \leq \frac{1}{4}$$

    for any random variable X with mean $\mu = 0$ and standard deviation $\sigma = 1$. To determine how sharp this inequality is in various cases, calculate the exact probability for the following random variables:

    a. X_1 is discrete, with $p_{X_1}(0) = \frac{3}{4}$ and $p_{X_1}(\pm a) = \frac{1}{8}$.

    b. X_2 is uniform on the interval $(-b, b)$.

    c. X_3 is discrete, with $p_{X_3}(0) = .99$ and $p_{X_3}(\pm c) = 0.005$.

    d. X_4 is standard normal.

    All of these random variables have $\mu = 0$. To make the comparisons meaningful, first determine the positive constants a, b, and c so that $\text{Var}(X_i) = 1$ for $i = 1, 2, 3$.

20. Let Z be a standard normal random variable and let $t > 0$.

    a. Prove that

    $$P\{Z > t\} = \frac{1}{\sqrt{2\pi}} t^{-1} e^{-t^2/2} - R(t)$$

    where

    $$R(t) = \frac{1}{\sqrt{2\pi}} \int_t^\infty z^{-2} e^{-z^2/2}\, dz$$

    (HINT: Integrate by parts.)

    b. Prove that

    $$0 < R(t) < \frac{1}{\sqrt{2\pi}} t^{-3} e^{-t^2/2}$$

    (HINT: In the range of integration the inequality $z^{-2} \leq t^{-3} z$ holds.)

    c. Calculate $P\{ Z > 3.5 \}$, using the approximation from part (a) and neglecting $R(t)$; compare your result with Table 3.

    d. Use parts (a) and (b) to show that $P\{Z > 6\} < 10^{-9}$.

21. Use Exercise 20 to show that the probability of at least $8,000$ heads in $10,000$ tosses of a fair coin is approximately 10^{-1200} (contrast this estimate with the crude upper bound $5/8$ from Markov's inequality obtained in Example 4.23).

22. Suppose a pair of fair dice are rolled. Let X be the larger and let Y be the smaller of the number of points showing on the dice. (For example, if one die shows 3 points and the other shows 5, then $X = 5$ and $Y = 3$.)

    a. Find the joint probability mass function of X, Y.

    b. Find the marginal probability mass functions of X and Y.

    c. Find the probability mass function of the random variable $Z = X - Y$.

23. Let $X_1, X_2, \ldots, X_n$ be independent random variables uniformly distributed on $(0, 1)$. Define the random variable $Y = \min\{X_1, X_2, \ldots, X_n\}$.

    a. Find the cumulative distribution function of Y.

    (HINT: Express the event $\{ Y > y \}$ as an intersection of independent events.)

    b. Find the probability density function for Y.

    c. Calculate $E[Y]$.

24. Let $X_1, X_2, \ldots, X_n$ be independent random variables uniformly distributed on $(0, 1)$. Define the random variables $X = \max\{X_1, X_2, \ldots, X_n\}$ and $Y = \min\{X_1, X_2, \ldots, X_n\}$.

    a. Let $0 \leq y \leq x \leq 1$. Calculate $P\{ y \leq Y;\ X \leq x\}$ by expressing the event in terms of $X_1, X_2, \ldots, X_n$.

    b. Use the result in part (a) to find the joint probability density of X, Y.

    c. Calculate $\mathrm{Var}(X)$, $\mathrm{Var}(Y)$, and $\mathrm{Cov}(X, Y)$.

    d. Use the result in part (b) to find the probability density function of the random variable $Z = X - Y$.

25. Let X be the number of aces in a poker hand (5 cards selected out of a playing deck of 52). Number the suits 1, 2, 3, 4, and let $X_i = 1$ if the hand contains the ace from suit i, and otherwise $X_i = 0$.

    a. Calculate $E[X_i]$.

    b. Calculate $\mathrm{Var}(X_i)$ and $\mathrm{Cov}(X_i, X_j)$.

    c. Calculate $E[X]$ and $\mathrm{Var}(X)$.

    (HINT: Write $X = X_1 + X_2 + X_3 + X_4$.)

26. Suppose that the waiting time X until the next incoming telephone call is an exponential random variable with parameter $\lambda = 1$ per minute. If the line is busy the caller hangs up, while if the line is free, the caller transmits a message of length Y uniformly distributed between 0 and 2 minutes. Let $Z = X + Y$ be the total elapsed time for the system to receive a message (Z is the length of a *message cycle* in the system). Find the probability density function of Z under the assumption that X and Y are independent.

27. Let X, Y be mutually independent random variables. Assume that X is standard normal and Y^2 has a Gamma distribution with parameters $(\alpha/2, 2)$. Find the density function of $T = X/Y$ (this is called *Student's t-distribution*).

28. Suppose X, Y are continuous random variables (but Y is not assumed to be positive). Show that the density function for the ratio $Z = X/Y$ is

$$f_Z(z) = \int_{-\infty}^{\infty} f_{X,Y}(yz, y)\, |y|\, dy.$$

    (HINT: When $Y > 0$, then the event $\{Z < a\}$ is the same as $\{X < aY\}$, whereas when $Y < 0$, then the event $\{Z < a\}$ is the same as $\{X > aY\}$. Thus

$$P\{X/Y < a\} = P\{X < aY \text{ and } Y > 0\} + P\{X > aY \text{ and } Y < 0\}$$

    Now calculate these probabilities by integration.)

29. Suppose X, Y are independent random variables with standard normal distributions. Set $Z = X/Y$. Use the previous exercise to show that Z is a Cauchy random variable.

30. Let X be a random variable with mean μ and standard deviation σ. The **skewness** of X (lack of symmetry around the mean) is measured by the dimensionless quantity

$$\alpha = \frac{E[(X - \mu)^3]}{\sigma^3}$$

    Calculate α for the following random variables:

    a. geometric with parameter p.

    b. negative binomial, parameters r, p.

    c. exponential, parameter λ.

    d. gamma, parameters r, λ.

    (HINT: Show that $\alpha = (\mu_3 - 3\mu\sigma^2 - \mu^3)/\sigma^3$, where μ_3 is the third moment of X, and use the moment generating function.)

31. Let X be a random variable with mean μ and standard deviation σ. A higher order measure of the spread of X around μ is the positive dimensionless quantity

$$\beta = \frac{E[(X-\mu)^4]}{\sigma^4}$$

called the **kurtosis** of X. Calculate β for the following random variables:

a. Bernoulli with parameter p.

b. geometric with parameter p.

c. exponential with parameter λ.

d. standard normal.

(HINT: First show that $\beta = (\mu_4 - 4\mu_3\mu + 6\mu^2\sigma^2 - 3\mu^4)/\sigma^4$, where μ_3 and μ_4 are the third and fourth moments of X; then use the moment generating function.)

32. Let $X_1, X_2, \ldots, X_n$ be independent, identically distributed random variables having mean zero, variance σ^2 and finite fourth moment $\mu_4 = E[(X_1)^4]$. Set $S_n = X_1 + X_2 + \cdots + X_n$.

a. Show that the fourth moment of S_n is $n\mu_4 + 6n(n-1)\sigma^4$.

(HINT: Expand $(S_n)^4$ as the sum of all products of four random variables $X_iX_jX_kX_l$. Show that such a product has expectation zero except in two cases: either the product is of the form $(X_i)^4$ (n choices), or else of the form $(X_i)^2(X_j)^2$ with $i \neq j$ ($6n(n-1)$ choices).)

b. Use the result of part (a) to calculate the kurtosis parameter β of Exercise 31 for a binomial random variable X having parameters n, p.

(HINT: Write $X - np$ as the sum of n independent random variables of mean zero.)

33. Let X be a random variable having mean μ and standard deviation σ. Assume that X has a finite kurtosis parameter $\beta = E[(X-\mu)^4]/\sigma^4$.

a. Use Markov's inequality to show that for any number $\epsilon > 0$,

$$P\{|X-\mu| \geq \epsilon\} \leq \frac{\beta\sigma^4}{\epsilon^4}$$

(This is a generalization of Chebyshev's inequality.)

b. A pair of fair dice are rolled 600 times. Let X be the number of times that the sum of the points showing is 7. Use the result in part (a) to estimate $P\{70 < X < 130\}$.

c. Compare the estimate in part (b) with the estimate obtained by using Chebyshev's inequality.

d. Compare the estimate in part (b) with the estimate obtained by using the Central Limit Theorem approximation.

Chapter 5

Conditional Expectations

When the important random variables in a stochastic model are not mutually independent, then probability calculations regarding the model become more difficult. A useful technique in this case is to analyze the model in terms of partial information. That is, we first study a simplified model in which a particular random variable Y is replaced by an observed value y. In this model we can calculate the **conditional expectations** of other random variables. Then we reintroduce the randomness of Y to obtain results about the original model. This yields a basic formula for calculating expectations in terms of conditional expectations relative to a random variable.

5.1 Conditional Distributions (Discrete Case)

Suppose that X and Y are two random variables in a stochastic model. We observe that Y has the numerical value y. What can we say about the probability distribution of X, taking into account this partial information about Y?

Example 5.1

Let the experiment consist of rolling a fair die twice. Let Y be the number that shows on the first roll of the die, and let X be the sum of the numbers showing on the two rolls. If the outcome of the first roll is $Y = 3$, for example, then the possible values for X are $4, 5, \ldots, 9$, each occurring with (conditional) probability $\frac{1}{6}$. ■

Example 5.2

Suppose X and Y are indicator random variables for events A and B, respectively. Then $E[X] = P(A)$ and $E[Y] = P(B)$. Furthermore, in Section 2.3 we defined the **conditional probability** of the event A, given that the event B has occurred, to be

$$P(A \mid B) = \frac{P(A \cap B)}{P(B)} = \frac{P\{X = 1, Y = 1\}}{P\{Y = 1\}}$$

If we observe that $Y = 1$, then we can predict, based on this knowledge, that $X = 1$ with probability $P(A \mid B)$, and that $X = 0$ with the complementary probability

$P(A^c \mid B)$. Thus the **conditional expectation** of X, given this information, is naturally defined to be

$$E[X \mid Y = 1] = 1 \cdot P(A \mid B) + 0 \cdot P(A^c \mid B) = P(A \mid B)$$

■

In general, if X and Y are random variables with a joint discrete probability mass function $p(x, y)$, then the **conditional probability mass function** of X, given Y, is defined to be

$$p_{X|Y}(x \mid y) = \frac{p(x, y)}{p_Y(y)} = \frac{P\{X = x \text{ and } Y = y\}}{P\{Y = y\}} \tag{5.1}$$

Recall that the probability mass function $p_Y(y)$ is calculated from $p(x, y)$ by summing over all x values:

$$p_Y(y) = \sum_x p(x, y) \tag{5.2}$$

To avoid division by zero in definition (5.1), we set $p_{X|Y}(x \mid y) = 0$ at all values x, y where $p_Y(y) = 0$. With this definition, we can then rewrite definition (5.1) as the following factorization of the joint probability mass function:

$$p(x, y) = p_{X|Y}(x \mid y) p_Y(y) \tag{5.3}$$

Fix a value of y such that $p_Y(y) \neq 0$. The following calculation shows that the function $p_{X|Y}(x \mid y)$ is suitably normalized to define a probability mass distribution of x values.

$$\sum_x p_{X|Y}(x \mid y) = \sum_x \frac{p(x, y)}{p_Y(y)} = 1$$

by formula (5.2). We then define the **conditional expectation** of X, given $Y = y$, to be the average of the X values, weighted according to this probability mass distribution:

$$E[X \mid Y = y] = \sum_x x \, p_{X|Y}(x \mid y). \tag{5.4}$$

For Example 5.1, we have $E[X \mid Y = y] = y + 3.5$, since the expected value for the number showing on the second roll is 3.5.

Example 5.3

Suppose that X and Y are independent random variables. Then the joint probability mass function can be factored as

$$p(x, y) = p_X(x) p_Y(y)$$

We have $p_{X|Y}(x \mid y) = p_X(x)$ in this case, and

$$E[X \mid Y = y] = E[X]$$

a constant unaffected by whatever Y value is observed. ■

Example 5.4

At the opposite extreme from Example 5.3, suppose that X is a function of Y — say, $X = Y^2 + 3Y$. Then whenever Y is observed to take the value y, we can say with certainty that X takes on the value $y^2 + 3y$. Thus $E[X \mid Y = y] = y^2 + 3y$ in this case. ■

5.2 Conditional Expectations

We continue the situation of a pair X, Y of discrete random variables in a stochastic model. So far we have tried to simplify the model by treating Y as a known quantity, determined by partial information about the outcome of an experiment. Now let's take into account the random fluctuations of Y.

It is natural to view the conditional expectation of X, given Y, as a *random variable*—say, Z—in the following way. We first define a real-valued function g by $g(y) = E[X \mid Y = y]$ for each possible value y of Y. Then we define the random variable Z to be $g(Y)$. Thus Z is a *function* of the random variable Y (see Section 3.5) and we write $Z = E[X \mid Y]$. By definition, Z takes on the value $E[X \mid Y = y]$ with probablility $p_Y(y)$.

Example 5.5

Return to Examples 5.1, 5.3, and 5.4. In Example 5.1, $Z = Y + 3.5$ and depends linearly on Y. In example 5.3, $Z = E[X]$ is constant. In Example 5.4, $Z = Y^2 + 3Y$ is a quadratic polynomial in Y. ■

In general, the conditional expectation of X, given Y, is a random variable with the following two key properties:

Theorem 5.1. (i) *The random variables $E[X \mid Y]$ and X have the same expected value:*

$$E[E[X \mid Y]] = E[X] \tag{5.5}$$

Thererore, $E[X \mid Y]$ is an **unbiased estimator** *of X.*

(ii) *The random variable $X - E[X \mid Y]$ has* **minimum variance** *among all random variables of the form $X - g(Y)$, for any function $g(y)$. Thus $E[X \mid Y]$ is at a minimum distance from X, in the probabilistic sense.*

Proof of Property (i) By definition of conditional expectation, we have

$$\begin{aligned} E[E[X \mid Y]] &= \sum_y p_Y(y) E[X \mid Y = y] = \sum_{x,y} x\, p_Y(y) p_{X \mid Y}(x \mid y) \\ &= \sum_x x \Big(\sum_y p(x, y) \Big) = \sum_x x p_X(x) \\ &= E[X] \end{aligned}$$

Here, we have used the factorization formula (5.3) and the analogue of formula (5.2) for $p_X(x)$.

Proof of Property (ii) Set $Z = E[X \mid Y]$. We must consider all random variables of the form $X - Z - g(Y)$, and show that the choice $g(y) = 0$ gives the smallest variance. We may assume that the random variable $g(Y)$ has mean zero; otherwise, we just subtract its mean, which doesn't change the variance. From Section 4.1, we know that

$$\mathrm{Var}(X - Z - g(Y)) = \mathrm{Var}(X - Z) + \mathrm{Var}(g(Y)) - 2\,\mathrm{Cov}(X - Z, g(Y)) \quad (5.6)$$

Since the random variables $X - Z$ and $g(Y)$ each have mean zero, we calculate the covariance as

$$\mathrm{Cov}(X - Z, g(Y)) = E[(X - Z)g(Y)]$$

To determine the expectation in this last equation, we condition on Y: if $Y = y$, then

$$Z = z = E[X \mid Y = y]$$

and $g(Y) = g(y)$ is a constant. Hence,

$$E[(X - Z)g(Y) \mid Y = y] = g(y) \cdot E[X - z] = g(y) \cdot (z - z) = 0$$

Now we use formula (5.5) applied to the random variable $(X - Z)g(Y)$:

$$E[(X - Z)g(Y)] = E[\,E[(X - Z)g(Y) \mid Y]] = E[0] = 0$$

Thus formula (5.6) becomes

$$\mathrm{Var}(X - Z - g(Y)) = \mathrm{Var}(X - Z) + \mathrm{Var}(g(Y))$$

Since X and Z are fixed, the only way to minimize the left side of this equation is to choose $g(y)$ so that $\mathrm{Var}(g(Y)) = 0$. But in this case $g(Y) = 0$, since we are already assuming that $g(Y)$ has zero expectation (see Example 3.19). This proves property (ii). ■

Notice that, in the course of proving property (ii) of Theorem 5.1, we also calculated that

$$E[g(Y)X \mid Y] = g(Y)E[X \mid Y] \quad (5.7)$$

for any function $g(Y)$ of the random variable Y. That is, in calculating conditional expectations relative to Y, any random quantity that depends only on Y behaves like a constant.

Example 5.6

Let X and Y be mutually independent Poisson random variables with means λ and μ. From Example 4.8, we know that $Z = X + Y$ is again Poisson, with mean $\lambda + \mu$. Moreover, the event $\{ X = k \text{ and } Z = n \}$ can be described by $\{ X = k \text{ and } Y = n - k \}$. Thus, using the independence of X and Y, we can calculate the *conditional* probability mass distribution of X, given Z, as

$$\begin{aligned} p_{X|Z}(k \mid n) &= \frac{P\{ X = k \text{ and } Y = n - k \}}{P\{ Z = n \}} \\ &= \frac{n!}{k!(n-k)!}\left(\frac{\lambda^k \mu^{n-k}}{(\lambda + \mu)^n}\right) \end{aligned}$$

where k and n are nonnegative integers. (The exponential factors e^{λ}, e^{μ}, and $e^{\lambda+\mu}$ associated with X, Y, and Z cancel.) This shows that *the conditional distribution of X, given $X + Y = n$, is binomial with parameters n, p, where $p = \lambda/(\lambda + \mu)$.* We will use this result often later in the book.

By the formula for the mean of a binomial random variable, we have

$$E[X \mid X + Y = n] = np$$

Expressed in terms of random variables, this formula says that

$$E[X \mid X + Y] = (X + Y)p$$

The expected value of the right side is then $(\lambda + \mu)p = \lambda = E[X]$, in agreement with property (i) of the Theorem 5.1. ■

5.3 Conditional Expectations (Continuous Case)

Consider two random variables X, Y in a stochastic model that have a joint continuous probability density $f(x, y)$. Recall that, just as in the discrete case, we can calculate the individual probability density functions for X or Y by integrating out the other variable,

$$f_Y(y) = \int_{-\infty}^{\infty} f(x, y)\, dx \tag{5.8}$$

with a similar formula for $f_X(x)$.

Suppose that the joint density f is continuous and nonzero at the point (x, y). Then the probability that a pair of observed X, Y values lies in a small rectangle $I \times J$ of dimensions Δx by Δy centered at (x, y) is approximately $f(x, y)\, \Delta x \Delta y$. The probability that the Y value lies in the interval J of width Δy centered at y is approximately $f_Y(y)\, \Delta y$, which is non-zero by formula (5.8). If we know that the observed Y value is in J, then the *conditional probability* that the observed X value falls within the interval I can be calculated approximately by the formula

$$P\{X \in I \mid Y \in J\} \approx \frac{f(x, y)\Delta x \Delta y}{f_Y(y)\Delta y} = \frac{f(x, y)}{f_Y(y)}\, \Delta x$$

Thus, it is natural to define the **conditional probability density** of X, given $Y = y$, to be the ratio

$$f_{X|Y}(x \mid y) = \frac{f(x, y)}{f_Y(y)} \tag{5.9}$$

for all values of y such that $f_Y(y) \neq 0$. We define the left side of formula (5.9) to be zero for all values of y where $f_Y(y) = 0$.

Notice that the event $\{Y = y\}$ has probability zero when Y is a continuous random variable, so this definition really is a new use of the term *conditional probability*. The approximation argument just given shows the reasonableness of this new definition. We also get directly from formula (5.9) the same factorization of the joint probability density function as in the discrete case—see Equation (5.3):

$$f(x, y) = f_{X|Y}(x \mid y) \cdot f_Y(y) \tag{5.10}$$

The same calculation as was performed in Section 5.1, but with integrals replacing sums, shows that the function $f_{X|Y}(x \mid y)$ is a probability density of x values:

$$\int_{-\infty}^{\infty} f_{X|Y}(x \mid y)\, dx = 1$$

for all y such that $f_Y(y) \neq 0$.

Example 5.7

Suppose an experiment consists of first picking a number Y at random uniformly between 0 and 10. If y is the number picked, we next pick a number X at random uniformly between 0 and y. We can use formula (5.10) to determine the joint probability density of X and Y. From the information given,

$$f_Y(y) = \frac{1}{10} \quad \text{for } 0 < y < 10$$

and is zero elsewhere, while the conditional density

$$f_{X|Y}(x \mid y) = \frac{1}{y} \quad \text{for } 0 < x < y < 10$$

and is zero otherwise. Thus, by formula (5.10), the joint density is

$$f(x, y) = \frac{1}{10y} \quad \text{for } 0 < x < y < 10$$

and is zero otherwise.

Now we reverse the roles of X and Y. We assume that a value of X is observed; what does this tell us about Y? The marginal density of X is

$$f_X(x) = \int_x^{10} \frac{1}{10y}\, dy = \frac{1}{10} \log \frac{10}{x}$$

(notice carefully the limits of integration) for $0 < x < 10$, and 0 otherwise. Hence, the conditional density of Y, given $X = x$, is

$$f_{Y|X}(y \mid x) = \frac{f(x, y)}{f_X(x)} = \left(\log \frac{10}{x}\right)^{-1} y^{-1}$$

for $x < y < 10$, and is zero otherwise. This density follows a hyperbolic arc when it is nonzero, with Y values near x more likely than values near 10. In particular, the conditional distribution of Y is *not* uniform on $(x, 10)$. ■

By analogy with the discrete case, we define the **conditional expectation** of X, given $Y = y$, to be the average of the X values, weighted according to the probability density function $f_{X|Y}(x \mid y)$.

$$E[X \mid Y = y] = \int_{-\infty}^{\infty} x f_{X|Y}(x \mid y)\, dx \tag{5.11}$$

(Notice that $E[X \mid Y = y] = 0$ if $f_Y(y) = 0$.) We can then view the conditional expectation of X, given Y, as a random variable, as in Section 5.2. Replacing sums by integrals in the proof of formula (5.5), we get the fundamental formula

$$E[E[X \mid Y]] = E[X] \tag{5.12}$$

for calculating expectations of continuous random variables by conditioning. It is clear from the proof that formula (5.12) also holds when one of the random variables is discrete and the other is continuous.

Example 5.8

Take X and Y to be defined as in Example 5.7. Then, for $0 < y < 10$,

$$E[X \mid Y = y] = \frac{1}{y}\int_0^y x\,dx = \frac{y}{2}$$

So as a random variable, $E[X \mid Y] = \frac{1}{2}Y$, which is also obvious from the verbal description of the experiment. Now take the expectation of this random variable to get

$$E[X] = E[E[X \mid Y]] = E\left[\frac{1}{2}Y\right] = \frac{5}{2}$$

(Remember that Y is uniform on $(0, 10)$.)

As a calculus exercise, try calculating $E[X]$ directly using the marginal density f_X that we calculated above:

$$E[X] = \frac{1}{10}\int_0^{10} x\,(\log 10 - \log x)\,dx$$

You will get the same answer, but with more work and less insight. So it's often easier to analyze a probabilistic situation by conditioning on the value of some random variable. ■

Example 5.9 Bivariate Normal Distribution

Consider a stochastic model for a system with a randomly varying input variable X and an output variable Y. We assume that the output is a linear function of the input plus a random noise W,

$$Y = \alpha X + \beta + W$$

where α and β are constants and $\alpha \neq 0$. Schematically, we can view such a system as shown in Figure 5.1.

Suppose that the input X and the noise W are independent normal random variables. Then Y is also a normal random variable (see Example 4.22). By making linear scaling transformations on X and W, we can arrange that $E[X] = E[W] = 0$ and $\mathrm{Var}(X) = 1$. Write $\sigma^2 = \mathrm{Var}(W)$. Then

$$\begin{aligned} E[Y] &= \alpha E[X] + \beta = \beta \\ \mathrm{Var}(Y) &= \alpha^2\,\mathrm{Var}(X) + \mathrm{Var}(W) = \alpha^2 + \sigma^2 \end{aligned}$$

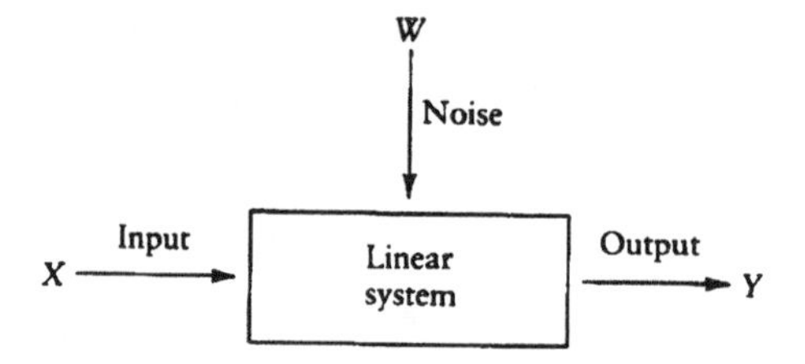

Figure 5.1: Random Noisy Linear System

by the independence of X and W (recall that adding a constant to a random variable doesn't change the variance). By independence, we also have $E[W \mid X] = E[W] = 0$. Thus,

$$E[Y \mid X] = \alpha X + \beta$$

Using Theorem 5.1 (with the roles of X and Y interchanged), we see that, if the input X and the coefficients α and β are known, then the linear function $\alpha X + \beta$ is the best predictor of the output Y. The **mean square prediction error** when we use this linear function of X to predict Y is defined to be

$$E\left[(Y - \alpha X - \beta)^2\right] = E\left[W^2\right] = \sigma^2$$

Notice that, since $Y - \alpha X - \beta$ has mean zero, the mean square prediction error is also the variance of the random variable $Y - \alpha X - \beta$. Suppose we take some other function $g(X)$ of the input X with $E[g(X)] = E[Y]$. Then $g(Y)$ also gives an unbiased prediction of the output Y. Theorem 5.1 guarantees that the mean square prediction error obtained by using $g(X)$ can't be any smaller than σ^2:

$$E\left[(Y - g(X))^2\right] = \mathrm{Var}(Y - g(X)) \geq \sigma^2$$

(See Exercise 10 for the inverse problem of predicting X, given Y.)

We shall use formula (5.10) (with x and y interchanged) to determine the joint density $f(x, y)$ of X, Y. We first observe that, for a given value $X = x$, we have $Y = \alpha x + \beta + W$, so that the *conditional distribution* of Y is normal, with mean $\alpha x + \beta$ and variance σ^2. Thus,

$$f_{Y|X}(y \mid x) = \frac{1}{\sigma(2\pi)^{1/2}} \exp\left(-\frac{(y - \alpha x - \beta)^2}{2\sigma^2}\right) \tag{5.13}$$

Since X has standard normal density $\phi(x)$, it follows by using formula (5.10)—with x and y interchanged–together with formula (5.13) that

$$f(x, y) = \frac{1}{2\pi\sigma} \exp\left(-\frac{(y - \alpha x - \beta)^2}{2\sigma^2} - \frac{x^2}{2}\right) \tag{5.14}$$

The joint density in formula (5.14) is an example of a **bivariate normal distribution**. ■

5.4 Sum of a Random Number of Random Variables

In setting up models for repeated trials of an experiment, we have always assumed that the n, the number of trials, was specified in advance. Given a (nonrandom) value for n, we calculated the expectation and variance of the sample mean for n trials. We also described the large-sample behavior of the model using the Law of Large Numbers (Section 4.6) and the Central Limit Theorem (Section 4.7).

In many situations, however, the number of trials is not fixed in advance; instead, it varies in a random way, as in the following example.

Example 5.10 Random Number of Customers

Consider a stochastic model for the operation of a fast-food restaurant. Let N be the number of customers served between 12 noon and 1 P.M., and let X_i be the amount spent by the ith customer, $i = 1, 2, \ldots$. In this situation, both N and X_i have to be treated as random variables. The random variable

$$S = \sum_{i=1}^{N} X_i \tag{5.15}$$

represents the gross receipts for this time period. Thus S is the sum of a random number of random variables. ■

Consider a random variable S, as in equation (5.15). Suppose that we assume that $X_1, X_2, \ldots$ are mutually independent identically distributed random variables, each having mean μ and variance σ^2. We assume that N is a random variable that takes on only integer values $n = 0, 1, 2, \ldots$ and has finite mean and variance. (When $N = 0$, then there aren't any X_i, so we define the "empty sum" S to be zero in this case.) Let's also assume that N is independent of the X_i. (These assumptions may be reasonable for the situation of Example 5.10, since fast-food restaurants are set up so that large orders don't slow down service very much. In general, such assumptions about the model can be tested by statistical methods.)

Given the assumptions just stated, we want to calculate the basic parameters for S—namely, $E[S]$ and Var$[S]$. We condition on the event $N = n$, which brings us back to the situation studied in Chapter 4; that is,

$$E[S \mid N = n] = E\Big[\sum_{i=1}^{n} X_i\Big] = \sum_{i=1}^{n} E[X_i] = n\mu$$

since X_i and N are mutually independent. Thus, as a random variable,

$$E[S \mid N] = \mu N$$

Taking the expectation of this random variable and using formula (5.12), we get

$$E[S] = E[\mu N] = \mu E[N] \tag{5.16}$$

This is just the formula you would guess.

Calculating the variance of S is somewhat trickier, and the answer is more interesting. Write ν for $E[N]$. We have just calculated that the mean of S is $\mu\nu$, so

$$\text{Var}(S) = E[(S - \mu\nu)^2]$$

We calculate this expectation by conditioning on the random variable N. When $N = n$, S has conditional expectation $n\mu$ and is the sum of n independent random variables, each with variance σ^2. Hence by formula (4.14)

$$E[(S - n\mu)^2 \mid N = n] = n\sigma^2$$

Notice on the left we have $n\mu$ (not $\nu\mu$), so we can't just take the expectation with respect to N in this formula to get the variance of S. But we can add and subtract $n\mu$:

$$\begin{aligned} (S - \mu\nu)^2 &= (S - n\mu + n\mu - \mu\nu)^2 \\ &= (S - n\mu)^2 + 2(S - n\mu)(n\mu - \nu\mu) + \mu^2(n - \nu)^2 \end{aligned}$$

Now calculate conditional expectations, given that $N = n$. The middle term on the right side of this last formula has mean 0, and the last term is a constant. So using the previous calculation, we get

$$E[(S - \mu\nu)^2 \mid N = \nu] = n\sigma^2 + \mu^2(n - \nu)^2$$

Thus, as a random variable,

$$E[(S - \mu\nu)^2 \mid N] = \sigma^2 N + \mu^2(N - \nu)^2$$

To get $\text{Var}(S)$, we take the expectation of this last equation. This yields the important formula

$$\text{Var}(S) = \sigma^2 E[N] + \mu^2 \,\text{Var}(N) \tag{5.17}$$

The first term in (5.17) is just what is expected and is consistent with formula (4.14). (In that case, N was nonrandom and $\text{Var}(N) = 0$.) The second term, however, shows that the randomness of N introduces bigger fluctuations into S than would occur with constant N.

Example 5.11 Random Number of Customers (Revisited)

In the situation of Example 5.10, suppose that the customers spend \$3.00, on average, with standard deviation $\sigma = \$1.00$, and that the number of customers is Poisson, with mean $\nu = 100$. Then $E[S] = \$300$, with standard deviation

$$\sigma_S = \left((1.00)^2 \cdot 100 + (3.00)^2 \cdot 100\right)^{1/2} = (1,000)^{1/2} \approx \$31.62$$

By contrast, suppose that you know on a particular day that the number of customers during this time period will be *exactly* 100. Then the (conditional) standard deviation of the receipts will be $\$10 \cdot \sigma = \$10$. ■

Exercises

1. An unbiased die is successively rolled. Let X and Y denote respectively the number of rolls necessary to obtain a 6 and a 5.

   a. Find $E[X]$.

   b. Find $E[X \mid Y = 1]$. Are X and Y independent random variables?

2. Suppose that the number of cars C and the number of trucks T which pass through Cactus Junction in a 1-hour midday period are mutually independent Poisson random variables with means 4 and 2, respectively.

   a. What is the probability that $C = 4$ and $T = 0$?

   b. On a particular day, $C + T$ was observed to be four vehicles. What is the (conditional) probability that $C = 4$ on that day? What is the (conditional) expectation of C on that day?

3. Suppose that Y is an exponential random variable with mean 1 and that X is a random variable whose conditional distribution, given $Y = y$, is exponential with mean y.

   a. Calculate $E[X]$.

   b. What is the joint probability density $f_{X,Y}(x, y)$?

4. Suppose that X and Y are continuous random variables. Assume that Y has an exponential distribution with parameter $\lambda = 1$. Assume that the *conditional distribution* of X, given $Y = y > 0$, is uniform on the interval $(y, 3y)$.

   a. Calculate $E[X]$. (HINT: First calculate $E[X \mid Y = y]$ and then use formula (5.12).)

   b. Calculate $E[X^2]$ and $\text{Var}(X)$. (HINT: Use the same method as in part (a).)

5. Tom and Huck are painting a fence together. Tom and Huck both start painting at the same time, but Lazy Huck stops working after Y hours, where the random variable Y is uniform on $(0, 1)$. Diligent Tom then finishes the job at time X, where the conditional distribution of the random variable X is uniform on $(Y, 1)$.

   a. What is the expected length of time that Tom works?

   b. If Huck and Tom are paid a total of \$ 10 to do the job, and they split the money according to the expected length of time each one works, what is the expected amount that Tom gets?

6. The joint density of X and Y is given by

$$f(x, y) = \begin{cases} y^{-1}e^{-y} & \text{when } 0 < x < y < \infty \\ 0 & \text{otherwise} \end{cases}$$

   a. What is the marginal distribution of Y?

   b. What is the conditional distribution of X, given $Y = y$?

   c. Calculate $E[X^2 \mid Y]$.

7. In a classroom there are 10 microcomputers which share a single disc storage device. Assume that during a typical class period the number of students using the computers is a binomial random variable with parameters $n = 10$ and $p = 0.7$. At the end of the period the students store their work on the disc, with the amount of disc storage used by each student uniformly distributed on $(0, 10)$. Let S be the total amount of data stored on the disc by the students from one period. Calculate $E[S]$ and $\text{Var}(S)$. What independence assumptions are you making?

8. Suppose that X and Y are continuous random variables. Assume that Y is uniformly distributed on the interval $(0, 2)$. Assume that the *conditional distribution* of X, given $Y = y$, is exponential with parameter $\lambda = y$.

   a. What is the joint probability density function for X and Y?

   b. Calculate $P\{X > 5 \mid Y = y\}$.

   c. Calculate $P\{X > 5\}$.

9. Let X and Y be as in Example 5.9 (bivariate normal distribution).

   a. Calculate $\text{Cov}(X, Y)$.

   b. Calculate the *correlation coefficient* $\rho = \text{Cov}(X, Y)/\sqrt{\text{Var}(X)\,\text{Var}(Y)}$.

10. Let X and Y be as in Example 5.9 (bivariate normal distribution).

    a. Calculate the conditional probability density of X, given $Y = y$.

    b. Calculate the conditional expectation $E[X \mid Y]$. This is the best estimate of the input X, given the output Y. Compare this with $E[Y \mid X]$.

Chapter 6

Markov Chains

In this chapter, we begin to emphasize the *dynamic* aspect of probability, as opposed to the *static* aspect. We use these words in the sense of physics: we will be studying a system that is in a particular state at each moment in time; as a result of various random influences, the state changes from one moment to the next. In this chapter, we study stochastic models that have a discrete set of states and discrete time steps. (In subsequent chapters we will study stochastic models with a continuous time variable.) The key probabilistic aspect of these systems is the *no-memory* property. Stochastic models having this property are called **Markov chains**. We shall concentrate on Markov chains that are also *time homogeneous*.

6.1 State Spaces and Transition Diagrams

We begin with an intuitive description of a stochastic system in terms of a transition diagram made up of nodes and one-way paths, where each node corresponds to a state of the system. To see how the system evolves in time, we can play a game using the transition diagram as the board. In this game, we choose a starting node and place a marker on it. At each discrete time step we move the marker one step along one of the paths leaving the current node (these are the allowed moves). The path is chosen by a random mechanism (spinning a pointer) with prescribed probabilities at each node (see Fig. 6.1).

We will call the system a **Markov chain** if the following two rules hold in the game.

No-memory Rule: At each time step, the choice of move can only depend on the current position of the marker and the random result of spinning the pointer. The past random history of moves cannot influence the choice of next move.

Time-homogeneous Rule: At each node, the probabilities for the allowed moves don't change as the game proceeds.

Here are some simple examples of Markov chains. In each example, p is a fixed number with $0 < p < 1$ and $q = 1 - p$.

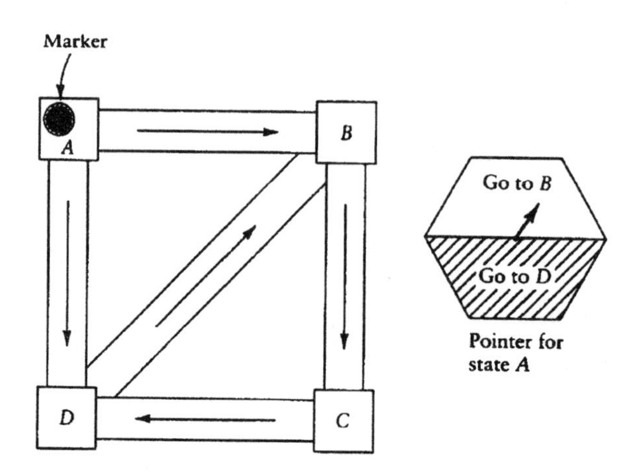

Figure 6.1: Markov Chain as a Board Game

Example 6.1 Waiting Game

Consider a system that can be in either of two states, which we label as 0 and 1. Suppose that the system changes randomly in time according to the following rules: When the marker is at state 0, it can move to state 1 with probability p or remain at state 0 with probability q; when the marker is at state 1, it cannot move.

Figure 6.2: Waiting Game

If we start the game with the marker at state 0, this game can be played by repeated independent flips of a coin (where the probability of heads in each flip is p). When the first outcome of heads occurs, we move the marker to state 1, and it stays there for all further trials. The independence of successive flips implies the no-memory property, and the constancy of p implies the time-homogeneity property. The transition diagram for this system is shown in Figure 6.2. Notice that, at each state, the probabilities of leaving from the state sum to 1. In Example 2.2 we gave some simulation data for this game when $p = \frac{1}{6}$. ■

Example 6.2 Gambler's Ruin

Suppose that the states of the system are $0, 1, \ldots, N$, and that the moves between the states are governed by the following rules: once the marker reaches state 0 or state N, no further move is possible; when the marker is at state k, with $0 < k < N$, then the next move is either to state $k-1$ (with probability q) or to state $k+1$ (with probability p). The transition diagram is shown in Figure 6.3. The movement of the marker in this system is called a *random walk with absorbing barriers at* 0 *and* N.

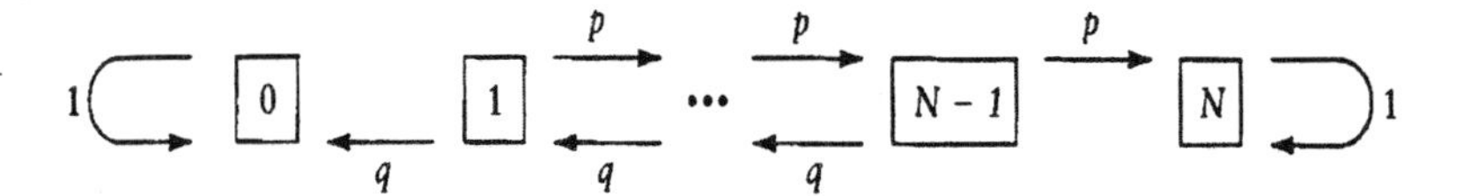

Figure 6.3: Gambler's Ruin

Just as in Example 6.1, we can describe this game in the context of Bernoulli trials. Namely, it is the gambler's ruin model, with N the total fortune of the two players (see Example 2.27). Starting the game with the marker in state k corresponds to starting with player A, say, having an initial fortune of $\$k$; A's win probability in each round is p.

We know from our study of gambler's ruin that, no matter where the marker is at the beginning of the game, it is certain to arrive at either state 0 or state N after a finite number of moves. Once it reaches either of these states, the game is over. Table 6.1 gives some sample data from a computer simulation of 1000 games, with $N = 5$ and the marker starting at state 1.

Table 6.1 Simulation of 1000 Games of Gambler's Ruin ($N = 5$, Initial State $= 1$)

p	0.2	0.3	0.4	0.5	0.6	0.7
WIN FREQUENCY	0.004	0.017	0.087	0.208	0.394	0.582
AVERAGE LENGTH	1.58	2.21	3.06	3.97	4.38	4.81

In this example, winning the game means reaching state 5, and the length of a game is the number of moves needed to reach either state 0 or state 5. ■

Example 6.3 Total Number of Heads

Suppose that we have a coin that shows heads with probability p. We flip the coin repeatedly, and keep track of the total number of heads that show. The possible states are $0, 1, 2, 3, \ldots$, and the transition diagram is as indicated in Figure 6.4.

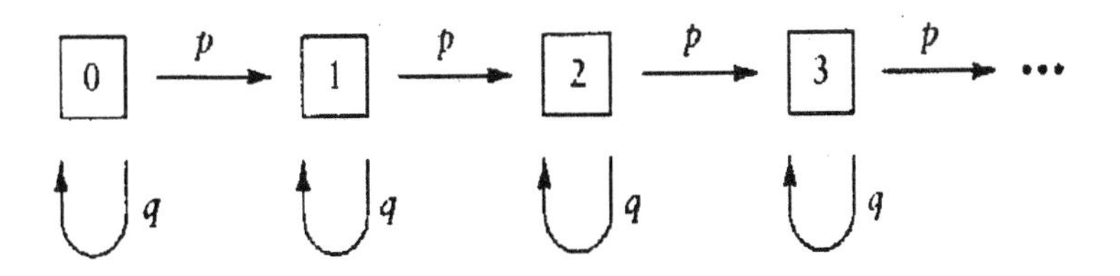

Figure 6.4: Total Number of Heads

In this model the no-memory property obviously holds, since we don't have to keep track of the particular past trials at which heads showed. ■

Example 6.4 Vending Machine

Suppose you use a food vending machine every day at lunchtime. Assume that the machine can be in two states:

$$\begin{aligned} 0 &= \text{Working} \\ 1 &= \text{Out of order} \end{aligned}$$

Suppose that, when the machine is working on a particular day, the probability is δ that it will be out of order the next day; conversely, suppose that, when the machine is out of order on a particular day, the probability is γ that it will be back in working order the next day.

With these assumptions, the one-day transition diagram for the operation of the machine is given by Figure 6.5. The two parameters δ and γ characterize the operation of the system. A reliable, well-maintained machine will be described by a value of δ near 0 and a value of γ near 1.

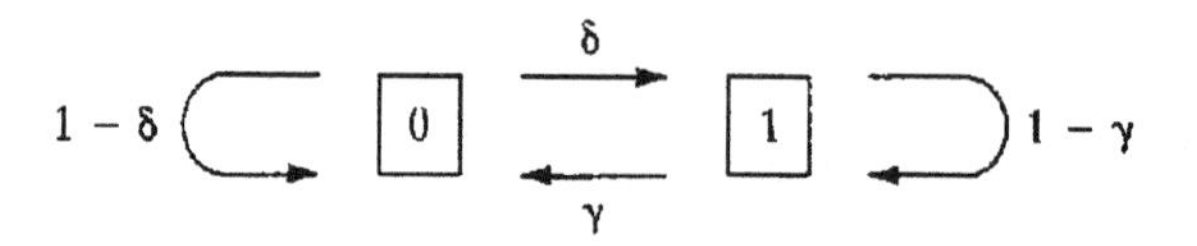

Figure 6.5: Vending Machine

Over a period of many days, the machine will alternate between *up cycles* (successive days in state 0) and *down cycles* (successive days in state 1). Table 6.2 gives some sample data from a computer simulation of this Markov chain, with $\delta = 0.2$ and $\gamma = 0.9$. The simulation data suggests that over a long time period the machine will

Table 6.2 Simulation of Vending Machine Operation

	NUMBER OF DAYS SIMULATED				
	10	50	100	500	1000
PROPORTION OF TIME MACHINE UP					
Machine up on day 0	0.90	0.82	0.84	0.842	0.815
Machine down on day 0	0.50	0.86	0.82	0.804	0.812

be up about 81% of the time. This long-term statistic doesn't seem to depend on the initial state of the machine. ■

Example 6.5 Bold Play

Suppose you have \$1, but desperately need \$5. You are offered the chance to play repeated rounds of a game, with win probability p each round and a double or nothing payoff (bets in multiples of \$1). You decide to follow the *bold play* strategy: at each round you bet all you have if winning the round would give you your goal of \$5 or less. Otherwise, you bet the amount that would make your total fortune \$5 if you win.

Thus, your initial bet is $1. If you win, then you have $2 and bet it all at the next round. If you win again, then you have $4, so now you only bet $1, and so on.

To model this game as a Markov chain, describe your state at any moment by the amount of money you have. Figure 6.6 shows the transition diagram for this strategy.

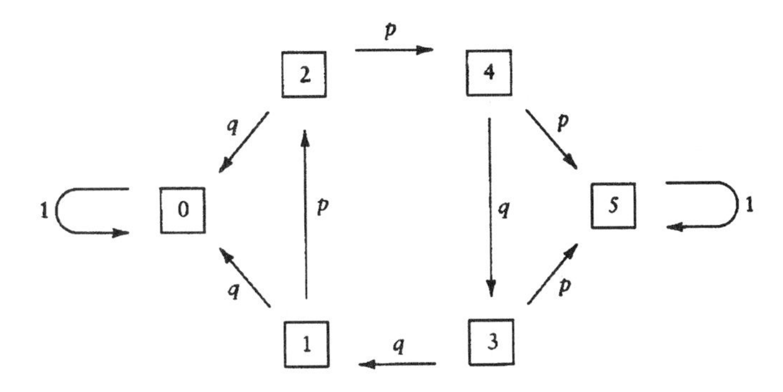

Figure 6.6: Bold Play

The starting point for the game is node 1. Winning the game means getting from node 1 to node 5. The no-memory property for this model means that the probability of winning or losing at a particular round is independent of your gambling track record for the previous rounds. The time-homogeneity property means that for each state i, your probability of winning when you have i dollars remains constant.

Table 6.3 Simulation of 1000 Games of Bold Play

p	0.2	0.3	0.4	0.5	0.6	0.7
WIN FREQUENCY	0.010	0.042	0.105	0.202	0.342	0.483
AVERAGE LENGTH	1.30	1.52	1.78	2.05	2.25	2.47

Table 6.3 shows some sample data from a computer simulation of 1000 games. Comparing this data with the gambler's ruin simulation in Example 6.2, we see a striking difference: when the odds are unfavorable ($p < 0.5$), the chance of winning the game (going from state 1 to state 5) is much higher using the bold play strategy than using the constant $1 bet strategy (*timid play*). When the odds are favorable ($p > 0.5$), however, the bold play strategy is less likely to succeed than the timid play strategy. In all cases, the games played using bold play are shorter, on average, than the games using timid play. ■

Example 6.6 Success Runs

We are playing repeated independent rounds of a game and have probability p of winning a single round. To win the whole game, we must win three consecutive rounds (the game stops when this happens).

To model this game as a Markov chain, we use four states: 0, 1, 2, and 3. Here, state i corresponds to a run of i consecutive wins. The transition diagram is given in Figure 6.7. In this model the no-memory property holds, since we forget about the past

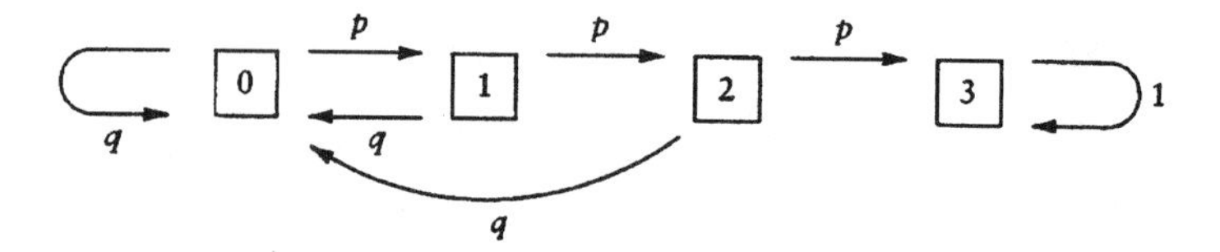

Figure 6.7: Success Runs

history of play whenever a success run (or failure run) begins. (This occurs whenever a loss is followed by a win, or vice versa.) ■

In general, we will describe a system by its state at any moment. If there are only a finite number of possible states, we have a *finite-state* system. The system goes from one state to another in discrete time steps by random jumps, as described above. We shall often refer to a single time step as a *trial*, just as we did for repeated trials of an experiment in Chapter 2.

For each pair of states (i, j), we let

$$p_{ij} = \text{probability of moving from state } i \text{ to state } j \text{ in one time step}$$

We call the numbers p_{ij} the **one-step transition probabilities** of the system. (We must include the case $i = j$ in this definition—see Example 6.1.) The transition diagram for the system consists of a node for each state, and a one-way path from node i to node j whenever $p_{ij} > 0$. We label each path by the probability of the corresponding move, as in the examples. To be consistent with the axioms of probability, the numbers p_{ij} must satisfy $0 \leq p_{ij} \leq 1$ and

$$\sum_j p_{ij} = 1 \tag{6.1}$$

Equation (6.1) says that, for each node, the probabilities of all the paths *leaving* this node (including a possible path from the node back to itself) sum to 1.

We can study such systems experimentally via computer-generated random simulation, using the game board description, and we can study them analytically by setting up an appropriate probability model (in the sense of Chapter 2). We shall be interested in answering two basic questions:

a. If winning the game requires that we move the marker from a designated starting node to a finishing node, then what is our chance of winning, and what is the expected length of time needed to win?

b. If the game can continue indefinitely, then what proportion of the time (over a long period of play) is the marker at each node?

In Examples 6.1, 6.2, and 6.5, we obtained experimental answers to question (a) using simulation data. In Example 6.4, we likewise answered question (b). Later in the chapter we will compare this simulation data with the corresponding theoretical probablities for these stochastic models.

6.2 Joint Probabilities

Now we relate the board-game version of the stochastic models described in Section 6.1 with the sample-space formulation of probability from Chapter 2.

Assume that we have a Markov chain whose states are labeled $0, 1, 2, \ldots$ (finite or infinite set of integers). In describing the time evolution of the system, as in Section 6.1, we follow the random moves of a marker starting at some particular node at time 0. Because of the no-memory property, the game effectively starts over at each move, with the starting node being the current position of the marker. This means that to analyze all the possible random paths of the marker, we should allow any node to be the starting node. Furthermore, we should allow the starting node to be chosen at random.

Fix some initial probability distribution $\boldsymbol{\pi} = \{\boldsymbol{\pi}_i\}$, where π_i is the probability of picking node i as the starting position. It must satisfy

$$\sum_i \pi_i = 1$$

where the sum ranges over all states. For example, we can choose a pair of states (i, j), set $\boldsymbol{\pi}_i = \frac{1}{2}$, $\boldsymbol{\pi}_j = \frac{1}{2}$, and let $\boldsymbol{\pi}_k = 0$ for all states $k \neq i, j$. This corresponds to flipping a fair coin before starting the game, and then starting at state i if the coin shows heads or at state j if the coin shows tails.

With the starting node chosen at random according to the distribution π, we proceed to play the Markov chain game as described above. For $n = 0, 1, 2, \ldots$ we define a discrete random variable $X(n)$ by setting $X(n) = i$ if the marker is at node i at the nth move. The sequence of nodes visited by the marker, starting at time 0, is thus

$$X(0), X(1), X(2), \ldots$$

So the whole set of random variables $\{X(n)\}$ describes the possible random paths that can be taken in playing the game.

The probability mass function of $X(0)$ is $\{\boldsymbol{\pi}_i\}$, by definition. To obtain the joint probability mass function of $X(0), \ldots, X(n)$ for all values of n, we restate the no-memory property in terms of these random variables. The marker is at the node labeled $X(n)$ at time n, and will move to the node labeled $X(n+1)$ at the next time step. The no-memory property for the choice of next move means that the conditional distribution of $X(n+1)$, given the complete history $X(n), X(n-1), \ldots, X(0)$ of the system up to the present, is the same as the conditional distribution of $X(n+1)$ given only information about the present state $X(n)$. We shall call this no-memory property the **Markov property**, after the mathematician who first studied this type of stochastic model.

Given a sequence of states $i_0, i_1, \ldots$, we define the event

$$A_k = \{X(k) = i_k\} \tag{6.2}$$

(at time k, the marker is at node i_k in the transition diagram) and the event

$$B_k = \{X(0) = i_0,\ X(1) = i_1, \ldots, X(k) = i_k\} \tag{6.3}$$

(starting at state i_0 at time 0, the marker arrives at state i_k by following the path $i_0 \to i_1 \to \cdots \to i_k$). The Markov property asserts that, for all values of n,

$$P\{A_{n+1} \mid B_n\} = P\{A_{n+1} \mid A_n\} \tag{6.4}$$

Notice that the right side of formula (6.4) is just the one-step transition probability $p_{i_n,i_{n+1}}$ from state i_n to state i_{n+1}. The time-homogeneity property that we have assumed holds for the transition probabilities implies that the probability in Equation (6.4) depends only on the choice of states i_{n+1} and i_n—not on the time n.

By iterating formula (6.4), we shall obtain the **joint probability mass function** (p.m.f.) of all the random variables $\{X(n)\}$. We start with the chain rule for conditional probabilities:

$$P(A \cap B) = P(A \mid B)\, P(B)$$

We apply this to the events A_k and B_{k-1}, defined in formulas (6.2) and (6.3), and we observe that $A_k \cap B_{k-1} = B_k$. Thus the chain rule yields

$$P(B_k) = P(A_k \mid B_{k-1})\, P(B_{k-1})$$

Now apply the same calculation to $P(B_{k-1})$, and continue in this way back to trial 0 (notice that $A_0 = B_0$):

$$P(B_k) = P(A_k \mid B_{k-1}) P(A_{k-1} \mid B_{k-2}) \cdots P(A_1 \mid A_0) P(A_0) \tag{6.5}$$

By the Markov property, every event B_j on the right side of (6.5) can be replaced by A_j. Thus,

$$P(B_k) = P(A_k \mid A_{k-1}) P(A_{k-1} \mid A_{k-2}) \cdots P(A_1 \mid A_0) P(A_0) \tag{6.6}$$

Equation (6.6) is the desired formula for the joint p.m.f. of the random variables $X(0), \ldots, X(k)$ in terms of the p.m.f. π of the initial random variable $X(0)$ and the one-step transition probabilities p_{ij}. We summarize this as follows:

Theorem 6.1 (Multiplication Rule for Path Probabilities). *Given a transition diagram with one-step transition probabilities p_{ij} and an initial distribution π for $X(0)$, formula* (6.6) *defines the joint p.m.f. of the family of random variables*

$$\{X(0), X(1), X(2), \ldots, X(k)\}$$

for every positive integer k. The probability of the event

$$\{X(0) = i_0,\ X(1) = i_1, \ldots, X(k) = i_k\}$$

is the product of the probability π_{i_0} for starting at node i_0 multiplied by the probabilities $p_{i_s,i_{s+1}}$ for each step $i_s \to i_{s+1}$ in the specified path from i_0 to i_k.

Example 6.7 Waiting Game

Consider Example 6.1. Suppose we start with the marker at 0. What is the probability that, after k flips of the coin, the marker will be at 1?

Solution We have already solved this problem in Example 3.4 in connection with the geometric random variable. It is instructive to rederive the result, however, using the multiplication rule for path probabilities.

Consider the paths of length k in the transition diagram that start at 0 and end at 1. For example, when $k = 2$, the possible paths are

$$(0) \to (1) \to (1) \quad \text{and} \quad (0) \to (0) \to (1)$$

By the multiplication rule, the probability of each of these paths is the product of the one-step probabilities, namely p for the first path and qp for the second. (Notice that the step $1 \to 1$ has probability 1.)

It is obvious from the transition diagram that every path starting at 0 and ending at 1 has exactly one segment $0 \to 1$ (the first outcome of heads). If the first outcome of heads appears on the mth flip (where $1 \le m \le k$), then the corresponding path is

$$\underbrace{(0) \to \cdots \to (0)}_{m} \to \underbrace{(1) \to \cdots \to (1)}_{k+1-m}$$

The probability of this particular path is $q^{m-1}p$. But any game of k rounds that starts at node 0 and ends at node 1 corresponds to exactly one such path. Hence the desired probability is the sum of all these path probabilities, for $m = 1, 2, \ldots, k$:

$$\begin{aligned} P\{X(k) = 1 \mid X(0) = 0\} &= p + pq + pq^2 + \cdots + pq^{k-1} \\ &= p\big(1 + q + q^2 + \cdots + q^{k-1}\big) \\ &= \frac{p(1 - q^k)}{1 - q} \\ &= 1 - q^k \end{aligned}$$

Here we have used the formula for the sum of a finite geometric series.

A easier way to obtain this answer is to consider the complementary event

$$\{X(k) = 0, \text{ given } X(0) = 0\}$$

This event corresponds to the single path $0 \to 0 \to \cdots \to 0$ of length k, which has probability q^k. ■

Example 6.8 Bold Play

Recall the stochastic model in Example 6.5: we start with \$1 and try to get to \$5 by following the bold play strategy. What are our chances of winning in k or fewer rounds?

Solution Following the same method as in Example 6.7, we look for paths of length k or less from state 1 to state 5. Since the transition probability from state 5 to itself is

1, we only need to consider paths that first reach node 5 at the last step. (Call such paths *essential*.) For each essential path, the probability of following it is the product of the one-step probabilities along the path, by formula (6.6). The desired probability is then the sum (over the set of essential paths of length $\leq k$) of these path probabilities.

To carry out this analysis, we see from the transition diagram that the shortest essential path is

$$(1) \to (2) \to (4) \to (5)$$

Call this path S. Its length is 3 steps, and the probability of following it is p^3 (win 3 consecutive rounds). This is also the probability that we will win in at most three rounds. The only essential path of length 4 steps is

$$(1) \to (2) \to (4) \to (3) \to (5)$$

and its probability is $p^3 q$ (win 3 rounds, lose 1 round). Call this path R. There are no essential paths of length 5 or 6 steps. Hence, the probability of winning in six or fewer rounds is $p^3 + p^3 q$.

There is an essential path of length 7 steps that consists of the loop

$$L: \quad (1) \to (2) \to (4) \to (3) \to (1)$$

followed by the path S. Write this path as $L \cdot S$ and define the number

$$\lambda = p^2 q^2$$

Since the path L has probability λ, the path $L \cdot S$ has probability λp^3. Hence the probability of winning in seven or fewer rounds is

$$p^3(1 + \lambda) + p^3 q$$

Similarly, we see that there is an essential path of length 8 steps, namely $L \cdot R$, with probability $\lambda p^3 q$. Adding this to the preceding probability, we find that the chance of winning in at most eight rounds is

$$p^3(1 + q)(1 + \lambda)$$

In general, the essential paths are obviously of two forms:

$$\underbrace{L \cdots L}_{n} \cdot S \quad \text{or} \quad \underbrace{L \cdots L}_{n} \cdot R \tag{6.7}$$

where $n = 0, 1, 2, \ldots$. These paths have respective probabilities $\lambda^n p^3$ and $\lambda^n p^3 q$. Note that $pq \leq 0.25$, so $\lambda \leq 0.0625$. Hence the probabilities of the paths (6.7) go to zero at a geometric rate as $n \to \infty$. Thus for an unlimited number of trials, the win probability is the sum of the infinite series

$$p^3(1 + q)(1 + \lambda + \lambda^2 + \cdots) = \frac{p^3(1 + q)}{1 - \lambda}$$

■

Example 6.9 Timid Play

Consider gambler's ruin (Example 6.2), where we use the timid play strategy of betting the minimum amount \$1 at each round. Suppose the total stake $N = 5$ and we start at state 1. What is the probability that we get to state 5 in k or fewer trials?

Solution We can use the same method as in Example 6.8, but the path enumeration is much more complicated. The shortest essential path in the transition diagram from node 1 to node 5 is of length 4 steps:

$$(1) \to (2) \to (3) \to (4) \to (5)$$

and it has probability p^4. This is the probability of winning in four or fewer trials, and requires winning four consecutive trials. There is no essential path of length 5 steps, but there are three different paths of length 6 steps, corresponding to winning five trials and losing one trial (with the single loss occurring at state 2, state 3, or state 4). For example,

$$(1) \to (2) \to (1) \to (2) \to (3) \to (4) \to (5)$$

is one of these three essential paths. Each such path has probability $p^5 q$. Thus the probability of winning the game in six or fewer trials is

$$p^4(1 + 3pq) \tag{6.8}$$

Enumerating all essential paths of length n can be done recursively, and the number of such paths can be determined using a generating function. The formulas are rather complicated and we omit the details. ■

Example 6.10 Bold Play versus Timid Play

In Example 6.5, we compared the bold play and the timid play strategies by computer simulation. Now let's see what the path analysis tells us about the advantage of bold play versus timid play when we face unfavorable odds.

For example, consider the choice of black or red in roulette, with the win probability $p = \frac{18}{38} = 0.4737$. Our chances of winning (going from \$1 to \$5 and stopping as soon as we get to \$5) in six or fewer trials using the timid play strategy can be calculated by formula (6.8) to be 0.0880. From the results of Example 6.8, the corresponding win probability using bold play is $p^3 + p^3 q = 0.1622$, which is about twice as high as the timid play odds.

When the number of roulette trials is unlimited, we know the probablity of winning using timid play because of our solution to the gambler's ruin problem (Example 2.27). Starting with stake $a = 1$, goal $N = 5$, and win/lose ratio $r = 18/20$, our ruin probability is

$$\frac{1 - r^4}{1 - r^5} = 0.8398$$

So, when play continues until either our fortune is either \$ 0 or \$5, our win probablity using timid play is 0.1602. By comparison, our win probablity for bold play was calculated in Example 6.8 to be

$$\frac{p^3(1 + q)}{1 - p^2 q^2} = 0.1730$$

We already calculated how advantageous bold play is in the first few rounds. This last comparison indicates the continuing superiority of bold play and is consistent with the simulation results in Tables 6.1 and 6.3 for the cases $p = 0.2$, 0.3 and 0.4. ■

6.3 More Examples

Here are some more examples of Markov Chains. There are two aspects to each example:

(i) Calculate the one-step transition probabilities p_{ij}.

(ii) Verify that the no-memory and time-homogeneity properties are valid.

Once we have carried out step (i), then we know by Theorem 6.1 that there is a Markov chain with the transition probabilities p_{ij}. The point of step (ii) is to make sure that the assignment of joint probabilities in this Markov chain is consistent with the situation described in the example.

Example 6.11 Repeated Independent Trials

Suppose we perform repeated independent trials of an experiment, where the possible outcomes at each trial are labeled $0, 1, 2, \ldots$ and outcome i has probability π_i. Let $X(n)$ be the outcome of the nth trial. Then $X(0), X(1), X(2), \ldots$ is a sequence of independent random variables, all having the p.m.f. π. For any pair of states i, j, we have

$$\begin{aligned} P\{X(n+1) = j \mid X(n) = i\} &= P\{X(n+1) = j\} \\ &= \pi_j \end{aligned}$$

by the independence of the successive trials. Thus the transition probabilities $p_{ij} = \pi_j$ don't depend on i. In terms of the board-game description, this means that current position of the marker has no influence at all on the choice of next position. This is clearly a very special type of Markov chain. ■

Example 6.12 Machine Maintenance

Consider a machine that can be in three possible states each day:

$$\begin{aligned} 0 &= \text{Working in good condition} \\ 1 &= \text{Working in poor condition} \\ 2 &= \text{Out of order} \end{aligned}$$

(This is a more realistic version of Example 6.4.) We assume that, on any particular day when the machine is working in good condition, the probability is δ that it will develop a defect. Suppose that, when this happens, the probability that the defect is major and will put the machine out of order is ϵ. From this description, we obtain the one-day transition probabilities from state 0:

$$p_{00} = 1 - \delta \qquad p_{01} = \delta(1 - \epsilon) \qquad p_{02} = \delta\epsilon$$

If the machine is currently working in poor condition, we assume that the probability is ϕ that on the next day it will fail completely, and that the probability is 0 that on the next day it will be working in good condition. This furnishes the one-day transition probabilities from state 1:

$$p_{10} = 0 \qquad p_{11} = 1 - \phi \qquad p_{12} = \phi$$

Finally, when the machine is out of order today, the probability that it will be repaired by tomorrow is ρ. However, the repair only has probability γ of restoring the machine to good working order. Thus the one-day transition probabilities from state 2 are

$$p_{20} = \gamma\rho \qquad p_{21} = (1 - \gamma)\rho \qquad p_{22} = 1 - \rho$$

The transition diagram for this Markov chain is shown in Figure 6.8.

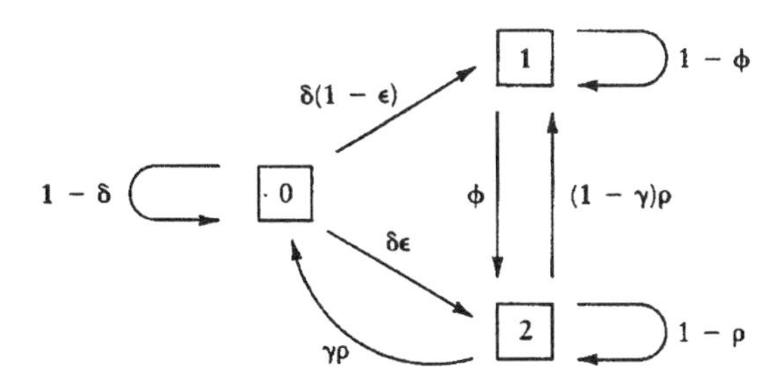

Figure 6.8: Machine Maintenance

In using this model, we would choose the initial probability distribution $\boldsymbol{\pi}$ to describe the condition of the machine when first installed. For example, a new machine in good working order would have initial distribution $\boldsymbol{\pi} = \begin{bmatrix} 1 & 0 & 0 \end{bmatrix}$. ■

Example 6.13 Genetic Chain

Consider the following genetic model: A gene is made up of N subgenes, each of which can be normal or abnormal. We say that a gene is in state i when it contains i normal subgenes (so the possible states are $0, 1, \ldots, N$). Assume that when a gene replicates, it first doubles itself so that it now has $2N$ subgenes. Then a new gene, composed of a set of N subgenes randomly selected from the $2N$, splits off. If we assume that normal subgenes replicate to normal, and abnormal to abnormal, then the replicated gene has $2i$ normal and $2N - 2i$ abnormal subgenes. Thus the probability that the new gene is in state j (that is, contains j normal subgenes and $N - j$ abnormal subgenes) is

$$p_{ij} = \frac{\binom{2i}{j}\binom{2N-2i}{N-j}}{\binom{2N}{N}} \tag{6.9}$$

This follows because the denominator in formula (6.9) is the number of subsets of size N chosen from $2N$ (all assumed to be equally likely), and the numerator is the number of subsets containing j normal subgenes and $N - j$ abnormal subgenes.

The combinatorial derivation of formula (6.9) makes it obvious that

$$\sum_{j=0}^{N} p_{ij} = 1$$

so we can define a Markov chain with p_{ij} as the transition probabilities. Using this chain, we can follow a randomly chosen line of descent of the original gene. The initial probability distribution π is determined by the distribution of normal and abnormal subgenes in the original gene. To say the chain is in state i at time n means that the nth gene in the line of descent has i normal subgenes. ■

Example 6.14 Inventory Model

A wholesale distributor has a randomly fluctuating inventory of a certain item. Each business day, orders for this item come from many retail stores, and are shipped out at the end of the day, provided that a sufficient supply of the item is on hand. (Whenever the stock is depleted, the unfilled demand is carried over to the next day as a back order). Let the state of the inventory system be the number of units in stock at the end of the day (after the day's orders are shipped). Assume that the full inventory size is S, so that the possible states are $S, S-1, \ldots, 1, 0, -1, \ldots$, where negative values correspond to back orders.

Suppose that the management policy for maintaining the inventory is to fix a *critical level* $s < S$, and then wait until the inventory level at the end of a day drops to s or lower before restocking. When this happens, an overnight delivery to the warehouse brings enough items to fill all back orders the following day and to raise the inventory up to size S again.

To model this system as a Markov chain, we assume that the retail stores order a total of $D(n)$ items on the nth day, where $D(n)$ is a random variable. If $X(n)$ as usual denotes the state of the system at the end of day n, then the inventory at the beginning of day $n+1$ is $X(n)$ if $X(n) > s$. If $X(n) \leq s$, then the inventory is S plus the number of back orders. After the back orders and the new orders on day $n+1$ are filled, the remaining inventory at the end of the day is

$$X(n+1) = \begin{cases} X(n) - D(n+1) & \text{if } X(n) > s \\ S - D(n+1) & \text{if } X(n) \leq s \end{cases} \tag{6.10}$$

If we assume that the demand sequence $D(1), D(2), D(3), \ldots$ is mutually independent, then we see from formula (6.10) that $X(n+1)$ depends on $X(n)$ and $D(n+1)$ but is independent of $X(n-1), X(n-2), \ldots$. Hence the no-memory property is satisfied. If we also assume that the random variables $\{D(n)\}$ are identically distributed, the time-homogeneity property is satisfied, too, and we have a Markov chain.

As a simple example of this model, suppose the daily demand is for either one item (probability p) or zero items (probability $q = 1 - p$). Thus $\{D(n)\}$ is a sequence of Bernoulli random variables in this case. Take the full inventory level to be $S = 2$ and the critical level $s = 0$. The states are then 2, 1, and 0. The one-day transitions $2 \to 1$ and $1 \to 0$ corresponding to an incoming order each occur with probability p.

To analyze the transitions from state 0, suppose the inventory is exhausted at the end of day n. At the beginning of day $n+1$, the inventory is back up to 2 because of restocking. At the end of day $n+1$, the inventory is either 1 (if $D(n+1)=1$) or 2 (if $D(n+1)=0$). Thus, the one-day transition $0 \to 1$ occurs with probability p, and the one-day transition $0 \to 2$ occurs with probability q. With this reasoning, we obtain the transition diagram shown in Figure 6.9. ■

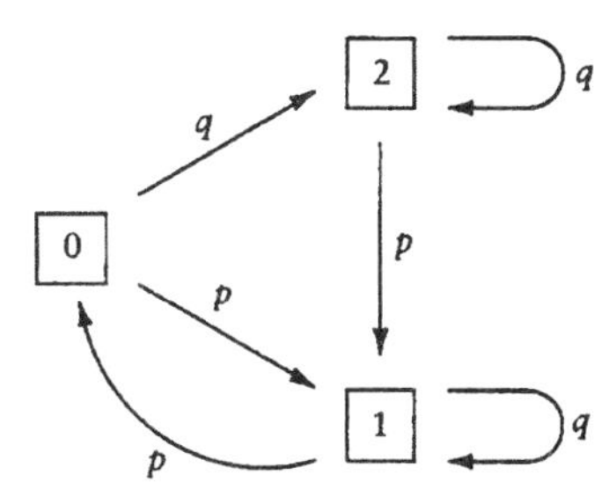

Figure 6.9: Inventory Model

Example 6.15 Simulation of Inventory Model

More complicated versions of the inventory model in Example 6.14 can be simulated on a computer, using random numbers to generate the random daily demands $D(n)$.

For example, suppose $S = 3$ and $s = 0$. (This means that no restocking is done until the current inventory is depleted.) Suppose that the possible values for $D(n)$ are 0, 1, 2, 3, or 4, with respective probabilities 0.1, 0.2, 0.3, 0.25, and 0.15. To simulate the demands, we use the method of Example 3.20. First, define the function

$$g(u) = \begin{cases} 0 & \text{for } 0 < u \le 0.1 \\ 1 & \text{for } 0.1 < u \le 0.3 \\ 2 & \text{for } 0.3 < u \le 0.6 \\ 3 & \text{for } 0.6 < u \le 0.85 \\ 4 & \text{for } 0.85 < u < 1 \end{cases}$$

Now use a random-number generator to obtain a sequence $u_1, u_2, \ldots$ uniformly distributed on the interval $0 < u < 1$. The corresponding simulated demand sequence is $d_1 = g(u_1)$, $d_2 = g(u_2)$, and so on.

If operations start with $X(0) = 3$ (full inventory), then the simulated values of $X(n)$ corresponding to the demand sequence are $x_1 = 3 - d_1$,

$$x_2 = \begin{cases} x_1 - d_2 & \text{if } x_1 \ge 1 \\ 3 - d_2 & \text{if } x_1 \le 0 \end{cases}$$

and so forth.

Numerical Example To simulate five days of operations of the inventory system, we use the random-number generator to get values

$$u_1 = 0.405 \quad u_2 = 0.829 \quad u_3 = 0.152 \quad u_4 = 0.927 \quad u_5 = 0.556$$

The corresponding sample demand values $d_n = g(u_n)$ are then

$$d_1 = 2 \quad d_2 = 3 \quad d_3 = 1 \quad d_4 = 4 \quad d_5 = 2$$

The sequence of end-of-day inventory levels is

$$\begin{aligned} x_1 &= 3 - d_1 = 1 \\ x_2 &= 1 - d_2 = -2 \quad \text{(This requires overnight restocking of 5 items)} \\ x_3 &= 3 - d_3 = 2 \\ x_4 &= 2 - d_4 = -2 \quad \text{(Another overnight restocking order of 5 items)} \\ x_5 &= 3 - d_5 = 1 \end{aligned}$$

In this five-day simulation, the inventory fell below the critical level at the end of days 2 and 4. The average end-of-day unfilled demand was 0.8 orders per day.

By changing the values of S and the critical level s, we can analyze the behavior of the system with these same random demands but different inventory management policies. Taking into account the restocking costs and the costs of unfilled demand, we can compare the effectiveness of different policies. ■

Example 6.16 Ehrenfest Urn Model

A famous Markov chain in physics is the Ehrenfest urn model for diffusion of a gas through a membrane separating two containers. In this discrete model, each container is represented by an urn. The gas molecules are represented by balls numbered $1, 2, \ldots, N$ distributed between the two urns.

Label the urns A and B, and say that the system is in state k when there are k balls in urn A. Thus, the possible states are $0, 1, \ldots, N$. Assume that the random mechanism for changing the state of the system at each time step is as follows. Pick an integer at random between 1 and N, and move the ball that bears this number from its current urn to the other urn. If the system is currently in state i, then the chance that the ball picked is in urn A is i/N. When this happens, the state changes to $i - 1$. Thus,

$$p_{i,i-1} = \frac{i}{N} \quad \text{if } 1 \leq i \leq N$$

By the same reasoning,

$$p_{i,i+1} = 1 - \left(\frac{i}{N}\right) \quad \text{if } 0 \leq i < N$$

All the other transition probabilities are zero. The transition diagram is shown in Fig. 6.10. From the diagram, we would expect that, when A starts out having most of the balls, it tends to empty and B tends to fill up. Then the roles of A and B are reversed, and the balls flow back into A. This model is an example of a random walk with *reflecting barriers* at 0 and N (compare this with the gambler's ruin model in Section 6.1). ■

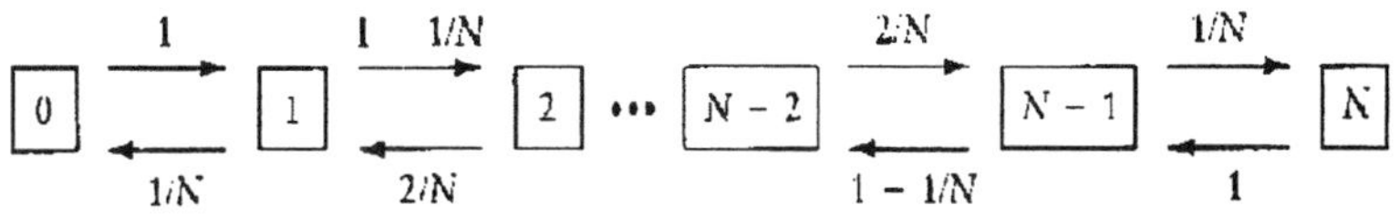

Figure 6.10: Ehrenfest Urn Model

Example 6.17 Non-Markov Chain

Consider a sequence of Bernoulli trials, with outcomes $Y(0), Y(1), Y(2), \ldots$. Here the $Y(j)$ are mutually independent and take the values 0 or 1 with probabilities q and p, respectively. By Example 6.11, $\{Y(n)\}$ forms a Markov chain.

We set

$$X(n) = Y(n) + Y(n+1) \quad \text{for } n = 0, 1, 2, \ldots$$

Then each $X(n)$ is a binomial random variable with parameters $(2, p)$. However, the sequence $\{X(n)\}$ does *not* form a Markov chain with states $\{0, 1, 2\}$, because the no-memory property in Equation (6.4) does not always hold. For example,

$$\begin{aligned} P\{X(2) = 0 \mid X(1) = 1\} &= P\{Y(3) = 0 \text{ and } Y(2) = 0 \mid X(1) = 1\} \\ &= P\{Y(3) = 0\} \cdot P\{Y(2) = 0 \mid X(1) = 1\} \end{aligned} \tag{6.11}$$

by the independence of $Y(3)$ and $X(1)$. But the event $\{Y(2) = 0 \text{ and } X(1) = 1\}$ can also be described as $\{Y(2) = 0 \text{ and } Y(1) = 1\}$. Hence by definition of conditional probability,

$$\begin{aligned} P\{Y(2) = 0 \mid X(1) = 1\} &= \frac{P\{Y(2) = 0 \text{ and } Y(1) = 1\}}{P\{X(1) = 1\}} \\ &= \frac{pq}{2pq} = \frac{1}{2} \end{aligned}$$

Substituting this in equation (6.11), we get

$$P\{X(2) = 0 \mid X(1) = 1\} = \frac{q}{2}$$

On the other hand, if we include the past history in this calculation, we can get a different value for the transition probability. For example, the event $\{X(1) = 1 \text{ and } X(0) = 0\}$ is the same as the event

$$B = \{Y(2) = 1 \text{ and } Y(1) = 0 \text{ and } Y(0) = 0\}$$

Hence,

$$\begin{aligned} P\{X(2) = 0 \mid X(1) = 1 \text{ and } X(0) = 0\} &= P\{Y(3) = 0 \text{ and } Y(2) = 0 \mid B\} \\ &= 0 \end{aligned}$$

This violates the Markov property. ■

Example 6.18 Enlarging the State Space

Let the random variables $X(n)$ and $Y(n)$ be as defined in Example 6.17. The intuitive reason why the sequence $\{X(n)\}$ does not form a Markov chain is that each $X(n)$ depends on *two* time steps relative to the Markov chain $\{Y(n)\}$. This creates an overlapping effect that allows $X(n-1)$ to influence the transition from $X(n)$ to $X(n+1)$, as we verified in Example 6.17.

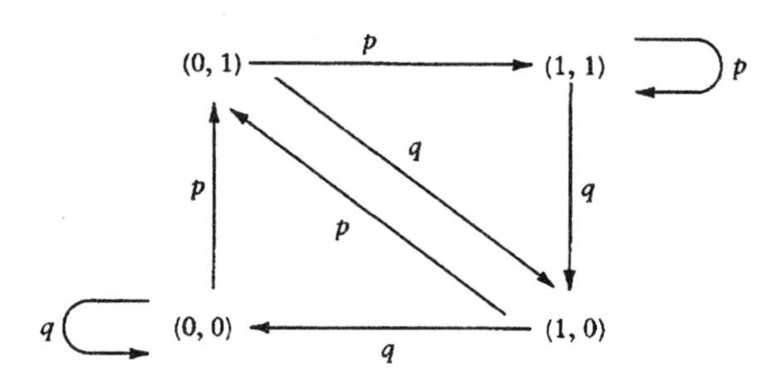

Figure 6.11: Enlarged State Space

We can turn the model in Example 6.17 into a Markov chain by enlarging the state space. Define a new state space, using the four possible pairs of values $Y(n)$, $Y(n+1)$. It is natural to label these new states by the ordered pairs $(0,0)$, $(0,1)$, $(1,0)$, and $(1,1)$, rather than by the integers $0, 1, 2, 3$. The one-step transition probability $p_{\alpha\beta}$ from state $\alpha = (i, j)$ to state $\beta = (k, m)$ is obviously 0 if $j \neq k$. If $j = k$, then

$$p_{\alpha\beta} = P\{Y(n+1) = j \text{ and } Y(n+2) = m \mid Y(n) = i \text{ and } Y(n+1) = j\} \quad (6.12)$$

By the independence of the sequence $\{Y(n)\}$, it is obvious that the probability in (6.12) is simply $P\{Y(n+2) = m\}$. Hence the transition diagram is as shown in Figure 6.11.

From formula (6.12) it is easy to check that the no-memory property is satisfied for this system. The past history before time n involves $Y(n)$, $Y(n-1), \ldots, Y(0)$. By the independence of the Y's, the only aspect of the past that influences the move at time $n+1$ is the value of $Y(n)$. But we have included this value in the present state $(Y(n), Y(n+1))$, so we may forget the past in calculating the probabilities for the next move. This is the Markov property.

Notice that the sequence $\{X(n)\}$ from Example 6.17 can be analyzed using the enlarged state space of this new Markov chain, since $X(n)$ is the sum of the coordinates of the state at time n. ■

6.4 n-Step Transition Probabilities

Let $\{X(n)\}$ be a Markov chain with one-step transition probabilities p_{ij}. Define the **n-step transition probabilities**

$$p_{ij}(n) = P\{X(n) = j \mid X(0) = i\}$$

In particular, $p_{ij}(1) = p_{ij}$. In terms of playing a game (treating the transition diagram as the game board described in Section 6.1), $p_{ij}(n)$ is the probability of moving the marker from node i to node j in n trials. (Don't confuse this with moving the marker n times. Whenever the transition $i \to i$ is randomly chosen at a trial, the marker doesn't move. This choice still counts as one trial, however, and thus contributes to the total number of trials.) In particular, for $i = j$, $p_{ii}(n)$ is the probability that, when the marker starts at node i, it is back at node i after n trials. The time homogeneity of the chain implies that, for any integer $k \geq 0$,

$$p_{ij}(n) = P\{X(k+n) = j \mid X(k) = i\}$$

Lemma 6.1. *For all states i, j and all integers $n \geq 0$,*

$$p_{ij}(n+1) = \sum_k p_{ik}(n) p_{kj}(1) = \sum_k p_{ik}(1) p_{kj}(n)$$

Proof By conditioning on the value of $X(n)$ and using the Rule of Total Causes (equation (2.24)), we can calculate $p_{ij}(n+1)$ as the sum

$$\sum_k P\{X(n) = k \mid X(0) = i\} P\{X(n+1) = j \mid X(n) = k, X(0) = i\} \qquad (6.13)$$

But by the no-memory and time-homogeneity properties

$$\begin{aligned} P\{X(n+1) = j \mid X(n) = k \text{ and } X(0) = i\} &= P\{X(n+1) = j \mid X(n) = k\} \\ &= p_{kj} \end{aligned}$$

Substituting this in formula (6.13) gives the first equation of Lemma 6.1. To obtain the second equation, condition on the value of $X(1)$ and use the same argument. ■

Chapman-Kolmogorov Equation
For all states i, j and all integers $m > 0, n \geq 0$,

$$p_{ij}(m+n) = \sum_k p_{ik}(m)\, p_{kj}(n) \qquad (6.14)$$

Proof For $m = 1$, this is true for all n, by Lemma 6.1. Now assume that equation (6.14) is true for a particular value of m and for all n. Apply the induction hypothesis for this value of m and for the value $n + 1$:

$$\begin{aligned} p_{ij}(m+n+1) &= \sum_k p_{ik}(m) p_{kj}(n+1) \\ &= \sum_k p_{ik}(m) \Big(\sum_s p_{ks} p_{sj}(n) \Big) && \text{(Lemma 6.1)} \\ &= \sum_s \Big(\sum_k p_{ik}(m)\, p_{ks} \Big) p_{sj}(n) && \text{(interchange order)} \\ &= \sum_s p_{is}(m+1)\, p_{sj}(n) && \text{(Lemma 6.1)} \end{aligned}$$

Hence, equation (6.14) is true, with m replaced by $m + 1$, for all positive integers n. This completes the inductive proof of equation (6.14). ■

One method of calculating n-step transition probabilities from state i to state j is to find all the paths of length n in the transition diagram that start at state i and end at state j. Here, by the *length* of a path we mean the number of number of time steps needed to go from the beginning to the end of the path. (This number is the same as the number of arrows that make up the path.) Denote by $T_{ij}(n)$ the set of these paths. The probability $P\{\gamma\}$ for a particular path γ is the product of the probabilities of each one-step segment. Since each different path correponds to a different way of going from state i to state j, it follows that

$$p_{ij}(n) = \sum_{\gamma \in T_{ij}(n)} P\{\gamma\}$$

We illustrated this method in Examples 6.7, 6.8, and 6.9.

An alternate method for calculating $p_{ij}(n)$ is to use matrix multiplication. This method avoids the combinatorics of path-enumeration. For a Markov chain with a finite number of states this method is well suited to computer calculations and to investigations of the long-term behavior of the model.

Let $\mathbf{P}$ be the matrix with entry p_{ij} in row i and column j, as i and j range over all states of the chain. If the states are labeled $1, 2, \ldots,$ then

$$\mathbf{P} = \begin{bmatrix} p_{11} & p_{12} & p_{13} & \cdots \\ p_{21} & p_{22} & p_{23} & \cdots \\ p_{31} & p_{32} & p_{33} & \cdots \\ \vdots & \vdots & \vdots & \ddots \end{bmatrix}$$

We call $\mathbf{P}$ the **one-step transition matrix** for the Markov chain. If the chain has a finite number N of states, then $\mathbf{P}$ is an $N \times N$ matrix. In this chapter we emphasize this case, although some of the examples (such as Bernoulli trials) require an infinite number of states.

Definition
For each integer $n = 1, 2, \ldots,$ let $\mathbf{P}(n)$ denote the n-step transition matrix, with entry $p_{ij}(n)$ in row i and column j.

Observe that the ith row of $\mathbf{P}(n)$ is the conditional probability mass function of $X(n)$, given that $X(0) = i$. Thus the entries in $\mathbf{P}(n)$ satisfy

$$0 \leq p_{ij}(n) \leq 1 \quad \text{and} \quad \sum_j p_{ij}(n) = 1$$

Any matrix possessing these two properties is called a **stochastic matrix**.

Theorem 6.2. *For all integers $n \geq 1$ the n-step transition matrix is the matrix product $\mathbf{P}(n) = \mathbf{P}^n$.*

Proof Theorem 6.2 is true for $n = 1$, by definition. If we assume by induction that it is true for a particular value of n, then, by Lemma 6.1,

$$p_{ij}(n+1) = \sum_k (\mathbf{P}^n)_{ik}\, p_{kj}(1) = (\mathbf{P}^{n+1})_{ij}$$

by definition of the product of matrices. Hence, the formula is true for $n+1$. ■

Remark The Chapman–Kolmogorov equation 6.14 is an immediate consequence of Theorem 6.2 and the associativity of matrix multiplication:

$$\mathbf{P}(m+n) = \mathbf{P}^{m+n} = \mathbf{P}^m\ \mathbf{P}^n = \mathbf{P}(m)\ \mathbf{P}(n)$$

(In fact, once we know that $\mathbf{P}(m+1) = \mathbf{P}(m)\ \mathbf{P}$, the rest of the proof of the Chapman-Kolmogorov equation is simply the proof of the associativity of matrix multiplication.) ■

Now that we have calculated the n-step transition probabilities using matrix multiplication, let's do the same thing for the initial probabilities. Write the probability mass function of $X(0)$ as a row vector:

$$\pi = [\ \boldsymbol{\pi}_1 \quad \boldsymbol{\pi}_2 \quad \dots\ ]$$

To find the p.m.f. for $X(n)$, condition on $X(0)$ and use Theorem 6.2:

$$\begin{aligned} P\{X(n)=j\} &= \sum_i P\{X(0)=i\} \cdot P\{X(n)=j \mid X(0)=i\} \\ &= \sum_i \pi_i (P^n)_{ij} \\ &= \left(\pi \cdot P^n\right)_j \end{aligned}$$

Thus, the row vector corresponding to the (unconditional) p.m.f. for $X(n)$ is simply the matrix product $\boldsymbol{\pi} \cdot \mathbf{P}^n$. In particular, if $P\{X(0)=i\} = 1$, then the vector $\boldsymbol{\pi}$ has a 1 in the ith position, and 0's elswhere. Hence, $\boldsymbol{\pi} \cdot \mathbf{P}^n$ is the ith row of the matrix $\mathbf{P}^n$ in this case.

Example 6.19 Waiting Game

Consider Example 6.1. The matrix $\mathbf{P}$ in this case is

$$\mathbf{P} = \begin{bmatrix} q & p \\ 0 & 1 \end{bmatrix}$$

which is upper triangular. (Here the first row corresponds to state 0, and the second row to state 1.) Thus, the diagonal elements of $\mathbf{P}^n$ are q^n and 1. (Recall that when two upper-triangular matrices are multiplied, the diagonal elements of the product are the products of the correponding diagonal elements.) Since the row-sums of $\mathbf{P}^n$ are all 1, it follows that

$$\mathbf{P}^n = \begin{bmatrix} q^n & 1-q^n \\ 0 & 1 \end{bmatrix}$$

This agrees with the calculations made via path probabilities in Example 6.7. ■

Example 6.20 Ehrenfest Urn Model

Consider Example 6.16, when there are two balls to distribute between the urns. There are three states in this case. The transistion matrix **P** satisfies

$$\mathbf{P} = \begin{bmatrix} 0 & 1 & 0 \\ 0.5 & 0 & 0.5 \\ 0 & 1 & 0 \end{bmatrix} \qquad \mathbf{P}^2 = \begin{bmatrix} 0.5 & 0 & 0.5 \\ 0 & 1 & 0 \\ 0.5 & 0 & 0.5 \end{bmatrix} \qquad \mathbf{P}^3 = \mathbf{P}$$

Thus, this chain is **periodic**, with period 2:

$$\begin{aligned} \mathbf{P}^4 &= (\mathbf{P}^3)\mathbf{P} = \mathbf{P}^2 \\ \mathbf{P}^5 &= (\mathbf{P}^4)\mathbf{P} = (\mathbf{P}^2)\mathbf{P} = \mathbf{P}^3 = \mathbf{P} \end{aligned}$$

All the odd-step transition matrices are the same as **P**, and the even-step transition matrices are the same as $\mathbf{P}^2$. ■

Example 6.21 Successes in Bernoulli Trials

Consider Example 6.3. The matrix **P** has infinitely many rows and columns; it starts as

$$\mathbf{P} = \begin{bmatrix} q & p & 0 & 0 & \cdots \\ 0 & q & p & 0 & \cdots \\ 0 & 0 & q & p & \cdots \\ 0 & 0 & 0 & q & \cdots \\ \vdots & \vdots & \vdots & \vdots & \ddots \end{bmatrix}$$

Notice that **P** is upper triangular. This makes it easy to calculate the powers of **P**, which will also be upper triangular:

$$\mathbf{P}^2 = \begin{bmatrix} q^2 & 2pq & p^2 & 0 & \cdots \\ 0 & q^2 & 2pq & p^2 & \cdots \\ 0 & 0 & q^2 & 2pq & \cdots \\ 0 & 0 & 0 & q^2 & \cdots \\ \vdots & \vdots & \vdots & \vdots & \ddots \end{bmatrix}$$

(binomial probabilities with parameters $(2, p)$ on each row, starting at the diagonal); and

$$\mathbf{P}^2 = \begin{bmatrix} q^3 & 3q^2p & 3qp^2 & p^3 & 0 & \cdots \\ 0 & q^3 & 3q^2p & 3qp^2 & p^3 & \cdots \\ 0 & 0 & q^3 & 3q^2p & 3qp^2 & \cdots \\ 0 & 0 & 0 & q^3 & 3q^2p & \cdots \\ \vdots & \vdots & \vdots & \vdots & \vdots & \ddots \end{bmatrix}$$

(binomial probabilities with parameters $(3, p)$ on each row, starting at the diagonal).

In general, the entries of $\mathbf{P}^n$ below the diagonal are zero. For $j = i + k$, with $0 \le k \le n$, we have

$$p(n)_{ij} = \binom{n}{k} q^{n-k} p^k$$

and $p(n)_{ij} = 0$ when $j > n + i$.

A curious thing happens in this model as $n \to \infty$. If we take a fixed row of $\mathbf{P}^n$, then the individual elements in this row get smaller and smaller as n increases, even though their sum remains 1. For example, if $p = q = 0.5$ and $n = 2m$ is even, then the binomial coefficients are symmetric about the midpoint, and the largest element of the first row is

$$\binom{2m}{m} 2^{-2m} \tag{6.15}$$

Replace the factorials in the binomial coefficient by Stirling's approximation

$$k! \sim \left(\frac{k}{e}\right)^k \sqrt{2\pi k}$$

Then the matrix element (6.15) is approximately $1/\sqrt{\pi m}$, which goes to zero as $m \to \infty$.

By such calculations we find that for each fixed pair of states i and j, the n-step probability $p(n)_{ij} \to 0$ as $n \to \infty$. Thus the limit of the infinite stochastic matrix $\mathbf{P}(n)$ as $n \to \infty$ is the zero matrix. This seems to contradict the Normalization Axiom of probability. The Law of Large Numbers explains what is happening: since the expected number of successes in n trials is np (which is tending toward ∞), the observed number of successes is also tending toward ∞ as n increases. The limiting matrix $\mathbf{P}(\infty)$ should have another column filled with 1's, corresponding to a state labeled ∞. This shows that results that are valid for $N \times N$ stochastic matrices with N finite do not automatically extend to infinite matrices.

6.5 Classification of States

With many examples now available as a guide, we shall analyze finite Markov chains in a systematic way. Many of the ideas also apply to infinite chains, such as the one in Example 6.3, but such chains raise subtle convergence questions involving the escape of the marker to infinity (see Example 6.21) that we shall not treat here. The gambler's ruin chain and the vending machine chain turn out to be the extreme cases, and a general finite-state chain is more or less a mixture of these two.

As a first step toward obtaining general results about finite Markov chains, let's go back to the board-game description and ask the following question:

> If the game starts with the marker at state i, then which states is it possible for the marker to visit during the course of the game? (Assume that the number of moves is unlimited.)

We write $i \to j$ if j can be reached from i with positive probability. In terms of the transition diagram, this means that there is at least one path from i to j. (Remember the paths are one-way!) It is useful to allow the path of length 0 from i to i in this definition, so that we always have $i \to i$.

To clarify this notion, let's review the examples in Section 6.1. For the waiting game, we have $0 \to 1$, but there is no path from 1 to 0. For the vending machine, we have $0 \to 1$ and $1 \to 0$. For bold play, we can get to any state starting at one of the states $1, 2, 3, 4$; but if we start at state 0 or at state 5, we can't go to any other states.

For any chain with a finite number N of states, it is a finite problem to determine whether $i \to j$. Indeed, if there exists a path from i to j, then either it has length less than N (with or without loops) or else it has length N or greater and it passes through some state k twice (see the example of bold play). By omitting the loop from k to k in a path with length $\geq N$, we obtain a shorter path. Now repeat this process a finite number of times, until we get a path with no repetitions. This final path must have length less than N.

The path approach to Markov chains is quite accessible intuitively, but it can be cumbersome computationally. Let $\mathbf{P}$ be the one-step transition matrix. The following matrix criterion is useful, especially for complicated transition diagrams.

Lemma 6.2. *Suppose a Markov chain has $N < \infty$ states. Let i and j be a pair of states. Then $i \to j$ if and only if there is an integer $0 \leq n < N$ such that the i, j entry of $\mathbf{P}^n$ is positive.*

Proof By Theorem 6.2 we know that the i, j entry of $\mathbf{P}^n$ is the probability of starting in state i and reaching state j after n trials. (Note that for $n = 0$ we have $\mathbf{P}^0 = \mathbf{I}$, the identity matrix, so the i, i entry of $\mathbf{P}^0$ is 1. This is consistent with the definition above.). By the path analysis made above, we can restrict n to be less than N. ■

Remark Define the matrix

$$\mathbf{G} = \mathbf{I} + \mathbf{P} + \mathbf{P}^2 + \cdots + \mathbf{P}^{N-1}$$

where $\mathbf{I}$ is the $N \times N$ identity matrix. Since all the entries in $\mathbf{P}^k$ are nonnegative, it follows from Lemma 6.2 that $i \to j$ if and only if $\mathbf{G}_{ij} > 0$. ■

Definition
States i and j **communicate** *if $i \to j$ and also $j \to i$. Thus there is at least one path in the transition diagram from i to j and vice versa. We write $i \leftrightarrow j$ when this is the case.*

In this definition, the lengths of the paths leading from i to j and from j back to i are not restricted. For example, in bold play, a path of length one exits from state 1 to state 2, but the shortest path from state 2 back to state 1 has length three.

Definition
A Markov chain is **irreducible** *if every pair of states communicate.*

Of the Markov chains in Section 6.1, only the vending machine is irreducible. In Section 6.3, the machine maintenance model, the Ehrenfest model, and Example 6.18 are irreducible.

For each state i, let

$$C(i) = \{\text{All states } j \text{ such that } i \leftrightarrow j\}.$$

Call $C(i)$ the **class** of the state i. If the chain is irreducible, then, by definition, there is only one class of states. On the other hand, consider the bold play example in Section 6.1. Here, the classes are

$$C(0) = \{0\} \qquad C(5) = \{5\} \qquad C(1) = \{1, 2, 3, 4\}$$

This grouping of states into classes has three general properties:

(i) $i \in C(i)$ for every state i.

(ii) If $j \in C(i)$ then $i \in C(j)$.

(iii) For any two states i and j, either $C(i) = C(j)$ or $C(i)$ is disjoint from $C(j)$.

Properties (i) and (ii) are obvious from the definitions. To prove (iii), suppose that $C(i)$ and $C(j)$ have some state k in common. Then $i \leftrightarrow k$ and $j \leftrightarrow k$. Hence $i \leftrightarrow j$ (use paths passing through k).

By properties (i), (ii), and (iii), we can always break up the state space S into a union of classes, with the classes mutually disjoint:

$$S = C_1 \cup C_2 \cup \cdots$$

For the bold play example, this decomposition is

$$S = \{0\} \cup \{1, 2, 3, 4\} \cup \{5\}$$

In defining the classes of a Markov chain, we require a two-way communication between all the states in a single class. On the other hand, when we view the chain as a board game, using the transition diagram, each path is one-way, and the marker moves along the paths in a specific direction. (Time is not reversible—we cannot make the clock run backward.) For chains such as bold play or gambler's ruin that have more than one class of states, it is possible to leave one class of states and enter another class. When this happens, it is then impossible to reenter the first class at some future move (otherwise, the two classes would coincide).

We shall call the state i **absorbing** if $p_{ii} = 1$. Since this forces $p_{ij} = 0$ for $i \neq j$, such a state does not communicate with any of the other states. The class containing it is simply $\{i\}$. For gambler's ruin (Example 6.2), states 0 and N are absorbing. Notice that, when a transition into an absorbing state occurs, the game is over.

A class C is called **ergodic** if every path that starts in C remains in C. That is,

$$P\{X(n) \in C \text{ for all } n \geq 1 \mid X(0) \in C\} = 1.$$

In matrix terms, a class C is ergodic if the partial row sum

$$\sum_{j \in C} p_{ij} = 1$$

for every $i \in C$. The individual states in an ergodic class will also be called *ergodic*. For example, a single absorbing state is obviously an ergodic class. An irreducible chain consists of a single ergodic class. In the bold play example, the only ergodic classes are the absorbing states $\{0\}$ and $\{5\}$.

A class C is called **transient** if there is a path out of C. In matrix terms, this can be described as follows: choose $i \in C$ and $k \notin C$ so that $i \to k$. Then the partial row sum

$$\sum_{j \in C} p_{ij} < 1 \tag{6.16}$$

Conversely, if inequality (6.16) holds for some $i \in C$, then there must be a state $k \notin C$ with $p_{ik} > 0$, since every row sum of $\mathbf{P}$ must equal 1. Hence, $i \to k$, and C

is transient. The individual states in a transient class are also be called *transient*. For example, in bold play, the class $\{1, 2, 3, 4\}$ is transient, since there are paths from this class to classes $\{0\}$ and $\{5\}$.

Standard Form for Transition Matrix

When a Markov chain includes more than one class of states, the powers of the transition matrix $\mathbf{P}$ are easier to analyze if we group the states in the following order: group the states in classes, as defined previously, making sure to list the ergodic classes before the transient classes.

For example, in gambler's ruin with combined stakes $N = 4$, we would list the classes in the order $\{0\}$, $\{4\}$, $\{1, 2, 3\}$. The matrix $\mathbf{P}$ would then become

$$\begin{array}{c} \text{State} \\ 0 \\ 4 \\ 1 \\ 2 \\ 3 \end{array} \quad \begin{array}{c} \begin{array}{ccccc} 0 & 4 & 1 & 2 & 3 \end{array} \\ \left[\begin{array}{ccccc} 1 & 0 & 0 & 0 & 0 \\ 0 & 1 & 0 & 0 & 0 \\ q & 0 & 0 & p & 0 \\ 0 & 0 & q & 0 & p \\ 0 & p & 0 & q & 0 \end{array}\right] \end{array}$$

Notice that each row sums to 1 ($\mathbf{P}$ is a stochastic matrix). We can write this matrix in block form as

$$\mathbf{P} = \begin{bmatrix} \mathbf{S} & \mathbf{0} \\ \mathbf{R} & \mathbf{Q} \end{bmatrix} \tag{6.17}$$

where

$$\mathbf{S} = \begin{bmatrix} 1 & 0 \\ 0 & 1 \end{bmatrix} \qquad \mathbf{R} = \begin{bmatrix} q & 0 \\ 0 & 0 \\ 0 & p \end{bmatrix} \qquad \mathbf{S} = \begin{bmatrix} 0 & p & 0 \\ q & 0 & p \\ 0 & q & 0 \end{bmatrix}$$

The matrices $\mathbf{R}$ and $\mathbf{Q}$ are **substochastic**: each row sum is ≤ 1.

In general, the ordering of the states we have just described will result in $\mathbf{P}$ having the shape shown in formula (6.17). Here, the block $\mathbf{Q}$ is of size $t \times t$ (with t the number of transient states). The matrix $\mathbf{Q}$ describes the transient $\to$ transient movements in the chain. For each class of transient states, at least one row in $\mathbf{Q}$ for this class will have sum < 1. The block $\mathbf{R}$ is of size $t \times e$ (with e the number of ergodic states). The matrix $\mathbf{R}$ describes the transient $\to$ ergodic movements in the chain. For each class of transient states, at least one row in $\mathbf{R}$ for this class will have a nonzero entry. Finally, the $e \times e$ block $\mathbf{S}$ describes the movements within each ergodic class in the chain. Suppose that the chain has d ergodic classes. Since it is impossible to leave an ergodic class, $\mathbf{S}$ has the block-diagonal shape

$$\mathbf{S} = \begin{bmatrix} \mathbf{S}_1 & \mathbf{0} & \cdots & \mathbf{0} \\ \mathbf{0} & \mathbf{S}_2 & \cdots & \mathbf{0} \\ \vdots & \vdots & \ddots & \vdots \\ \mathbf{0} & \mathbf{0} & \cdots & \mathbf{S}_d \end{bmatrix}$$

where $\mathbf{S}_i$ is the transition matrix within the ith ergodic class.

Periodic and Aperiodic Classes

Thus far in our analysis of the possible paths in a Markov chain, we have usually ignored the length of the paths. Because time is measured discretely in these models, a special periodicity phenomenon plays a role in the long-term behavior of the chain.

Consider gambler's ruin (Example 6.2), for example. Every path that starts and ends at state 1 has even length of $2n$. This is true because every move to the right must eventually be balanced by a move to the left, in order to return to the starting point. If $\mathbf{P}$ is the transition matrix for this example, then

$$p_{11}(k) = 0 \quad \text{when } k \text{ is odd}$$

In this example, the same property obviously holds for all of the states $2, 3, \ldots, N-1$ that are in the same class as state 1.

In general, we say that a state is **periodic**, with period $d > 1$, if every path that starts and ends at this state has length nd, and n can be any integer $0, 1, 2, \ldots$. (Taking $n = 1$, we see that d is uniquely defined by the state, if the state is periodic.)

Lemma 6.3. *Suppose state i is periodic and $i \leftrightarrow j$. Then j is periodic and has the same period as i.*

Proof Let d be the period of state i. We assume that there is a path α from i to j and a path β from j back to i in the transition diagram. Suppose γ is any path in the transition diagram that starts and ends at j. We must show that the length of γ is a multiple of d. Denote by $\alpha \cdot \beta$ the path from i to i that we get by first following α from i to j, and then following β from j back to i. Denote by $\alpha \cdot \gamma \cdot \beta$ the path from i to i that we get by first following α from i to j, then following the loop γ from j to j, and finally following β from j back to i. Obviously, we have

$$\begin{aligned} \text{length}(\alpha \cdot \gamma \cdot \beta) &= \text{length}(\alpha) + \text{length}(\gamma) + \text{length}(\beta) \\ &= \text{length}(\alpha \cdot \beta) + \text{length}(\gamma) \end{aligned}$$

By the periodicity of state i, we know that d divides $\text{length}(\alpha \cdot \gamma \cdot \beta)$ and $\text{length}(\alpha \cdot \beta)$. Hence d divides $\text{length}(\gamma)$. ■

Example 6.22 Bold Play

In Example 6.5, the states $1, 2, 3, 4$ are all periodic of period 4. This periodicity arises from the loop $1 \to 2 \to 4 \to 3 \to 1$ in the transition diagram. ■

Definition

A state is **aperiodic** *if it is not periodic.*

By Lemma 6.3, aperiodicity is a class property. If a Markov chain is irreducible, we say that it is *aperiodic* if every state is aperiodic.

Example 6.23 Machine Maintenance

Consider Examples 6.4 and 6.12. Because there is a path of length 1 from state 0 to state 0, this state is not periodic. Since these chains are irreducible, all states are aperiodic. ■

The following lemma plays an important role in Section 6.7 where we investigate the long-term behavior of chains such as those in Example 6.23.

Lemma 6.4. *Given an irreducible, aperiodic chain with a finite number of states, we can find an integer $n^* > 0$ with the property that, for every integer $n \geq n^*$ and every pair (i, j) of states, there exists a path of length n between i and j.*

Proof We may assume that the chain has more than one state. First consider the case $i = j$. If $p_{jj} > 0$, then the path $j \to j$ occurs in the transition diagram. Repeating this path n times, we get a path of length n from j to j. If $p_{jj} = 0$, then the assumption that j is aperiodic means that there are two different paths—say, α and β—that start and end at j and whose lengths are not multiples of a common period $d > 1$. In arithmetic terms, this means that 1 is the only integer which divides both

$$a = \text{length}(\alpha) \quad \text{and} \quad b = \text{length}(\beta)$$

The Euclidean division algorithm says in this case that we can find integers r and s such that

$$1 = ar + bs \tag{6.18}$$

Observe that r and s must be of opposite signs, since $a > 1$ and $b > 1$. Starting from the arithmetic relation in formula (6.18), we will show that, when n is sufficiently large, we can write it as

$$n = au + bv \quad \text{with integers } u \geq 0 \text{ and } v \geq 0 \tag{6.19}$$

Once we have found u and v, the path α repeated u times followed by the path β repeated v times will furnish the required path of length n from j back to j.

Now we prove formula (6.19). The key requirement is that both u and v be positive. (Multiplying equation (6.18) by n doesn't work, since either rn or sn is negative.) We may assume that $r > 0$ and $s < 0$. Take a positive integer n and use the Euclidean algorithm to write

$$n = bq + k \quad \text{with } 0 \leq k < b \tag{6.20}$$

Then multiply equation (6.18) by the remainder k and substitute in formula (6.20):

$$n = au + bv, \quad \text{with } u = kr \text{ and } v = q + ks \tag{6.21}$$

For any positive n, the coefficient $u \geq 0$ in (6.21). The coefficient v will be nonnegative, provided that $bv \geq 0$. To determine when this holds, notice that $bv = bq + bks$ and $bq = n - k$, from formula (6.20). Hence, we can write

$$bv = n - k + bks = n - k(1 - bs)$$

Now, from formula (6.18) we have $1 - bs = ar$, so that

$$bv = n - kar. \tag{6.22}$$

But $0 \leq k < b$ and $ar \geq 0$, so from formula (6.22) we have the lower bound $bv \geq n - abr$. In particular, if

$$n \geq abr \tag{6.23}$$

then $v \geq 0$. This proves formula (6.19).

We have now shown that we can construct a path of length n from j back to j, provided n satisfies inequality (6.23). Repeat this process for all the states of the chain, and let m_0 be the maximum of the numbers abr that arise in inequality (6.23). Since there are only a finite number of states, $m_0 < \infty$.

Now consider the case $i \neq j$. The chain is irreducible, so we can choose a particular path from i to j for each pair of states i, j. Let n_0 be the maximum of the lengths in this finite set of paths. Define $n^* = m_0 + n_0$. We can then construct paths from i to j of any prescribed length $n \geq n^*$, as follows. First take the chosen path from i to j. It has length $k \leq n_0$. Since $n - k \geq n - n_0 \geq m_0$, there is a path of length $n - k$ from j to j, by the definition of m_0 in the previous paragraph. Combine these two paths to get a path from i to j of length n, as required. ■

6.6 Absorbing Chains

We say that a Markov chain is **absorbing** if every state in it is either absorbing or transient. This means that each ergodic class consists of a single absorbing state. We have already studied several examples of absorbing chains (waiting game, gambler's ruin, and bold play)—both by computer simulation and by path analysis.

We now show that every absorbing chain behaves like gambler's ruin (see Example 2.27): if we start in a transient state, we are certain of reaching an absorbing state after a finite number of trials. The key step in proving this is to determine the behavior of the powers $\mathbf{P}^n$ of the transition matrix $\mathbf{P}$ as $n \to \infty$.

We begin our analysis by listing the states of an absorbing chain in the order given in Section 6.5. Let a be the number of absorbing states and let t be the number of transient states in the chain. Since each ergodic class reduces to a single state, the matrix $\mathbf{S}$ that describes transitions within the ergodic classes is the $a \times a$ identity matrix $\mathbf{I}$, with 1's on the diagonal and 0's elsewhere. Thus, the transition matrix for the whole chain has the block form

$$\mathbf{P} = \begin{bmatrix} \mathbf{I} & \mathbf{0} \\ \mathbf{R} & \mathbf{Q} \end{bmatrix}$$

Here $\mathbf{R}$ is a $t \times a$ matrix describing the movement from transient to absorbing states, and $\mathbf{Q}$ is a $t \times t$ matrix describing the movement among transient states.

Since the formula for matrix multiplication also applies to matrices written in block form (provided that we are careful to preserve the order of multiplication of the blocks), we can calculate the powers of $\mathbf{P}$ in terms of the matrices $\mathbf{R}$ and $\mathbf{Q}$:

$$\begin{aligned} \mathbf{P}^2 &= \begin{bmatrix} \mathbf{I} & \mathbf{0} \\ \mathbf{R} + \mathbf{QR} & \mathbf{Q}^2 \end{bmatrix} \\ \mathbf{P}^3 &= \begin{bmatrix} \mathbf{I} & \mathbf{0} \\ \mathbf{R} + \mathbf{QR} + \mathbf{Q}^2\mathbf{R} & \mathbf{Q}^3 \end{bmatrix} \end{aligned}$$

The general formula is clear:

$$\mathbf{P}^n = \begin{bmatrix} \mathbf{I} & \mathbf{0} \\ \mathbf{N}_n\mathbf{R} & \mathbf{Q}^n \end{bmatrix} \tag{6.24}$$

where $\mathbf{N}_n = \mathbf{I} + \mathbf{Q} + \mathbf{Q}^2 + \cdots + \mathbf{Q}^{n-1}$. The fundamental results about the long-term behavior of an absorbing chain will follow from Theorem 6.3.

Theorem 6.3. *When $n \to \infty$, then $\mathbf{Q}^n \to 0$. The matrix $\mathbf{I} - \mathbf{Q}$ is invertible and $\mathbf{N}_n \to (\mathbf{I} - \mathbf{Q})^{-1}$.*

To prove Theorem 6.3, we need to estimate the magnitude of $\mathbf{Q}^n$. For this purpose, we define the following **norm** of a matrix. If $\mathbf{A} = [a_{ij}]$ is an $m \times n$ matrix, then we define its norm to be

$$||\mathbf{A}|| = \max_{1 \le i \le m} \sum_{j=1}^{n} |a_{ij}|$$

(this is also called the ∞ norm of the matrix and is often written as $||\mathbf{A}||_\infty$). This function of matrices has the properties

(i) $|a_{ij}| \le ||\mathbf{A}||$

(ii) $||\mathbf{A} + \mathbf{B}|| \le ||\mathbf{A}|| + ||\mathbf{B}||$

(iii) $||\mathbf{AB}|| \le ||\mathbf{A}|| \cdot ||\mathbf{B}||$

for any matrices $\mathbf{A}$ and $\mathbf{B}$ (we assume in (ii) that $\mathbf{A}$ and $\mathbf{B}$ are the same size, and in (iii) that $\mathbf{A}$ has size $m \times n$ and $\mathbf{B}$ has size $n \times r$ so that the product $\mathbf{AB}$ is defined.) The first and second properties are obvious. To prove the third property, let $\mathbf{C} = \mathbf{AB}$. Then, by the definition of matrix product,

$$|c_{ij}| = \Big|\sum_{k=1}^{n} a_{ik}b_{kj}\Big| \le \sum_{k=1}^{n} |a_{ik}||b_{kj}|$$

Thus

$$\begin{aligned} \sum_{j=1}^{n} |c_{ij}| &\le \sum_{j=1}^{n}\Big(\sum_{k=1}^{n} |a_{ik}||b_{kj}|\Big) = \sum_{k=1}^{n} |a_{ik}|\Big(\sum_{j=1}^{r} |b_{kj}|\Big) \\ &\le \sum_{k=1}^{n} |a_{ik}|\, ||\mathbf{B}|| \le ||\mathbf{A}|| \cdot ||\mathbf{B}|| \end{aligned}$$

The norm $||\mathbf{C}||$ is the maximum (over i) of the sum appearing at the beginning of this string of inequalities, so we have proved (iii).

The key point in establishing Theorem 6.3 is the inequality in Lemma 6.5.

Lemma 6.5. *There exist constants $b > 0$ and $0 < r < 1$ such that $||\mathbf{Q}^n|| \le br^n$ for all positive integers n.*

Proof The main idea of the proof is well illustrated by the example of gambler's ruin with total stakes $N = 4$. As we already observed in equation (6.17),

$$\mathbf{Q} = \begin{bmatrix} 0 & p & 0 \\ q & 0 & p \\ 0 & q & 0 \end{bmatrix}$$

in this case, where $p + q = 1$. In particular, $||\mathbf{Q}|| = 1$, because of the second row of $\mathbf{Q}$. But now calculate

$$\mathbf{Q}^2 = \begin{bmatrix} pq & 0 & p^2 \\ 0 & 2pq & 0 \\ q^2 & 0 & pq \end{bmatrix}$$

To determine the norm of $\mathbf{Q}^2$, we calculate the row sums:

$$\begin{aligned} pq + p^2 &= p(p+q) = p < 1 \\ pq + q^2 &= q(p+q) = q < 1 \\ 2pq &= 2p(1-p) \le \frac{1}{2} \quad \text{(the maximum occurs at } p = \tfrac{1}{2}) \end{aligned}$$

Since either p or q is at least $\frac{1}{2}$, it follows that $||\mathbf{Q}^2|| = \max\{p, q\} < 1$. This estimate means that there is a positive probability of going from a transient state to an absorbing state in at most two moves. Thus, the shortest path from any transient state to some absorbing state has length at most 2.

Let w be the larger of p and q, and set $r = w^{1/2} < 1$. Then $||\mathbf{Q}^2|| = r^2$. Once we have this estimate, we can use it to estimate the norm of any even power of $\mathbf{Q}$:

$$||\mathbf{Q}^{2n}|| = ||(\mathbf{Q}^2)^n|| \le ||\mathbf{Q}^2||^n = r^{2n} \tag{6.25}$$

Here, we have used property (iii) of the norm to obtain the inequality. For odd powers of $\mathbf{Q}$, we use property (iii) of the norm again, together with the estimate for even powers:

$$||\mathbf{Q}^{2n+1}|| = ||\mathbf{Q} \cdot \mathbf{Q}^{2n}|| \le ||\mathbf{Q}|| \cdot ||\mathbf{Q}^{2n}|| \le br^{2n+1} \tag{6.26}$$

(where $b = ||Q||/r$). This proves Lemma 6.5 for this example.

The general case uses essentially the same argument: there is some integer n_0 so that the shortest path from any transient state to some absorbing state has length at most n_0. This implies that the norm of $\mathbf{Q}^{n_0}$ is strictly less than 1, because the row sums of $\mathbf{Q}^{n_0}$ give the probabilities of staying within the class of transient states. The rest of the argument follows the pattern of inequalities (6.25) and (6.26) above, with the multiples of n_0 replacing the even integers. ■

Proof of Theorem 6.3
Since $0 < r < 1$ in Lemma 6.5, we conclude that $||\mathbf{Q}||^n \to 0$ as $n \to \infty$. To show that the matrix $\mathbf{I} - \mathbf{Q}$ is invertible we use the geometric series:

$$\begin{aligned} (\mathbf{I} - \mathbf{Q})\mathbf{N}_n &= \mathbf{I} + \mathbf{Q} + \mathbf{Q}^2 + \cdots + \mathbf{Q}^{n-1} - (\mathbf{Q} + \mathbf{Q}^2 + \cdots + \mathbf{Q}^n) \\ &= \mathbf{I} - \mathbf{Q}^n. \end{aligned}$$

But $\mathbf{Q}^n \to \mathbf{0}$ as $n \to \infty$, by property (i) of the norm and by the estimate in Lemma 6.5. Thus $(\mathbf{I}-\mathbf{Q})\mathbf{N}_n \to \mathbf{I}$ as $n \to \infty$. This shows that the matrix $\mathbf{I}-\mathbf{Q}$ is invertible, and its inverse is the limit of the matrices $\mathbf{N}_n$ as $n \to \infty$. ■

We now know, by formula (6.24) and Theorem 6.3, that the limiting form of the transition matrix for an absorbing Markov chain (relative to the standard ordering of the states) is

$$\lim_{n\to\infty} \mathbf{P}^n = \begin{bmatrix} \mathbf{I} & \mathbf{0} \\ \mathbf{NR} & \mathbf{0} \end{bmatrix} \tag{6.27}$$

where $\mathbf{N} = (\mathbf{I}-\mathbf{Q})^{-1}$. We shall see that the matrices $\mathbf{N} = [n_{ij}]$ and $\mathbf{NR}$ contain important information about what happens in the chain in the long run.

As usual, we view the Markov chain as a game played on the transition diagram. Suppose the game starts at state i. If A is an event or Y is a random variable associated with the chain, we denote the conditional probabilties and conditional expectations relative to this starting state as follows:

$$\begin{aligned} P_i(A) &= P\{A \mid X(0) = i\} \\ E_i[Y] &= E[Y \mid X(0) = i] \end{aligned}$$

For absorbing chains, the only interesting starting states are the transient ones. Let T be the set of transient states. Fix an initial state $i \in T$. For each state $j \in T$, define the random variable

$$V_j = \text{Total number of visits to state } j \text{ during the entire game}$$

In the case $j = i$, the game starts in state j and we count this as a visit to state j.

It can be shown that $V_j < \infty$ for every transient state j, and that V_j has finite expectation. Assuming these properties, we now prove that the matrix $\mathbf{N}$ gives the expected number of visits to each transient state.

Theorem 6.4. *For every pair of transient states i, j,*

$$E_i[V_j] = n_{ij}$$

Proof We use a starting-over argument. In the first trial of the game, we move from state i to state k, with probability p_{ik}. If k is an absorbing state, we can never get to state j. If k is a transient state, we are in the same situation as at the beginning of the game, except that the starting position is now state k. Thus, by the Markov property,

$$E_i[V_j] = \delta_{ij} + \sum_{k\in T} p_{ik}\, E_k[V_j] \tag{6.28}$$

(recall that $\delta_{ii} = 1$ and $\delta_{ij} = 0$ if $i \neq j$). The term δ_{ij} on the right side counts the initial visit to state j in case the starting position is j.

We denote by $\mathbf{M}$ the matrix with i, j entry $E_i[V_j]$, as i, j range over T. Then (6.28) can be written as the matrix equation

$$\mathbf{M} = \mathbf{I} + \mathbf{QM}.$$

Hence, $(\mathbf{I}-\mathbf{Q})\mathbf{M} = \mathbf{I}$, so that $\mathbf{M} = (\mathbf{I}-\mathbf{Q})^{-1} = \mathbf{N}$, as claimed. ■

Corollary 6.5 (Mean Absorbtion Time). *Let the random variable W be the number of steps until an absorbing state is reached. Then $W < \infty$ and*

$$E_i[W] = \sum_{j \in T} n_{ij}.$$

Proof Clearly $W = \sum_{j \in T} V_j$, so this follows from Theorem 6.4. ■

Corollary 6.5 gives the expected length of the game when we start at a transient state i. We can also determine the probabilities of reaching each absorbing state, as indicated in Theorem 6.6:

Theorem 6.6. *Let i be a transient state, and let j be an absorbing state. Then*

$$P_i\{\text{Game ends at state } j\} = (\mathbf{NR})_{ij}$$

Proof Let A_j be the event "The game ends at the absorbing state j", and set $b_{ij} = P_i[A_j]$. To get an equation for the $t \times a$ matrix $\mathbf{B} = [b_{ij}]$, we condition on the outcome of the first step of the game. There are three mutually exclusive possibilities:

(i) The marker moves to state j at first step, ending the game.

(ii) The marker moves to an absorbing state other than j at first step, ending the game.

(iii) The marker moves to a transient state k at the first step, enabling us to continue the game from state k.

Now use the Rule of Total Causes to calculate $P_i[A_j]$, noting that in case (ii) the conditional probability of A_j is zero and that in case (iii) the Markov property allows us to forget the first step:

$$P_i[A_j] = p_{ij} + \sum_{k \in T} p_{ik} P_k[A_j]$$

In matrix form, this equation reads

$$\mathbf{B} = \mathbf{R} + \mathbf{QB}$$

Hence $(\mathbf{I} - \mathbf{Q})\mathbf{B} = \mathbf{R}$, so we obtain

$$\mathbf{B} = (\mathbf{I} - \mathbf{Q})^{-1}\mathbf{R} = \mathbf{NR}$$

as claimed. ■

To use Theorems 6.4 and 6.6, we need to invert the matrix $\mathbf{I} - \mathbf{Q}$. Theorem 6.3 guarantees that this matrix is invertible, and provides an algorithm (infinite geometric series in powers of $\mathbf{Q}$) for calculating the inverse. In practice, we can pick whatever matrix inversion algorithm is most convenient for the problem at hand.

Example 6.24

For the waiting game (Example 6.1), state 0 is transient and state 1 is absorbing. The matrices $\mathbf{Q} = [q]$ and $\mathbf{R} = [p]$ are 1×1, and

$$\mathbf{N} = \left[\frac{1}{1-q}\right] = \left[\frac{1}{p}\right] \qquad \mathbf{NR} = [1]$$

By Theorem 6.4, if the game starts in state 0, then the expected number of trials until state 1 is reached is $1/p$. This agrees with our calculation of the mean of a geometric random variable in Section 3.4. By Theorem 6.6, the equation $\mathbf{NR} = [1]$ corresponds to the fact that only one absorbing state exists, and the game must end at this state. ■

Example 6.25 Duration of Gambler's Ruin

Consider gambler's ruin, with total fortune $N = 3$. The states 0 and 3 are absorbing, and states 1 and 2 transient. We have

$$\mathbf{Q} = \begin{array}{c} \text{State} \\ 1 \\ 2 \end{array} \begin{array}{c} \begin{array}{cc} 1 & 2 \end{array} \\ \left[\begin{array}{cc} 0 & p \\ q & 0 \end{array}\right] \end{array} \qquad \mathbf{R} = \begin{array}{c} \text{State} \\ 1 \\ 2 \end{array} \begin{array}{c} \begin{array}{cc} 0 & 3 \end{array} \\ \left[\begin{array}{cc} q & 0 \\ 0 & p \end{array}\right] \end{array}$$

We can invert the matrix $\mathbf{I} - \mathbf{Q}$ using cofactors (Cramer's rule), and get

$$(1-pq)\mathbf{N} = \begin{array}{c} \text{State} \\ 1 \\ 2 \end{array} \begin{array}{c} \begin{array}{cc} 1 & 2 \end{array} \\ \left[\begin{array}{cc} 1 & p \\ q & 1 \end{array}\right] \end{array}$$

$$(1-pq)\mathbf{NR} = \begin{array}{c} \text{State} \\ 1 \\ 2 \end{array} \begin{array}{c} \begin{array}{cc} 0 & 3 \end{array} \\ \left[\begin{array}{cc} q & p^2 \\ q^2 & p \end{array}\right] \end{array}$$

Let's check that the probabilities for absorption at state 0 (*ruin*) given by Theorem 6.6 agree with our earlier analysis of gambler's ruin (Example 2.27). For example, if the starting point (*initial stake*) is state 1, then the probability of the game ending at state 0 is the matrix entry

$$(\mathbf{NR})_{\text{state 1, state 0}} = \frac{q}{1-pq} \tag{6.29}$$

On the other hand, by formula (2.42), with $r = p/q$, total stake $N = 3$, and $a = 1$, this probability is

$$\frac{1-r^2}{1-r^3} \tag{6.30}$$

After some algebraic manipulations, we find that matrix entry (6.29) and probability (6.30) are equal. (Factor $1 - r$ from numerator and denominator of expression (6.30), multiply by q^2, and use the relation $p + q = 1$.)

Observe that the row sums of the matrix $\mathbf{NR}$ are 1. (We know this even without calculation, since the limit of a finite stochastic matrix is again a stochastic matrix.) This property agrees with our calculations in Chapter 2, where we concluded that one of the players is certain to be ruined.

If we start in state 1, then expected duration of the game, by Corollary 6.5, is the sum of the elements of the first row of **N**:

$$E_1[W] = \frac{1+p}{1-pq} \tag{6.31}$$

This is information that we didn't get by the method used in Chapter 2. ■

6.7 Regular Chains

The absorbing chains arise in models for *terminating* processes. We now turn to the study of *recurrent* processes.

Definition
A finite Markov chain is called **regular** *if it is irreducible and aperiodic.*

Lemma 6.6. *Let* $\mathbf{P}$ *be the one-step transition matrix for a finite Markov chain. Then the chain is regular if and only if there exists an integer* $n > 0$ *such that every entry of* $\mathbf{P}^n$ *is strictly positive. In this case, every entry of* $\mathbf{P}^m$ *is strictly positive for all integers* $m \geq n$.

Proof If the chain is regular, then the the existence of such an integer n follows from Lemma 6.4. Conversely, if every entry of $\mathbf{P}^n$ is positive, then the same is true for $\mathbf{P}^{n+1}$. To see this, start with the Chapman-Kolmogorov equation

$$p_{ij}(n+1) = \sum_k p_{ik}\, p_{kj}(n) \tag{6.32}$$

and notice that for each state i, there is a state k with $p_{ik} > 0$ (the row sums of $\mathbf{P}$ are 1). Since we are assuming that $p_{kj}(n) > 0$ for *all* states k and j, it follows that the right side of (6.32) is positive.

Iterating this argument, we conclude that $p_{ij}(m) > 0$ for all $m \geq n$. Hence $i \leftrightarrow j$ for all states i and j, and there are paths of any length $m \geq n$ from i to j. This proves that the chain is irreducible and aperiodic. ■

We shall call the transition matrix $\mathbf{P}$ **regular** if it satisfies the condition of Lemma 6.6. The obvious example of a regular $\mathbf{P}$ is a stochastic matrix in which *all* entries are positive. A larger class of examples consists of all stochastic matrices $\mathbf{P}$ that describe irreducible chains and have at least one positive diagonal entry:

$$p_{ii} > 0 \quad \text{for some } i \tag{6.33}$$

(Notice that condition (6.33) means that there is a path $i \to i$ of length 1; hence, state i is aperiodic. Since the chain is irreducible, it follows that all of its states are aperiodic.) The vending machine example of Section 6.1 and the machine maintenance and inventory models of Section 6.3 are of this type.

The long-term behavior of a Markov chain is determined by limiting behavior of $\mathbf{P}^n$ as $n \to \infty$. We already determined this behavior for absorbing chains in Section 6.6, where we found that gambler's ruin was a typical model. Now let's look at the vending machine example in more detail to get an idea of the long-term behavior of a regular chain.

Example 6.26 Vending Machine

Consider the vending machine model (Example 6.4). Assume that the one-day probability that the machine will break down is 0.2, and the probability of its being fixed in one day is 0.9. Then

$$\mathbf{P} = \begin{array}{c} \text{State} \\ 0 \\ 1 \end{array} \begin{array}{c} \begin{array}{cc} 0 & 1 \end{array} \\ \begin{bmatrix} 0.8 & 0.2 \\ 0.9 & 0.1 \end{bmatrix} \end{array}$$

By matrix multiplication, we find that

$$\mathbf{P}^2 = \begin{bmatrix} 0.82 & 0.18 \\ 0.81 & 0.19 \end{bmatrix} \qquad \mathbf{P}^3 = \begin{bmatrix} 0.818 & 0.182 \\ 0.819 & 0.181 \end{bmatrix}$$

Note that the rows of $\mathbf{P}^3$ are almost equal. If we calculate higher powers of $\mathbf{P}$, this pattern becomes unmistakable:

$$\mathbf{P}^4 = \begin{bmatrix} 0.8182 & 0.1818 \\ 0.8181 & 0.1819 \end{bmatrix} \approx \mathbf{P}^5$$

(with equality when the entries are rounded to four decimal places).

Recall that the first row of $\mathbf{P}^n$ gives the (conditional) probabilities of finding the machine in state 0 (working) or state 1 (broken) on day n, given that the machine was working on day 0. Likewise, the second row of $\mathbf{P}^n$ gives the conditional probabilities for state 0 or state 1 on day n, given that the machine was broken on day 0. The fact that these two rows are essentially the same when $n \geq 4$ shows that the influence of the initial state of the machine disappears after several days.

If the initial probability vector for $X(0)$ happens to be

$$\boldsymbol{\alpha} = \begin{bmatrix} 0.818 & 0.182 \end{bmatrix}$$

then this vector also gives (approximately) the p.m.f. of $X(1)$, since

$$\begin{bmatrix} 0.818 & 0.182 \end{bmatrix} \begin{bmatrix} 0.8 & 0.2 \\ 0.9 & 0.1 \end{bmatrix} = \begin{bmatrix} 0.818 & 0.182 \end{bmatrix}$$

(rounded to three decimal places). Under these starting conditions, all of the random variables $X(n)$ will have approximately the same probability distribution. ■

In the case of regular chains, we have the following fundamental result, which confirms the numerical calculations made in Example 6.26.

Theorem 6.7 (Limiting Probabilities). *If* $\mathbf{P}$ *is a regular transition matrix, then*

(i) *The powers* $\mathbf{P}^n$ *converge to a stochastic matrix* $\mathbf{A}$.

(ii) *Each row of* $\mathbf{A}$ *is the same vector* $\boldsymbol{\alpha}$, *and every component of* $\boldsymbol{\alpha}$ *is positive.*

Call $\mathbf{A}$ *the* **limiting matrix** *and* $\boldsymbol{\alpha}$ *the* **limiting vector** *for the chain.*

Before proving Theorem 6.7, let's consider its assertions in more detail. It claims that, when n is large, every row of $\mathbf{P}^n$ is essentially the same. This means that all the entries in any particular column of $\mathbf{P}^n$ are essentially equal. Also, it claims that all the entries of $\mathbf{P}^n$ are strictly positive. Thus they are all at least β, for some fixed number $\beta > 0$. The one-step matrix $\mathbf{P}$ generally won't have these properties, so there must be some mechanism in the iterative process $\mathbf{P} \to \mathbf{P}^2 \to \cdots \to \mathbf{P}^n$ that produces this effect. Let's see if we can find it.

Assume for the moment that every entry of $\mathbf{P}$ is strictly positive, and let $\beta > 0$ be the smallest entry. Assume the chain has d states, labeled $1, 2, \ldots, d$. We focus on a particular column of $\mathbf{P}^n$—say, the jth. Denoting this column by $\mathbf{v}$, we observe that, by the definition of matrix multiplication, the jth column of $\mathbf{P}^{n+1}$ is the vector $\mathbf{Pv}$. But the kth entry of $\mathbf{Pv}$ is the weighted average

$$\sum_{j=1}^{d} v_j w_j \tag{6.34}$$

of the components of $\mathbf{v}$, with the weights $w_j = p_{kj}$. We know that $w_1 + \cdots + w_d = 1$, since $\mathbf{P}$ is a stochastic matrix, and $w_j \geq \beta$ by assumption. The key to the proof of Theorem 6.7 is the following leveling effect produced by taking averages.

Lemma 6.7 (Averaging Lemma). *Let $w_1, \ldots, w_d$ be positive numbers with $w_1 + \cdots + w_d = 1$. Suppose that $w_i \geq \beta > 0$ for all i. Let $v_1, \ldots, v_d$ be any set of nonnegative real numbers. Define*

$$m = \min_{1 \leq i \leq d} \{v_i\} \qquad M = \max_{1 \leq i \leq d} \{v_i\} \tag{6.35}$$

Then the average (6.34) of the numbers $\{v_i\}$, relative to the weights w_i, satisfies

$$(1-\beta)m + \beta M \leq \sum_{i=1}^{d} v_i w_i \leq \beta m + (1-\beta)M \tag{6.36}$$

In particular, if the numbers v_i are not all equal, then the weighted average (6.34) is strictly bigger than the smallest and strictly less than the largest of the v_i.

Proof of Averaging Lemma Assume for the moment that $m = 0$ and $M = 1$, so that the inequality to be proved is

$$\beta \leq \sum_{i=1}^{d} v_i w_i \leq 1 - \beta \tag{6.37}$$

We may assume that the weights w_i are numbered in increasing magnitude, so that w_1 is the smallest. Thus, $\beta \leq w_1$.

If v_1 happens to be the smallest of the v_i, then $v_1 = 0$, since $m = 0$. Hence, we can bound the average (6.34) from above by

$$v_2 w_2 + \cdots + v_d w_d \leq w_2 + \cdots + w_d \leq 1 - \beta$$

in this case, since each $v_i \leq 1$ and $w_2 + \cdots + w_d = 1 - w_1$.

If v_1 happens to be the largest of the v_i, then $v_1 = 1$ and we can bound the average (6.34) from below by

$$\sum_{i=1}^{d} v_i w_i \geq v_1 w_1 = w_1 \geq \beta$$

In general, when v_1 is neither the largest nor the smallest of the v_i, we can see what happens to the average (6.34) when we interchange v_1 and v_i in the sum. Note that, for $i > 1$,

$$v_1 w_1 + v_i w_i = v_i w_1 + v_1 w_i + (w_i - w_1)(v_i - v_1). \tag{6.38}$$

Recall that we have arranged the weights so that $w_i - w_1 \geq 0$. If we choose i so that v_i is the smallest, then $v_i - v_1 \leq 0$. Therefore, by formula (6.38), we see that

$$v_1 w_1 + v_i w_i \leq v_i w_1 + v_1 w_i$$

Thus average (6.34) increases when v_1 and v_i are interchanged. We have already proved that the upper inequality in (6.36) holds when the first element is the smallest. Hence this inequality holds for any labeling of the v's. Similarly, if we choose i so that v_i is the largest, then $v_i - v_1 \geq 0$. Therefore, by formula (6.38), we have

$$v_1 w_1 + v_i w_i \geq v_i w_1 + v_1 w_i$$

Hence in this case average (6.34) decreases when v_1 and v_i are interchanged. We have already proved that the lower inequality in (6.36) holds when the first element is the largest. Hence this inequality also holds for any labeling of the v's. This proves Lemma 6.7 in the special case that $m = 0$ and $M = 1$.

Given a general sequence $\{v_i\}$ having maximum term M and smallest term m, we make a scaling transformation: we replace v_i by the normalized sequence $v_i' = (v_i - m)/(M - m)$. Here we may assume that $m < M$. (If $m = M$, then the sequence is constant and inequalities (6.36) reduce to the statement $m = M$.)

For the normalized sequence we have $\min\{v_i'\} = 0$ and $\max\{v_i'\} = 1$. Hence by (6.37) we conclude that

$$\beta \leq \sum_{i=1}^{d} w_i \,(v_i - m)/(M - m) \leq 1 - \beta$$

To get the inequalities (6.36), we now multiply through by $M - m$ and use the condition $w_1 + \cdots + w_d = 1$. ■

Proof of Limit Theorem We first treat the case where *all* the elements of $\mathbf{P}$ are strictly positive: $p_{ij} \geq \beta > 0$. We may assume that the chain has at least two states. Since the row sums of $\mathbf{P}$ are 1, this implies that $\beta \leq \frac{1}{2}$. Fix a column j and define two sequences $\{m_n\}$ and $\{M_n\}$ of positive numbers:

$$m_n = \min_{1 \leq i \leq d} p_{ij}(n) \qquad M_n = \max_{1 \leq i \leq d} p_{ij}(n)$$

By the Averaging Lemma (Lemma 6.7) and the remarks preceding it, we have the lower and upper bounds

$$(1 - \beta) m_n + \beta M_n \leq p_{ij}(n+1) \leq \beta m_n + (1 - \beta) M_n$$

for all values of n and $1 \le i \le d$. Hence, taking the values of i that give the minimum and maximum for each n, we obtain the inequalities

$$M_{n+1} \le \beta m_n + (1-\beta)M_n \tag{6.39}$$
$$m_{n+1} \ge (1-\beta)m_n + \beta M_n \tag{6.40}$$

In particular, replacing m_n with M_n in formula (6.39) shows that

$$M_{n+1} \le M_n \tag{6.41}$$

Likewise, replacing M_n with m_n in formula (6.40) shows that

$$m_{n+1} \ge m_n \tag{6.42}$$

Now subtract inequality (6.40) from inequality (6.39). This gives the bound

$$M_{n+1} - m_{n+1} \le (1-2\beta)(M_n - m_n) \tag{6.43}$$

for the difference between the largest and smallest elements in the jth column of $\mathbf{P}^{n+1}$, as compared to the corresponding difference for $\mathbf{P}^n$.

Set $C = M_1 - m_1$ and $r = 1 - 2\beta$. Then $C < 1$ and $0 \le r < 1$. Iterating the inequality (6.43), we find that the difference $M_n - m_n$ goes to zero at a geometric rate:

$$\begin{aligned} M_2 - m_2 &\le Cr \\ M_3 - m_3 \le r(M_2 - m_2) &\le Cr^2 \\ &\vdots \\ M_{n+1} - m_{n+1} \le r(M_n - m_n) \le \cdots &\le Cr^n \end{aligned}$$

This shows that the jth column of $\mathbf{P}^n$ converges to a vector with all components equal. The common value a_j of this component is the limit of the monotone increasing sequence m_n as $n \to \infty$, and hence a_j is strictly positive. Since the row sums of $\mathbf{P}^n$ are all 1, the same is true for the limiting matrix $\mathbf{A}$. This proves Theorem 6.7 when $\mathbf{P}$ is strictly positive.

In the general case of a regular $\mathbf{P}$, it is easy to see (without using the Averaging Lemma) that inequalities (6.41) and (6.42) still hold. Some power $\mathbf{P}^N$ is strictly positive, by Lemma 6.6, so the subsequence $\mathbf{P}^{kN} = (\mathbf{P}^N)^k$ converges as $k \to \infty$ by the proof just given. Together with the monotonicity estimates (6.41) and (6.42), this implies that the full sequence $\mathbf{P}^n$ converges as stated. ■

6.8 Stationary Distributions

Up to now we have emphasized Markov chains that begin in a specific state i at time 0. In studying the long-term behavior of a chain, it is essential to allow a random choice of initial state. We shall restrict attention to finite chains with d states labeled $1, 2, \ldots, d$, and we shall write the probability mass function of the state at time 0 as a row vector

$$\boldsymbol{\pi} = [\ \pi_1 \ \cdots \ \pi_d\ ]$$

where $\pi_i = P\{X(0) = i\}$. Recall that, given any initial probability distribution vector $\boldsymbol{\pi}$ for the starting position of the marker in the Markov chain game, the probability distribution for the position of the marker after n moves is the row vector $\boldsymbol{\pi}\, \mathbf{P}^n$, where $\mathbf{P}$ is the $d \times d$ transition matrix for the chain.

Definition
The vector $\boldsymbol{\pi}$ *is a* **stationary distribution** *for the chain if*

$$\boldsymbol{\pi}\, \mathbf{P} = \boldsymbol{\pi} \tag{6.44}$$

Suppose that $\boldsymbol{\pi}$ is a stationary distribution. Then $\boldsymbol{\pi}\, \mathbf{P}^2 = (\boldsymbol{\pi}\, \mathbf{P})\, \mathbf{P} = \boldsymbol{\pi}\, \mathbf{P} = \boldsymbol{\pi}$, and in general

$$\boldsymbol{\pi}\, \mathbf{P}^n = \boldsymbol{\pi} \quad \text{for all integers } n > 0 \tag{6.45}$$

If the game starts with the marker's position being generated from the initial probability distribution $\boldsymbol{\pi}$, then formula (6.45) shows that, at every step of the game, the same distribution $\boldsymbol{\pi}$ describes the random position of the marker. This is the meaning of the term *stationary*.

Theorem 6.8 (Ergodicity of Regular Markov Chain). *Suppose that* $\mathbf{P}$ *is the transition matrix for a regular chain, and that* $\boldsymbol{\alpha}$ *is the limiting vector for* $\mathbf{P}$*. Then*

(i) $\boldsymbol{\alpha}$ *is the unique stationary distribution for the chain.*

(ii) *For any initial probability distribution* $\boldsymbol{\beta}$*, the n-step distribution* $\boldsymbol{\beta}\, \mathbf{P}^n$ *converges to* $\boldsymbol{\alpha}$ *as* $n \to \infty$.

Proof Let $\mathbf{e}_i$ be the row vector with 1 in position i and 0 elsewhere. If $\mathbf{A}$ is the limiting matrix for the chain, then $\mathbf{e}_i\, \mathbf{A} = \boldsymbol{\alpha}$ for every value of i, since all the rows of $\mathbf{A}$ are $\boldsymbol{\alpha}$. Hence,

$$\boldsymbol{\alpha} = \lim_{n\to\infty} \mathbf{e}_i\, \mathbf{P}^n \tag{6.46}$$

for every value of i. In particular,

$$\boldsymbol{\alpha}\, \mathbf{P} = \lim_{n\to\infty} \mathbf{e}_1\, \mathbf{P}^{n+1} = \lim_{n\to\infty} \mathbf{e}_1\, \mathbf{P}^n = \boldsymbol{\alpha}$$

This shows that $\boldsymbol{\alpha}$ is a stationary distribution.

If $\boldsymbol{\beta}$ is an arbitrary initial distribution for the chain, we can write

$$\boldsymbol{\beta} = \sum_{i=1}^{d} b_i \mathbf{e}_i$$

where $b_i \geq 0$ and $b_1 + \cdots + b_d = 1$. Thus, the probability distribution for $X(n)$, starting with the distribution $\boldsymbol{\beta}$ for $X(0)$, is given by

$$\boldsymbol{\beta}\, \mathbf{P}^n = \sum_{i=1}^{d} b_i \mathbf{e}_i\, \mathbf{P}^n \tag{6.47}$$

Now let $n \to \infty$ in formula (6.47), and use formula (6.46):

$$\lim_{n\to\infty} \boldsymbol{\beta}\,\mathbf{P}^n = \sum_{i=1}^{d} c_i\,\boldsymbol{\alpha} = \boldsymbol{\alpha} \tag{6.48}$$

If $\boldsymbol{\beta}$ happens to be a stationary distribution for $\mathbf{P}$, then the left side of equation (6.48) is $\boldsymbol{\beta}$ (no limit needed). Hence $\boldsymbol{\beta} = \boldsymbol{\alpha}$, which shows that $\boldsymbol{\alpha}$ is the unique stationary vector. ■

To find the stationary distribution for a regular chain, we must find a row vector $\boldsymbol{\alpha}$ with positive entries summing to 1 which satisfies the system of homogeneous linear equations

$$\boldsymbol{\alpha}\,\mathbf{P} = \boldsymbol{\alpha} \tag{6.49}$$

(This system of equations means that $\boldsymbol{\alpha}$ is a *left eigenvector* for $\mathbf{P}$ with *eigenvalue* 1, in the terminology of linear algebra.) Theorem 6.8 guarantees that a unique vector with these properties exists. There are several ways to calculate $\boldsymbol{\alpha}$.

An iterative algorithm for determining $\boldsymbol{\alpha}$ is as follows: choose any initial distribution $\boldsymbol{\beta}$, set $\boldsymbol{\beta}(0) = \boldsymbol{\beta}$, and calculate the sequence of row vectors $\boldsymbol{\beta}(n+1) = \boldsymbol{\beta}(n)\,\mathbf{P}$ for successive values of n, until the sequence converges numerically (within some prescribed accuracy). This gives an approximation to the stationary vector $\boldsymbol{\alpha}$. We illustrated this method in Example 6.26.

An alternate method is to solve the equations (6.49) for $\boldsymbol{\alpha}$ directly using Gaussian elimination as in the following examples. We know that the solution is unique up to a scaling factor and that all its components have the same sign (by Theorem 6.8); hence, we can then normalize the solution vector to make it a probability distribution.

Example 6.27 Vending Machine

Consider the vending machine model (Example 6.4). The matrix $\mathbf{P}$ is

$$\mathbf{P} = \begin{array}{c} \text{State} \\ 0 \\ 1 \end{array} \begin{array}{c} \begin{array}{cc} 0 & \quad 1 \end{array} \\ \begin{bmatrix} 1-\delta & \delta \\ \gamma & 1-\gamma \end{bmatrix} \end{array}$$

where the parameters γ and δ are between 0 and 1. Let $\boldsymbol{\alpha} = [a\ b]$ be the limiting vector for $\mathbf{P}$. Its components satisfy the system of homogeneous equations

$$\begin{aligned} a(1-\delta) + b\gamma &= a \\ a\delta + b(1-\gamma) &= b \end{aligned}$$

and the normalizing condition $a + b = 1$. By elimination, we reduce the system of equations to the single equation $b = (\delta/\gamma)a$. Substituting this in the normalizing equation, we find

$$a = \frac{\gamma}{\delta+\gamma} \qquad b = \frac{\delta}{\delta+\gamma}$$

In particular, when $\delta = 0.2$ and $\gamma = 0.9$, as in Example 6.26, we have

$$a = \frac{0.9}{1.1} = 0.81818\ldots \qquad b = \frac{0.2}{1.1} = 0.18181\ldots$$

This confirms the numerical data of that example. ■

Example 6.28 Inventory Model

Consider the inventory model (Example 6.14), with transition diagram Figure 6.9. As in that example, we label the states 0, 1, 2. The transition matrix is

$$\mathbf{P} = \begin{array}{c} \text{State} \\ 0 \\ 1 \\ 2 \end{array} \begin{array}{c} \begin{array}{ccc} 0 & 1 & 2 \end{array} \\ \left[\begin{array}{ccc} 0 & p & q \\ p & q & 0 \\ 0 & p & q \end{array} \right] \end{array}$$

where $p+q = 1$. Let $\alpha = \left[\begin{array}{ccc} a_0 & a_1 & a_2 \end{array} \right]$ be the limiting vector for $\mathbf{P}$. The equation $\alpha \mathbf{P} = \alpha$ is the homogeneous system

$$\begin{aligned} pa_1 &= a_0 \\ pa_0 + qa_1 + pa_2 &= a_1 \\ qa_0 + qa_2 &= a_2 \end{aligned}$$

We solve these equations by elimination (we already know that there is a unique solution, up to a scalar multiple). From the first equation, we get $a_1 = (1/p)a_0$; from the third equation, we get $a_2 = (q/p)a_0$. Substituting these values in the normalizing equation $a_0 + a_1 + a_2 = 1$, we find

$$a_0 = \frac{p}{2} \qquad a_1 = \frac{1}{2} \qquad a_2 = \frac{q}{2}$$

■

6.9 Time Averages

Consider a finite-state regular Markov chain. Let $X(t)$ be the state of the chain at time t, where $t = 0, 1, 2, \ldots$. If we use random numbers to generate a simulation of the chain for a finite number of time steps, we get a **sample path** $x(t)$. After generating a simulation over a large number of time steps, we are interested in answering the following questions.

a. What proportion of the time is the chain in a particular state?

b. Does the proportion in (a) depend on the initial position at time 0?

Example 6.29 Inventory Model

Consider the inventory model (Examples 6.14 and 6.28). A simulation of this model yielded the data in Table 6.4. We see that the choice of initial state strongly influences the 10-day results, but its effect disapears in longer runs.

The data in the table indicate that, over a long time period, the inventory level will be 0 for about 10% of the time, 1 for about 50% of the time, and 2 for about 40% of the time. Recall from Example 6.28 that the stationary distribution for this chain (when $p = 0.2$) is

$$\alpha = \left[\begin{array}{ccc} 0.1 & 0.5 & 0.4 \end{array} \right]$$

Table 6.4 Simulations of Inventory Model ($p = 0.2$)

Simulation 1: Initial State 0 (Empty Inventory)

	NUMBER OF DAYS SIMULATED			
	10	100	1000	10,000
PROPORTION OF DAYS IN				
State 0	0.3	0.08	0.096	0.1031
State 1	0.3	0.54	0.514	0.5154
State 2	0.4	0.38	0.390	0.3815

Simulation 2: Initial State 1 (Full Inventory)

	NUMBER OF DAYS SIMULATED			
	10	100	1000	10,000
PROPORTION OF DAYS IN				
State 0	0.0	0.10	0.106	0.1019
State 1	0.3	0.56	0.494	0.5060
State 2	0.7	0.34	0.400	0.3921

Thus each component of $\boldsymbol{\alpha}$ gives the long-term proportion of time that the chain is in the corresponding state. ■

Example 6.29 illustrates the Law of Large Numbers for regular Markov chains: *the components of the stationary distribution give the long-term proportion of time that the chain spends in each state.* To prove this result, we introduce indicator random variables $I_j(t)$ for each state j:

$$I_j(t) = \begin{cases} 1 & \text{if } X(t) = j \\ 0 & \text{otherwise} \end{cases}$$

The proportion of time that the chain is at state j over the interval $1 \le t \le n$ is then

$$A_j(n) = \frac{1}{n} \sum_{t=1}^{n} I_j(t) \tag{6.50}$$

Notice that formula (6.50) expresses a time average, not a probability; consequently $A_j(n)$ is a *random variable* rather than just a number. Running repeated random simulations of the chain will give different sample values of $A_j(n)$.

The Law of Large Numbers (Section 4.6) asserts that the sample means of repeated independent observations of a random variable X converge (in the probabilistic sense) toward a constant value as the number of observations $n \to \infty$, and this limiting value is $E[X]$. For regular Markov chains, the same behavior occurs.

Theorem 6.9 (Law of Large Numbers for Markov Chains). *Let*

$$\boldsymbol{\alpha} = \begin{bmatrix} a_1 & a_2 & \dots & a_d \end{bmatrix}$$

be the stationary distribution of a regular finite Markov chain. Then, for every state j,

$$\lim_{n\to\infty} E[A_j(n)] = a_j \tag{6.51}$$

Furthermore, for every fixed $\epsilon > 0$

$$\lim_{n\to\infty} P\{|A_j(n) - a_j| \geq \epsilon\} = 0 \tag{6.52}$$

In particular, limits (6.51) *and* (6.52) *are not affected by the initial distribution of the chain at time* $t = 0$.

Proof We first observe that

$$\left|E[A_j] - a_j\right| = \left|E[A_j - a_j]\right| \leq E[|A_j - a_j|]$$

For any nonnegative random variable Y, we have

$$0 \leq \mathrm{Var}(Y) = E[Y^2] - E[Y]^2$$

Hence $E[Y] \leq E[Y^2]^{1/2}$. Applying this inequality for the random variable $Y = |A_j - a_j|$, we obtain the upper bound

$$\left|E[A_j] - a_j\right| \leq E\left[|A_j - a_j|^2\right]^{1/2} \tag{6.53}$$

We can now use the same argument as in the proof of the Law of Large Numbers in Section 4.6. By estimate (6.53) and the Chebyshev inequality (Section 4. 5), it will suffice to prove that

$$\lim_{n\to\infty} E\left[|A_j(n) - a_j|^2\right] = 0 \tag{6.54}$$

Suppose that the initial distribution is $\boldsymbol{\beta} = \begin{bmatrix} b_1 & b_2 & \dots & b_d \end{bmatrix}$. Then, for any random variable Y associated with the chain, we have

$$E[Y] = \sum_{i=1}^{d} b_i \, E_i[Y]$$

(Recall the notation $E_i[Y] = E[Y \mid X(0) = i]$.) Hence, if we prove limit (6.54) when the initial state is i (for every i), then limit (6.54) also holds for any initial distribution $\boldsymbol{\beta}$.

We start by writing

$$A_j(n) - a_j = \frac{1}{n} \sum_{t=1}^{n} \left(I_j(t) - a_j\right)$$

Squaring this expression, we have

$$|A_j(n) - a_j|^2 = \frac{1}{n^2} \sum_{s=1}^{n} \sum_{t=1}^{n} \left(I_j(s) - a_j\right)\left(I_j(t) - a_j\right)$$

Fix the states i and j, and define

$$m(s,t) = E_i\left[\left(I_j(s) - a_j\right)\left(I_j(t) - a_j\right)\right]$$

To prove limit (6.54) we must show that

$$\frac{1}{n^2} \sum_{s=1}^{n} \sum_{t=1}^{n} m(s,t) \to 0 \tag{6.55}$$

as $n \to \infty$. Since $m(s,t) = m(t,s)$, we may assume in making estimates that $s \le t$. Since $|I_j(t) - a_j| \le 2$, it is obvious that $m(s,t) \le 4$. However, this only shows that the left side of expression (6.55) is bounded by 4, since there are n^2 terms in the sum. To prove that the limit is 0, we must show that $m(s,t) \to 0$ sufficiently rapidly when $s, t \to \infty$. The precise estimate is given in Lemma 6.8.

Lemma 6.8. *There exist constants $C > 0$ and $0 < r < 1$ so that for all integers $0 \le s \le t$,*

$$|m(s,t)| \le C\left(r^{t-s} + r^t\right)$$

Proof We shall use the estimates for the n-step transition probabilities that we obtained in Section 6.7. We observe first that

$$E_i[I_j(t)] = p_{ij}(t) \tag{6.56}$$

by definition of $I_j(t)$. In addition,

$$I_j(s)I_j(t) = \begin{cases} 1 & \text{if } X(s) = j \text{ and } X(t) = j \\ 0 & \text{otherwise} \end{cases}$$

Hence,

$$\begin{aligned} E_i[I_j(s)I_j(t)] &= P_i\{X(s) = j \text{ and } X(t) = j\} \\ &= P_i\{X(t) = j \mid X(s) = j\} \cdot P_i\{X(s) = j\} \end{aligned}$$

by the chain rule for probabilities. But

$$\begin{aligned} P_i\{X(t) = j \mid X(s) = j\} &= P\{X(t) = j \mid X(s) = j \text{ and } X(0) = i\} \\ &= P\{X(t) = j \mid X(s) = j\} \qquad \text{(Markov property)} \\ &= P\{X(t-s) = j \mid X(0) = j\} \qquad \text{(shift invariance)} \\ &= p_{jj}(t-s) \end{aligned}$$

Combining these calculations, we find that

$$E_i\left[I_j(s)I_j(t)\right] = p_{jj}(t-s) \cdot p_{ij}(s) \tag{6.57}$$

Next, we expand the products in the formula for $m(s,t)$ and use formulas (6.56) and (6.57):

$$\begin{aligned} m(s,t) &= E_i\left[I_j(s)I_j(t)\right] - a_j E_i\left[I_j(t)\right] - a_j E_i\left[I_j(s)\right] + a_j^2 \\ &= p_{jj}(t-s) \cdot p_{ij}(s) - a_j p_{ij}(t) - a_j p_{ij}(s) + a_j^2 \end{aligned}$$

By the estimates in Section 6.7, we know that

$$p_{ij}(t) = a_j + e_{ij}(t) \tag{6.58}$$

where the term $e_{ij}(t)$ goes to zero at a geometric rate

$$|e_{ij}(t)| \le br^t \tag{6.59}$$

for some constants $b > 0$ and $r < 1$. Substituting formula (6.58) into the formula for $m(s,t)$, we find that the a_j^2 terms cancel, and we have

$$m(s,t) = a_j\big(e_{jj}(t-s) - e_{ij}(t)\big) + e_{jj}(t-s) \cdot e_{ij}(s)$$

Finally, we use estimate (6.59) and the fact that $a_j \leq 1$ to obtain the bound

$$|m(s,t)| \leq b\big(r^{t-s} + r^t\big) + b^2 r^{t-s} r^s = b r^{t-s} + \big(b + b^2\big) r^t$$

Therefore, we may take the constant $C = b + b^2$ and get the inequality stated in Lemma 6.8. ∎

Completion of Proof of Theorem 6.9 Since $m(s,t) = m(t,s)$, we may assume that $s \leq t$. From the estimate in Lemma 6.8, we have

$$\begin{aligned}\sum_{s=1}^{n}\sum_{t=1}^{n} |m(s,t)| &\leq 2\sum_{t=1}^{n}\sum_{s=1}^{t} |m(s,t)| \\ &\leq 2C\sum_{t=1}^{n}\sum_{s=1}^{t} \left(r^{t-s} + r^t\right)\end{aligned}$$

Replacing t with ∞ in the upper limit of the s-summation, we can estimate

$$\sum_{s=1}^{t} r^{t-s} \leq \sum_{k=0}^{\infty} r^k = \frac{1}{1-r}$$

since $r < 1$. Thus,

$$\sum_{t=1}^{n}\sum_{s=1}^{t} r^{t-s} \leq \frac{n}{1-r}$$

since each of the n sums on s is bounded by $1/(1-r)$. Similarly,

$$\begin{aligned}\sum_{t=1}^{n}\sum_{s=1}^{t} r^t = \sum_{t=1}^{n} t r^t &\leq n\sum_{t=1}^{n} r^t \\ &\leq n\sum_{t=1}^{\infty} r^t = \frac{n}{1-r}\end{aligned}$$

Combining these two estimates, we find that

$$\sum_{s=1}^{n}\sum_{t=1}^{n} m(s,t) \leq \frac{4Cn}{1-r} \tag{6.60}$$

Recall that, to prove Theorem 6.9, we only need to show that, when the left side of formula (6.60) is divided by n^2, the ratio approaches zero as $n \to \infty$. But inequality (6.60) shows that this ratio is bounded by C'/n, where $C' = 4C/(1-r)$. Hence we have the same $1/n$ rate of convergence toward zero as in the proof of the Law of Large Numbers (see Section 4.6). ∎

Example 6.30 Vending Machine

Consider the vending machine model again. We found in Example 6.27 that, if n is large, the probability is approximately $\gamma/(\delta+\gamma)$ that the machine will be in state 0 (working) on day n. From Theorem 6.9, we can conclude that

$$\text{Long-term proportion of time that the machine is working} = \frac{\gamma}{\delta+\gamma}$$

In particular, when $\delta = 0.2$ and $\gamma = 0.9$, this proportion is $\frac{9}{11} = 0.81818\ldots$. This is consistent with the simulation data of Example 6.4. ■

6.10 Sojourn Times and First Entrance Times

We continue to analyze the dynamic behavior of a Markov chain, which we can visualize in terms of a marker moving around the transition diagram. Suppose that we observe at some fixed time that the marker is at state i. Because of the no-memory property of a Markov chain, we may assume that the moment of observation is $n = 0$. For each state j, we want to know

a. How long will the marker stay at state i?

b. How long will it take for the marker to get to state j?

Question (a) is easy to answer. Let S_i be the number of time steps that the marker stays at state i (the **sojourn time** for this state). If $p_{ii} = 0$, then $S_i = 1$ and it is not random. If $0 < p_{ii} < 1$, then S_i is a geometric random variable with parameter $p = 1 - p_{ii}$, since p is the probability of leaving state i at the next step (see the waiting game, Example 6.1). Thus

$$\text{Expected sojourn time in state } i = (1 - p_{ii})^{-1} \tag{6.61}$$

in this case. Finally, if $p_{ii} = 1$, then state i is absorbing, and the marker will never leave it. Thus $S_i = \infty$ in this case. Notice that, in all cases, formula (6.61) is true, with the understanding that both sides are ∞ when $p_{ii} = 1$.

Question (b) is more difficult. We have already answered it in Section 6.6 for an irreducible chain with a single absorbing state j (the expected duration of the game). From now on we restrict our attention to **regular** chains. For each state j, we define the random variable

$$T_j = \text{Number of steps to reach state } j \text{ for the first time}$$

Here we assume that the chain starts in some fixed state i, and we allow the case $j = i$. We call T_j the **first entrance time** into state j.

Notice the difference between sojourn times and first entrance times when $i = j$. For example, if $p_{ii} > 0$ and the first transition happens to be $i \to i$, then T_i takes the sample value 1; however, the sample value of S_i will be greater than 1, since the system is still in state i. (In fact, the sample value of S_i is not determined until the system jumps out of state i.)

Suppose the chain has d states. Define the $d \times d$ **recurrence matrix R** to have entries

$$\mathbf{R}_{ij} = E_i[T_j]$$

(It can be shown that all of these expectations are finite). In particular, the expected number of steps between visits to state i is the ith diagonal entry of the matrix **R**. We call this the **mean recurrence time** for state i. We shall first analyze these recurrence times.

Example 6.31 Vending Machine

Consider the vending machine model (Example 6.4). By formula (6.61), the sojourn times S_0 and S_1 are geometric random variables with expectations $1/\delta$ and $1/\gamma$, respectively. For example, if $\delta = 0.2$ and $\gamma = 0.9$, then

$$\begin{aligned}\text{Expected length of up cycle} &= \frac{1}{0.2} = 5 \text{ days} \\ \text{Expected length of down cycle} &= \frac{1}{0.9} = 1.11 \text{ days}\end{aligned}$$

We can get sample values of the sojourn and recurrence times by simulation. Suppose we start with the machine is in state 0 (working) on day 0. A computer simulation of this model (with $\delta = 0.2$, $\gamma = 0.9$) produced the data in Table 6.5.

Table 6.5 Simulations of Vending Machine Model

	NUMBER OF DAYS SIMULATED			
TIMES	100	1000	10,000	50,000
Mean Sojourn times				
State 0	5.25	5.33	4.84	4.99
State 1	1.07	1.13	1.10	1.11
Mean recurrence times				
State 0	1.19	1.21	1.23	1.22
State 1	5.88	5.74	5.42	5.49

The values for the mean sojourn times match the predictions. To understand the mean recurrence times, recall from Example 6.27 that the long-term proportions of time the system spends in states 0 and 1, respectively, are $\frac{9}{11}$ and $\frac{2}{11}$. The reciprocals of these proportions are 1.22 days and 5.5 days. From the simulation data, these appear to be the mean recurrence times. ■

The data of Example 6.23 suggest that *the mean recurrence time for a state is the reciprocal of the steady-state probability for the state*. This is true for every regular chain.

Theorem 6.10 (Mean Recurrence Times). *Let* **P** *be the* $d \times d$ *transition matrix for a regular finite Markov chain, and let*

$$\boldsymbol{\alpha} = \begin{bmatrix} a_1 & a_2 & \dots & a_d \end{bmatrix}$$

be the stationary distribution that satisfies $\boldsymbol{\alpha}\mathbf{P} = \boldsymbol{\alpha}$. *Then the mean recurrence time for state* i *is* $1/a_i$.

Proof Suppose that we follow the movements of the marker, starting in state i. What is the expected number of steps taken to reach state j? We calculate by conditioning on the outcome of the first step. If the first step is to state j, then this expectation is 1. If the first step is to state $k \neq j$, then we begin again (by the no-memory property) with the step count increased by 1. Taking into account the probabilities for the first step, we thus get the recursion

$$E_i[T_j] = p_{ij} + \sum_{k \neq j} p_{ik}\big(E_k[T_j] + 1\big) \tag{6.62}$$

Since

$$p_{ij} + \sum_{k \neq j} p_{ik} = 1$$

we can write equation (6.62) as

$$E_i[T_j] = 1 + \sum_{k \neq j} p_{ik} E_k[T_j] \tag{6.63}$$

Equation (6.63) can be written in matrix form as

$$\mathbf{R} = \mathbf{U} + \mathbf{P}(\mathbf{R} - \mathbf{D}) \tag{6.64}$$

Here $\mathbf{U}$ is the $d \times d$ matrix with all entries 1, and $\mathbf{D}$ is the $d \times d$ matrix with main diagonal $E_i[T_i]$ and all off-diagonal entries zero.

To find $\mathbf{D}$ from equation (6.64), we multiply that equation on the left by the stationary vector $\boldsymbol{\alpha}$ and use the equation $\boldsymbol{\alpha}\,\mathbf{P} = \boldsymbol{\alpha}$:

$$\boldsymbol{\alpha}\mathbf{R} = \boldsymbol{\alpha}\mathbf{U} + \boldsymbol{\alpha}\mathbf{R} - \boldsymbol{\alpha}\mathbf{D}$$

Since the entries of $\boldsymbol{\alpha}$ sum to 1, the entries in $\boldsymbol{\alpha}\mathbf{U}$ are all 1. Hence,

$$\boldsymbol{\alpha}\mathbf{D} = \begin{bmatrix} 1 & 1 & \cdots & 1 \end{bmatrix}$$

But we also know that the entries of $\boldsymbol{\alpha}\mathbf{D}$ are $a_i E_i[T_i]$. Hence,

$$a_i E_i[T_i] = 1$$

for each state i. This proves Theorem 6.10. ∎

Let $\mathbf{P}$ be the $d \times d$ transition matrix for a regular finite Markov chain, and let $\boldsymbol{\alpha}$ be the stationary distribution. We have just proved that the diagonal entry R_{ii} in the recurrence matrix $\mathbf{R}$ is given by $1/a_i$. To find the off-diagonal entry R_{ij}, which is the expected time to go from state i to state j when $i \neq j$, we use the matrices in the proof of Theorem 6.10. Let $\mathbf{D}$ be the $d \times d$ matrix with diagonal entries $1/a_i$ and 0 off-diagonal entries, and let $\mathbf{U}$ be the $d \times d$ matrix with all entries 1. Define the matrix $\mathbf{M} = \mathbf{R} - \mathbf{D}$. The following theorem shows how to find $\mathbf{M}$ and hence $\mathbf{R}$.

Theorem 6.11 (Mean First Entrance Times). *The matrix* $\mathbf{M}$ *is uniquely determined by the equation*

$$(\mathbf{I} - \mathbf{P})\mathbf{M} = \mathbf{U} - \mathbf{D} \tag{6.65}$$

and the condition that all the diagonal entries of $\mathbf{M}$ *be zero.*

Proof Since $\mathbf{R} = \mathbf{M} + \mathbf{D}$, we see that equation (6.65) is the same as equation (6.64). Thus, by the proof of Theorem 6.10, we know that $\mathbf{M}$ satisfies equation (6.64). We must show that, given $\mathbf{P}$, equation (6.65) always has a unique solution $\mathbf{M}$ with 0 diagonal entries. (Notice that $\mathbf{P}$ determines $\mathbf{D}$, via the stationary distribution $\boldsymbol{\alpha}$.)

The matrix equation (6.65) is the same as d sets of inhomogeneous linear equations for the d columns of $\mathbf{M}$. If we denote the jth column of $\mathbf{M}$ by $\mathbf{M}_j$ and the jth column of $\mathbf{U} - \mathbf{D}$ by $\mathbf{A}_j$, then the equations are

$$(\mathbf{I} - \mathbf{P})\mathbf{M}_j = \mathbf{A}_j \quad \text{for } j = 1, 2, \ldots, d. \tag{6.66}$$

By Theorem 6.8, we know that the matrix $\mathbf{I} - \mathbf{P}$ has rank $d - 1$, with left null space consisting of multiples of the row vector $\boldsymbol{\alpha}$ and right null space consisting of multiples of the column vector $\mathbf{u}$ having all entries 1. By the theory of inhomogeneous systems of linear equations, it follows that equation (6.66) will have a solution if and only if the right-hand side is orthogonal to $\boldsymbol{\alpha}$. Since

$$\boldsymbol{\alpha}\mathbf{U} = \begin{bmatrix} 1 & \cdots & 1 \end{bmatrix} = \boldsymbol{\alpha}\mathbf{D}$$

it follows that $\boldsymbol{\alpha}(\mathbf{U} - \mathbf{D}) = 0$. But $\boldsymbol{\alpha}\,\mathbf{A}_j$ is the jth column of $\boldsymbol{\alpha}(\mathbf{U} - \mathbf{D})$. Thus,

$$\boldsymbol{\alpha}\,\mathbf{A}_j = 0 \quad \text{for } j = 1, \ldots, d$$

so equation (6.66) always has solutions.

For each column index j, let $\mathbf{C}_j$ be some particular solution of equation (6.66). We can then write

$$\mathbf{M}_j = \mathbf{C}_j + \lambda_j \mathbf{u}$$

for an appropriate choice of the constant λ_j. The condition that the jth entry in $\mathbf{M}_j$ be 0 uniquely determines λ_j, since $\lambda_j \mathbf{u}$ is the column vector with all entries equal λ_j. This proves Theorem 6.11. ■

Example 6.32 Vending Machine

Consider the vending machine model again. We found in Section 6.8 that the stationary distribution is

$$\boldsymbol{\alpha} = \begin{bmatrix} \dfrac{\gamma}{\delta + \gamma} & \dfrac{\delta}{\delta + \gamma} \end{bmatrix}$$

Hence, the mean recurrence times are $R_{00} = 1 + (\delta/\gamma)$ for state 0 and $R_{11} = 1 + (\gamma/\delta)$ for state 1.

An obvious relation exists between recurrence times and sojourn times in a two-state chain. Suppose that we start observing the chain at the moment it leaves state 1 and enters state 0. Then $T_1 = S_0$, since the system goes to state 1 when it leaves state 0. Hence,

$$E_0[T_1] = E_0[S_0] = \frac{1}{\delta}$$

by formula (6.61). By the same argument,

$$E_1[T_0] = E_1[S_1] = \frac{1}{\gamma}$$

Let's check that the matrix equation of Theorem 6.11 yields the same values for the off-diagonal entries R_{01} and R_{10} in the recurrence matrix. In this case, we have

$$\mathbf{I}-\mathbf{P}=\begin{bmatrix} \delta & -\delta \\ -\gamma & \gamma \end{bmatrix} \qquad \mathbf{U}-\mathbf{D}=\begin{bmatrix} -\delta/\gamma & 1 \\ 1 & -\gamma/\delta \end{bmatrix}$$

Thus, equation (6.65) takes the form

$$\begin{bmatrix} \delta & -\delta \\ -\gamma & \gamma \end{bmatrix}\begin{bmatrix} 0 & -R_{01} \\ R_{10} & 0 \end{bmatrix}=\begin{bmatrix} -\delta/\gamma & 1 \\ 1 & -\gamma/\delta \end{bmatrix}$$

Carrying out the matrix multiplication on the left side of this equation, and comparing entries with the right side, we find that

$$R_{01}=\frac{1}{\delta} \qquad R_{10}=\frac{1}{\gamma}$$

which agrees with the previous calculation. ■

Example 6.33 Inventory Model

Consider the inventory model. From Example 6.28, the steady-state distribution is

$$\boldsymbol{\alpha}=\begin{bmatrix} \frac{p}{2} & \frac{1}{2} & \frac{q}{2} \end{bmatrix}$$

where p is the probability of an incoming order per day and $q = 1-p$. From Theorem 6.10, we conclude that the mean recurrence times for states 0, 1, and 2 are $2/p$, 2, and $2/q$ days, respectively.

Suppose, for example, that $p = 0.2$. Then the mean time between visits to state 0 is 10 days. From the description of the model (Example 6.14), this means that the average time between restocking orders is 10 days.

We can use the matrix equation of Theorem 6.11 to obtain the off-diagonal entries in the recurrence matrix. In this case, we have

$$\mathbf{I}-\mathbf{P}=\begin{bmatrix} 1 & -p & -q \\ -p & p & 0 \\ 0 & -p & p \end{bmatrix} \qquad \mathbf{U}-\mathbf{D}=\begin{bmatrix} 1-2/p & 1 & 1 \\ 1 & -1 & 1 \\ 1 & 1 & 1-2/q \end{bmatrix}$$

Thus, equation (6.65) for this example takes the form

$$\begin{bmatrix} 1 & -p & -q \\ -p & p & 0 \\ 0 & -p & p \end{bmatrix}\begin{bmatrix} 0 & R_{01} & R_{02} \\ R_{10} & 0 & R_{12} \\ R_{20} & R_{21} & 0 \end{bmatrix}=\begin{bmatrix} 1-2/p & 1 & 1 \\ 1 & -1 & 1 \\ 1 & 1 & 1-2/q \end{bmatrix}$$

Notice that this matrix equation is the same as nine inhomogeneous linear equations on the six unknown off-diagonal entries of **R**, which is an *over-determined* system. But Theorem 6.11 guarantees that this system of equations is consistent and has a unique solution. Thus we can take any independent set of six of the equations to determine

$R_{01}, \ldots, R_{21}$. Carrying out the matrix multiplication on the left side of the equation and comparing entries with the right side, we find that

$$\mathbf{R} = \begin{bmatrix} 2/p & 1/p & 2/p \\ 1/p & 2 & (1+p)/pq \\ 2/p & 1/p & 2/q \end{bmatrix}$$

As a numerical example, suppose that $p = 0.2$ orders per day. When the inventory level is 2 (full) at the end of a day, we can expect to wait $R_{20} = 2/p = 10$ days before the end-of-day inventory next reaches 0 and requires a restocking order. When the current inventory level is 1 at the end of a day, we can expect to wait $R_{12} = (1+p)/pq = 7.5$ days before the end-of-day inventory next reaches 2. ■

Exercises

1. In Example 6.3, suppose that the game starts in state 0. Let T_n be the number of flips of the coin needed to get to state n. What is the probability distribution of T_n?

2. In the vending machine model (Example 6.4), suppose that the one-day breakdown probability $\delta = 0.2$, and the one-day repair probability $\gamma = 0.9$. You observe on Monday that the machine is working.

   a. What is the probability that the machine will remain in working order Tuesday, Wednesday, and Thursday?

   b. What is the probability that the machine will be in working order on Thursday?

3. In Example 6.6 (success runs), suppose that your win probability per round is $p = \frac{18}{38}$ and that you have won two consecutive rounds. What is the probability that you will win the game (that is, have a success run of length 3) within the next five rounds?

4. In Example 6.6 (success runs), suppose that the rules are changed so that now you need a run of four consecutive successes in order to win the game.

   a. Set up this game as a Markov chain, and draw the transition diagram.

   b. Suppose that your win probability per round is $p = \frac{18}{38}$ and that you have won three consecutive rounds. What is the probability that you will win the game (that is, have a success run of length 4) within the next five rounds?

5. You are playing successive independent rounds of a game where your win probability in each round is p; the bets are in multiples of \$1, and the payoff is double or nothing. You decide to follow the bold play strategy (Example 6.5). Suppose that you start playing with \$1 and decide to stop playing either when you go broke or when your fortune reaches \$7.

   a. Set this game up as a Markov chain, and draw the transition diagram.

   b. Calculate the probability that you will have \$7 in six or fewer rounds of play.

c. Calculate the probability that you will get \$ 7 in an unlimited number of rounds of play.

6. For the repeated independent trials chain (Example 6.11), suppose each trial consists of tossing two fair coins and counting the number of heads that show. Draw the transition diagram, and write down the transition matrix for this chain.

7. In Exercise 2, let X and Y be the states of the machine on Tuesday and Wednesday, respectively. We assume that the state is 0 (working) on Monday.

   a. Calculate the joint probability mass function of X, Y.

   b. Calculate the marginal probability mass functions of X and Y. Are X and Y independent? Are they identically distributed?

8. Modify the vending machine model (Example 6.4) as follows. Suppose that the probability that the machine is working on any particular day only depends on the state of the machine for the previous *two* days. Specifically, assume that

$$P\{X(n+1) = 0 \mid X(n-1) = j \text{ and } X(n) = k\} = q_{jk}$$

   where $X(n)$ is the state of the machine on day n and $q_{00} = \frac{3}{4}, q_{10} = \frac{4}{5}, q_{01} = \frac{1}{2}$, and $q_{11} = \frac{2}{3}$.

   a. Show that $\{X(n)\}$ is *not* a Markov chain.

   b. We define a new state space for this model by taking the pairs (j, k), where j and k are 0 or 1. We say the system is in state (j, k) on day n if the machine was in state j on day $n-1$ and in state k on day n. Show that, with this enlarged state space, the system is a Markov chain, and draw the transition diagram.

   c. Suppose that the machine was working on Monday and on Tuesday. What is the probability that it will be working on Thursday?

   (HINT: Rephrase the question in terms of the enlarged state space.)

9. Consider bold Play versus timid play (Examples 6.8 and 6.9). Suppose that you start with \$ 1 and that your probability of winning a single round is p. Let $B_6(p)$ and $T_6(p)$ be the probability that you get \$ 5 in six or fewer rounds of play following the bold play and timid play strategies, respectively.

   a. Calculate $B_6(p)/T_6(p)$ for $p = 0.3, 0.4, 0.5, 0.6$, and 0.7.

   b. Is the ratio in part (a) always greater than 1 when $0 < p < 1$?

10. For the genetic chain (Example 6.13) with $N = 3$ subgenes, draw the transition diagram and write down the one-step transition matrix.

11. For the inventory model with daily random demand either 1 or 0 items (Example 6.14), assume that the restocking policy is $S = 2$ and $s = -1$ (that is, we allow the inventory to reach 0 and then wait for one order to come in before restocking). Draw the transition diagram and write down the one-step transition matrix in this case.

12. Using the same restocking policy as was described in the inventory model in Example 6.15, generate sample demands for 5 days by using the first five entries in column 1 of the random-number table (Table 4 at the back of the book). Determine the sample values $x_1, \ldots, x_5$ from this data, under the assumption that $x_0 = 3$.

13. Consider the following stochastic matrices:

$$\mathbf{P}_1 = \begin{bmatrix} 1/2 & 0 & 1/2 \\ 0 & 1 & 0 \\ 3/4 & 0 & 1/4 \end{bmatrix} \qquad \mathbf{P}_2 = \begin{bmatrix} 0 & 1/3 & 2/3 \\ 2/3 & 0 & 1/3 \\ 1/2 & 1/2 & 0 \end{bmatrix}$$

$$\mathbf{P}_3 = \begin{bmatrix} 1/2 & 1/4 & 1/8 & 0 & 1/8 \\ 1/3 & 1/3 & 1/3 & 0 & 0 \\ 1/4 & 1/4 & 1/2 & 0 & 0 \\ 0 & 0 & 1/2 & 0 & 1/2 \\ 0 & 0 & 0 & 0 & 1 \end{bmatrix} \quad \mathbf{P}_4 = \begin{bmatrix} 1/2 & 0 & 0 & 0 & 1/2 \\ 0 & 1/2 & 0 & 1/2 & 0 \\ 0 & 3/4 & 1/8 & 1/8 & 0 \\ 0 & 1/4 & 0 & 3/4 & 0 \\ 1/2 & 0 & 0 & 0 & 1/2 \end{bmatrix}$$

Carry out the following analyses of each of these matrices:

a. Draw the transition diagram for the associated Markov chain. Is the chain irreducible?

b. Determine the classes of states. Determine whether each class is transient or ergodic. Determine whether each ergodic class is periodic or aperiodic.

c. Reorder the states so that the transition matrix for the chain is in standard form.

14. Suppose that you are flipping a fair coin.

a. How many times do you expect tails to appear before you get three successive heads?

b. How many flips do you expect to make before you get three successive heads?

c. Suppose that the first flip comes up heads. Now what is the answer to part (a)?

d. Suppose that the first flip comes up heads. How many more flips do you expect to make before getting three successive heads? (The first flip may be included in your run of three heads.)

15. Consider the genetic chain (Example 6.13), with $N = 3$ subgenes.

a. Classify the states, and show that the descendants of the gene will eventually have only all normal or only all abnormal subgenes.

b. Write the transition matrix in standard form (see Exercise 10 and Section 6.6), and calculate the matrices $\mathbf{R}$, $\mathbf{Q}$, and $(\mathbf{I} - \mathbf{Q})^{-1}$.

16. In Exercise 15, suppose that we start with a gene having two normal subgenes and one abnormal subgene.

a. Calculate the probability that the descendants of the gene will eventually have all normal subgenes.

b. Calculate the expected number of generations before a descendent has either all normal or all abnormal subgenes.

17. Consider the bold play chain (Example 6.5).

    a. Classify the states, and write the transition matrix $\mathbf{P}$ in standard form (see Section 6.6).

    b. Show that the matrix $\mathbf{Q}$ in the block decomposition of $\mathbf{P}$ satisfies $Q^4 = p^2q^2\mathbf{I}$, where $\mathbf{I}$ is the 4×4 identity matrix. Use this to calculate $(\mathbf{I} - \mathbf{Q})^{-1}$

    (HINT: Use the geometric series.)

    c. Calculate the expected length of the game, starting in state 1.

    d. Calculate your probability of winning (getting to state 5), if you start in state 1. Compare your answer with the answer obtained by path analysis in Example 6.8.

    e. Compare your answers in part (c) and part (d) with the simulation data in Example 6.5.

18. Consider gambler's ruin with $N = 5$ (Example 6.2).

    a. Write the transition matrix in canonical form (see Section 6.6).

    b. Calculate the matrix $(\mathbf{I} - \mathbf{Q})^{-1}$ for $p = 0.3$ and $p = 0.7$.

    c. Calculate the win probability and average length of the game for $p = 0.3$ and $p = 0.7$ when the initial state is 1. Compare your answers with the simulation data in Example 6.2.

19. In Exercise 9, let $B(p)$ (respectively $T(p)$) be the probability that you win the game (get \$5) following the bold play (respectively, timid play) strategy.

    a. Calculate $B(p)/T(p)$ for $p = 0.3$ and $p = 0.7$.

    b. Compare your values with those obtained by simulation in Examples 6.2 and 6.5.

    c. In there a contradiction between the answers in part (a) and the answers in Exercise 9?

20. Suppose that you are managing a vending machine that generates an average daily profit of \$200 when it is working. The machine has a daily breakdown rate of $\delta = 0.2$. Assume that a repair company offers a service contract to restore the machine to working order with probability γ in one day whenever it breaks down. The cost of such a service contract is $\$10/(1-\gamma)$ per day (whether the machine is working or not). What value of γ (where $0 < \gamma < 1$) should you specify to obtain the maximum net profit from the machine? (Assume that the operation of the machine can be described by the model in Example 6.4.)

21. Consider the machine maintenance model (Example 6.12), with $\delta = 0.2$, $\epsilon = 0.7$, $\rho = 0.4$, $\phi = 0.6$, and $\gamma = 0.8$.

    a. Write down the transition matrix $\mathbf{P}$.

    b. Calculate the stationary distribution $\boldsymbol{\alpha}$ for the chain.

    c. On a long-term basis, what is the ratio of the number of days during which the machine is in good working order to the number of days during which the machine is in poor order working order? What is the ratio of the number of days during which the machine is in good working order to the number of days during which the machine is broken?

22. Refer to Exercise 21.

    a. What is the expected number of days between breakdowns of the machine?

    b. If the machine is in good working condition on a particular day, what is the expected number of days before it breaks down?

    c. If the machine is in poor working condition on a particular day, what is the expected number of days until it is back in good working condition?

    d. If the machine is not working on a particular day, what is the expected number of days until it is back in good working condition?

23. Consider the Markov chain with states 1, 2, and 3 and transition probability matrix

$$\mathbf{P} = \begin{bmatrix} 0 & 1/2 & 1/2 \\ 1/3 & 0 & 2/3 \\ 0 & 1 & 0 \end{bmatrix}$$

    a. Draw the transition diagram for this chain, and show that the chain is regular.

    b. Find the stationary distribution for the chain.

    c. Try to guess from the transition diagram which state has the shortest mean recurrence time and which state has the longest mean recurrence time. Check your guesses by calculating the mean recurrence times using Theorem 6.10.

24. Suppose that you are playing the following game on the transition diagram of the chain in Exercise 23: you choose either state 2 or state 3 as your goal, start with the marker on state 1, and move the marker randomly according to the transition probabilities until you reach your chosen state. If you win $ 1 for each move of the marker, which goal gives the higher expected payoff? Try to guess from the transition diagram, and then check your guess by calculating mean first entrance times using Theorem 6.11.

25. Consider the repeated independent trials chain (Example 6.11).

    a. Show that the chain is regular, and find the stationary distribution.

    b. Show that Theorem 6.9 implies the (weak) Law of Large Numbers (Theorem 4.2).

26. (Simulation) Write a program to simulate the genetic chain with $N = 3$ subgenes. Compare the long-term simulation data with the theoretical probabilities in parts (a) and (b) of Exercise 16. Obtain the empirical distributions and sample variances of the absorption times, starting in states 1 or 2.

27. (Simulation) Write a program to simulate the inventory model (Example 6.15). Use the long-term simulation data to estimate the stationary distribution and mean recurrence times for the chain and compare these estimates with the theoretical values. Obtain the empirical distributions and sample variances of the recurrence times for state 0.

Chapter 7

The Poisson Process

The probability models we have studied up to this point have all involved random events or random variables occurring in a *discrete* sequence, such as the outcomes of n successive Bernoulli trials. In this chapter, we introduce the basic stochastic model for events that occur at random moments in *continuous* time. This model, called the **Poisson process**, applies to such varied situations as telephone calls arriving at a switchboard and cosmic rays arriving from outer space.

Because of its importance, we shall describe the Poisson process in several different, but equivalent, ways in this chapter and in Chapter 8. The first description is based on a simple part-replacement model. This model is especially convenient for computer simulation of a Poisson process by random-number generation; it also easily leads to consideration of the characteristic properties of shift-invariance and independent increments for the Poisson process. At the end of the chapter, we obtain another description of the Poisson process by means of differential equations. This approach turns out to be very useful in analyzing complicated continuous-time stochastic models. We shall return to it in Chapter 9.

7.1 Memoryless Random Variables

Suppose we have a bin full of integrated circuit chips of a particular type. The total operating lifetime X of a single chip is a random variable. We assume that all the chips were manufactured under identical conditions, so that the probability distribution of X is the same for all the chips in the bin. We assume that X is a continuous random variable and that $P\{X > 0\} = 1$ (that is, there are no chips in the bin that are nonoperative).

Consider the following scenario: We select a chip at random from the bin, install it, and turn it on at time $t = 0$. The probability that it is still operating at time t is $P\{X > t\}$. After the chip has been operating successfully for t hours, we unplug it and place it back in the bin. Someone else comes along, takes this same chip and installs it. What is the probability that the chip will still be working after s hours of use in the second installation? This is obviously the *conditional* probability

$$P\{X > s + t \mid X > t\} \tag{7.1}$$

(Notice that we have conditioned on the past history of the chip: it has already survived t hours of use during the first installation.)

One surprising thing about most types of solid-state electronic components is that they do not "grow old." This means that it is normally impossible for this second user of the chip to detect that we had previously used it for t hours. If the chip is working when he first installs it, then it behaves as if it were newly manufactured. By equation (7.1), this property is the same as the probabilistic equation

$$P\{X > s\} = P\{X > s + t \mid X > t\} \quad \text{for all } s \geq 0 \text{ and } t \geq 0 \tag{7.2}$$

A random variable $X \geq 0$ that satisfies property (7.2) is said to have **no memory**.

Now let's reconsider the problem of setting up a probabilistic model for this situation. Suppose we assume, as a law of nature, that equation (7.2) is true for the lifetime of a certain type of component. Then the following theorem shows that our model is completely determined once we know the average lifetime. (Recall that, for an exponential random variable X with parameter λ, $E[X] = 1/\lambda$.)

Theorem 7.1. *An exponential random variable has no memory. Conversely, if $X > 0$ is a random variable that satisfies property* (7.2) *(that is, X has no memory), then X is an exponential random variable with parameter $\lambda = 1/\mu$, where $\mu = E[X]$.*

Proof Define the function

$$R(s) = P\{X > s\} = 1 - F_X(s)$$

The function $R(s)$, which gives the probability that the lifetime is greater than s, is called the **reliability function** of the random variable X. By the definition of conditional probability, equation (7.2) describing the no-memory property can be written in terms of the reliability function as

$$R(s + t) = R(s) \cdot R(t), \quad \text{for all } s \geq 0 \text{ and } t \geq 0 \tag{7.3}$$

Now suppose that we know that X is exponentially distributed, with parameter λ. We calculated in Example 3.8 that $F_X(t) = 1 - e^{-\lambda t}$, so the reliability function is

$$R(t) = e^{-\lambda t}$$

for $t \geq 0$ in this case. By the multiplicative property of the exponential function, we see that formula (7.3) is satisfied. Thus, every exponential random variable is memoryless.

In the other direction, if we start with formula (7.3), then for $t \geq 0$ and every positive integer m, we have

$$\begin{aligned} R(mt) &= R(t + (m-1)t) \\ &= R(t) \cdot R((m-1)t) = \cdots \\ &= R(t)^m \end{aligned}$$

Taking $t = 1/n$, with n a positive integer, we see that $R(1) = R(1/n)^n$, which we can also write as

$$R\left(\frac{1}{n}\right) = R(1)^{1/n}$$

Using this in the previous equation gives us the following formula for $R(t)$:

$$R\Big(\frac{m}{n}\Big) = \big[R(1)\big]^{m/n} \quad \text{for all positive integers } m, n \tag{7.4}$$

We claim that equation (7.4) implies that $0 < R(1) < 1$. Indeed, if to the contrary $R(1) = 1$, we would have $R(m) = 1$ for every positive integer m Taking the limit as $m \to \infty$, we would conclude that $R(\infty) = 1$, and hence $F_X(\infty) = 0$, contradicting the basic probability axiom that $F_X(\infty) = 1$. On the other hand, if $R(1) = 0$, then we would have $R(1/n) = 0$ for every positive integer n. Taking the limit as $n \to \infty$, we would conclude that $R(0) = 0$, and hence $F(0) = 1$. But this would contradict the assumption that all components are in working order when first installed.

Now that we know $R(1)$ falls strictly between 0 and 1, we can set $\lambda = -\log R(1)$. Then λ is in the range $0 < \lambda < \infty$. Since $R(1) = e^{-\lambda}$, equation (7.4) can be written as

$$R(t) = e^{-\lambda t} \tag{7.5}$$

for all *rational* numbers $t \geq 0$. However, each side of equation (7.5) is a right-continuous function of the real variable t. Since these functions are equal for all rational values of t, by continuity they must be equal for all real positive values of t. The calculation at the beginning of the proof then implies that $F_X(t) = 1 - R(t)$ is the cumulative distribution function of an exponential random variable with parameter λ. ■

Suppose the operating lifetime T of a certain type of component has the no-memory property, and is hence an exponential random variable with parameter λ. The average lifetime is then $1/\lambda$, a quantity with physical dimensions of time; so λ has dimensions 1/time. It has the following interpretation.

If the component is working at time t, then the probability that it fails in the time interval between t and $t + \delta t$ is

$$P\{T < t + \delta t \mid T > t\}$$

By the no-memory property, this probability is the same as

$$P\{T < \delta t\} = 1 - e^{-\lambda \delta t} \approx \lambda\, \delta t$$

if δt is small (here we are using the approximation $e^h \approx 1 + h$ for h near zero). We can write this relation as

$$\frac{P\{\text{Failure occurs between } t \text{ and } t + \delta t \mid \text{Lifetime} > t\}}{\delta t} \approx \lambda$$

Thus λ can be interpreted as the **failure rate** of a currently functioning component. (We will return to this interpretation in Section 7.4 and in Chapter 12.)

Example 7.1

Return to the probability model for the lifetime of an integrated circuit chip. Suppose we have an electronic device consisting of two different chips. Let X be the lifetime of the first chip, and Y the lifetime of the second chip. Assume that both lifetimes have the no memory property, and are hence exponential random variables with failure rates α and β, respectively. Assume, too, that X and Y are mutually independent. If both chips must be functioning for the device to work, what is the probability distribution of the total lifetime Z of the device?

Solution We have $Z = \min\{X, Y\}$. To calculate the probability distribution of Z, fix a real number $c \geq 0$. Then

$$\begin{aligned} P\{Z \geq c\} &= P\{X \geq c \text{ and } Y \geq c\} \\ &= P\{X \geq c\} \cdot P\{Y \geq c\} \qquad \text{(independence)} \\ &= e^{-ac} e^{-bc} = e^{-(a+b)c} \end{aligned}$$

Thus Z is again an exponential random variable, but with parameter $\alpha + \beta$. In particular, the expected lifetime of the instrument is

$$E[Z] = \frac{1}{\alpha + \beta}$$

■

A fundamental fact about pairs of independent exponential random variables— and one that we will use repeatedly—is stated in Lemma 7.1.

Lemma 7.1. *If X and Y are independent exponential random variables with parameters α and β respectively, then $Z = \min\{X, Y\}$ is an exponential random variable with parameter $\alpha + \beta$. Furthermore, if the random variable B is defined by*

$$B = \begin{cases} 1 & \text{when } X \leq Y \\ 0 & \text{when } X > Y \end{cases}$$

then

$$E[B] = \frac{\alpha}{\alpha + \beta} \tag{7.6}$$

and the random variables B and Z are mutually independent.

Proof We already know from Example 7.1 that Z is a exponential random variable with parameter $\alpha + \beta$. Obviously, $E[B] = P\{X \leq Y\}$. But we calculated in Example 4.9 that

$$P\{X \leq Y\} = \int_{-\infty}^{\infty} F_X(y)\, f_Y(y)\, dy$$

Since X is exponential with parameter α, and since Y is exponential with parameter β, the preceding integral is

$$\begin{aligned} \int_0^{\infty} \left(1 - e^{-\alpha y}\right) \beta e^{-\beta y}\, dy &= 1 - \beta \int_0^{\infty} e^{-(\alpha+\beta)y}\, dy \\ &= \frac{\alpha}{\alpha + \beta} \end{aligned}$$

It remains for us to prove that B and Z are mutually independent. Let c be any positive real number. Then

$$\begin{aligned} P\{B = 1 \text{ and } Z > c\} &= P\{X \leq Y \text{ and } Z > c\} \\ &= P\{c < X \leq Y\} \end{aligned}$$

We can calculate this last probability by conditioning on Y. If $Y = y$ and $y > c$, then

$$\begin{aligned} P\{c < X \le Y \mid Y = y\} &= P\{c < X \le y\} \\ &= e^{-\alpha c} - e^{-\alpha y} \end{aligned}$$

since X and Y are independent. On the other hand, if $Y = y$ and $y \le c$, then

$$P\{c < X \le Y \mid Y = y\} = 0$$

Now average these conditional probabilities, using the probability density of Y:

$$\begin{aligned} P\{B = 1 \text{ and } Z > c\} &= \int_0^\infty P\{c < X \le Y \mid Y = y\}\, f_Y(y)\, dy \\ &= \int_c^\infty \left(e^{-\alpha c} - e^{-\alpha y}\right)\beta e^{-\beta y}\, dy \\ &= \frac{\alpha}{\alpha+\beta}\, e^{-(\alpha+\beta)c} \end{aligned}$$

(the evaluation of the last integral is elementary). From this calculation and the facts already proved about the random variables B and Z, we conclude that

$$P\{\, B = 1 \text{ and } Z > c\,\} = P\{B = 1\} \cdot P\{Z > c\}$$

for every value of c. This proves that B and Z are mutually independent. ■

Example 7.2

A quick copy shop has two printers. Jobs come into the shop and are randomly assigned to printer A or printer B. Jobs done on Printer A require X minutes to print, where X is an exponential random variable with parameter $\alpha = 2$ jobs/minute. Jobs done on Printer B require Y minutes to print, where Y is an exponential random variable with parameter $\beta = 3$ jobs/minute. The printers function independently of each other.

Suppose that we walk into the shop and see that both printers are busy. The probability that printer Y will finish its current job before printer X does is $\beta/(\alpha+\beta) = 0.6$. On the other hand, the probability that at least one printer will finish its current job in the next 30 seconds is

$$P\{\,\min\{X, Y\} \le 0.5\,\} = 1 - e^{-5\cdot(0.5)} = 0.92$$

since $Z = \min\{X, Y\}$ is exponential, with parameter $\alpha + \beta = 5$. Finally, the probability that *both* printers will finish their current jobs in the next 30 seconds is

$$\begin{aligned} P\{X \le 0.5 \text{ and } Y \le 0.5\} &= P\{X \le 0.5\} \cdot P\{Y \le 0.5\} \\ &= (1 - e^{-1})(1 - e^{-1.5}) \\ &= 0.49 \end{aligned}$$

by the no-memory property and the independence of the printing times. ■

7.2 Replacement Models

Consider a machine containing a particular type of electronic component. As in Section 7.1, we assume that the operating lifetime of any such component is an exponential random variable T with parameter λ.

Suppose that we have a supply of these components, all manufactured under identical conditions. To build an appropriate stochastic model for this situation, we assume that the nth component has lifetime T_n— an exponential random variable with parameter λ—and that $T_1, T_2, \ldots, T_n$ are mutually independent for all values of n. Here the failure-rate parameter λ is the same for all the components.

Now consider the following **replacement model** scenario. We install the first component in the machine and turn on the machine at time $t = 0$. When the first component fails, we immediately replace it with the second component, and we restart the machine. We continue this mode of operation indefinitely. At any clock time t, we can then observe

$$N(t) = \text{The number of components that have failed in the period up to and including time } t \tag{7.7}$$

We are assuming that every component has a positive lifetime, so $N(0) = 0$. We want to determine the probabilistic properties of the family of random variables $N(t)$, for $t > 0$.

Notice that $N(t)$ is a *discrete* random variable that depends on a continuous parameter $t \geq 0$. To analyze $N(t)$, we consider a natural complementary *continuous* random variable that depends on a discrete parameter n. Define

$$S_n = T_1 + T_2 + \cdots + T_n \tag{7.8}$$

Thus, S_n is the sum of the lifetimes of the first n components that are used. Since each T_k is an exponential random variable, we know from Example 4.19 that S_n is a gamma random variable, with parameters n, λ (this property depends on the mutual independence of the T_k).

Observe that, for any $t \geq 0$, these events are the same:

$$\{S_n \leq t\} = \{N(t) \geq n\} \tag{7.9}$$

That is, if the sum of the lifetimes of the first n components was at most t, then at least n components have failed by time t, and vice versa. Notice that, on the left side of equation (7.9), we have a family of *continuous* random variables indexed by the discrete parameter n; meanwhile, on the right side, we have a family of *discrete* random variables indexed by the continuous variable t. This equation is the link between the discrete counting parameter n and the continuous time parameter t. We will use it often.

From (7.9) and the formula for the gamma density, we calculate that

$$\begin{aligned} P\{\,N(t) \geq n\,\} &= P\{\,S_n \leq t\,\} \\ &= \frac{\lambda}{(n-1)!} \int_0^t (\lambda x)^{n-1}\, e^{-\lambda x}\, dx \end{aligned}$$

To evaluate this, we integrate by parts, with

$$u = e^{-\lambda x} \qquad dv = \frac{\lambda}{(n-1)!}(\lambda x)^{n-1}\,dx$$
$$du = -\lambda e^{-\lambda x}\,dx \qquad v = \frac{(\lambda x)^n}{n!}$$

Evaluating uv at $x = 0$ and $x = t$, we get the formula

$$P\{\, N(t) \geq n \,\} = \frac{(\lambda t)^n}{n!} e^{-\lambda t} + \lambda \int_0^t \frac{(\lambda x)^n}{n!} e^{-\lambda x}\,dx$$

But the integral on the right is just $P\{\, N(t) \geq n + 1 \,\}$. Subtracting it from the left side thus gives the formula for the probability mass function of the random variable $N(t)$:

$$P\{\, N(t) = n \,\} = \frac{(\lambda t)^n}{n!} e^{-\lambda t} \tag{7.10}$$

This shows that $N(t)$ has a *Poisson* distribution, with parameter λt (see Example 3.6).

7.3 Simulation of a Poisson Process

Consider again the replacement model of Section 7.2. To study the random variables $N(t)$ experimentally, we don't have to take actual electronic components and operate them in real time. Instead, we can *simulate* the failure times of a set of components by first calling on a random-number generator to obtain mutually independent sample values $u_1, u_2, \ldots$ that are uniformly distributed on the interval $(0, 1)$. We then set

$$\begin{aligned} t_k &= -(1/\lambda)\log u_k \\ s_n &= t_1 + t_2 + \cdots + t_n \\ n(t) &= \max\{k \text{ such that } s_k \leq t\} \end{aligned}$$

By Lemma 3.1, we know that t_k is a sample value from an exponential random variable with parameter λ. The mutual independence of the u_k values implies the mutual independence of the t_k values. It thus follows that s_n is a sample value of the gamma random variable S_n, and that $n(t)$ is a sample value of the Poisson random variable $N(t)$.

Example 7.3

Suppose that $\lambda = 1$. Using the random number key on a hand calculator, we obtain the sequence of random numbers $u_1 = 0.405$, $u_2 = 0.829$, $u_3 = 0.499$, and $u_4 = 0.556$. The corresponding sample values of $t_k = -\log u_k$ are calculated to be $t_1 = 0.904$, $t_2 = 0.188$, $t_3 = 0.695$, $t_4 = 0.587$. Taking the partial sums of these numbers, we get $s_1 = 0.904$, $s_2 = 1.091$, $s_3 = 1.787$, $s_4 = 2.374$, and so on. From the relation between S_n and $N(t)$ (see equation (7.9), we have

$$n(t) = \begin{cases} 0 & \text{for} \quad 0 \le t < 0.904 \\ 1 & \text{for } 0.904 \le t < 1.091 \\ 2 & \text{for } 1.091 \le t < 1.787 \\ 3 & \text{for } 1.787 \le t < 2.374 \\ 4 & \text{for } 2.374 \le t < \ldots \end{cases}$$

(see Figure 7.1). ■

We shall call the family of random variables $\{N(t) : t > 0\}$ a **Poisson Process** (with **rate** λ). A **sample observation** of a Poisson process is a function $n(t)$ such as shown in Figure 7.1. The jumps in the graph of $n(t)$ occur at the moments when a component fails. Notice that we have defined $n(t)$ so that it is *right*-continuous.

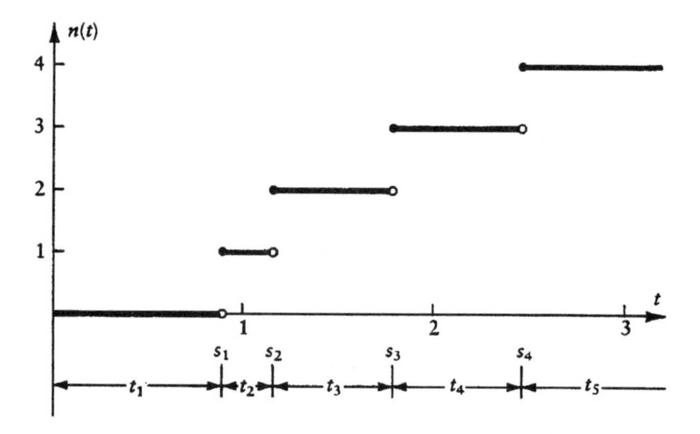

Figure 7.1: Simulation of Poisson Process

Of course, the graph in Figure 7.1 theoretically continues indefinitely to the right. (We assume that we have an unlimited stock of replacement parts in this model).

7.4 Fundamental Properties of a Poisson Process

In this section, we continue to study the replacement model of Section 7.2. We shall derive the following two fundamental properties of the Poisson process $\{N(t)\}$:

(i) **Shift Invariance:** For any $s \ge 0$ and $h > 0$, the random variable

$$N(s+h) - N(s) = \text{Number of components failing in time period } s < t \le s+h \tag{7.11}$$

is Poisson, with parameter λh.

(ii) **Independent Increments:** Fix any set of successive times $0 = \tau_0 < \tau_1 < \tau_2 < \tau_3 < \ldots < \tau_n$, and define

$$X_i = N(\tau_i) - N(\tau_{i-1})$$

Then the random variables $X_1, X_2, \ldots, X_n$ are mutually independent.

Proof of Property (i) This is a consequence of the no-memory property. If we start observing the system at time s, the component currently in use has number $N(s) + 1$ (number the components, in order of use; see Fig. 7.1). It hasn't failed at time $t = s$, so it behaves during time $t > s$ as if it were new, and as if the machine had been first turned on at $t = s$. The random variable in equation (7.11) therefore must have the same probability distribution as $N(h)$. We calculated in Section 7.2 that $N(h)$ is a Poisson random variable with parameter λh. (Notice carefully that, although the random variables $N(h)$ and $N(s+h) - N(s)$ are identically distributed, they are *different* random variables). ∎

Proof of Property (ii) To understand why this independence property is valid, take (for example) $\tau_1 = 1$ and $\tau_2 = 2$ in the numerical example of Section 7.3. The sample values of the random variables X_1 and X_2 can then be read off from the graph of $n(t)$:

$$x_1 = \{\text{Number of jumps of } n(t) \text{ in } 0 < t \leq 1\} = 1$$

and

$$x_2 = \{\text{Number of jumps of } n(t) \text{ in } 1 < t \leq 2\} = 2$$

The value of x_1 was determined from the lifetimes t_1 and t_2 of the first two components; however, the only information we needed about the second component was that it was still functioning at time $t = 1$. The value of x_2 was determined from the *remaining* lifetime of the second component, starting at time t = 1, and the lifetimes t_3 and t_4 of the third and fourth components. We assume in the replacement model that the random variables $T_1, \ldots, T_4$ are mutually independent, so the only possible lack of independence between X_1 and X_2 in this example results from the use of t_2 in determining both x_1 and x_2. But by the no-memory property of the exponential random variable, the remaining lifetime of the second component, starting at time $t = 1$, is independent of the length of time that it has been in operation before $t = 1$ (assuming that it is still functioning at time $t = 1$). The random variable X_1 depends only on the past (before time $t = 1$), while the random variable X_2 depends only on the future (after $t = 1$). The no-memory property means that events depending on the past are independent of events depending on the future. From this we conclude that X_1 and X_2 are independent. The proof of property (ii) in general follows along the same lines laid down in this example. ∎

From equation (7.11), we see that

$$E[N(s+h) - N(s)] = \lambda h$$

from the formula for the expectation of a Poisson random variable. Hence, the parameter λ is the **expected failure rate** of the components:

$$\lambda = \frac{\text{Expected number of failures during } s < t \leq s+h}{h} \tag{7.12}$$

This formula is similar to the one for λ obtained in Section 7.1. Here, however, we are dealing with an expected value, rather than a probability, and this formula is *exact* rather than being an approximation.

Remarks In terms of the replacement model, the independence property (ii) can be seen as follows. Suppose that, at a schedule of *fixed* times $\tau_1, \tau_2, \tau_3, \ldots$, we automatically replace the component currently in use by a new component. We continue as before to replace components when they fail at random times. By the mutual independence of the lifetimes of the different components, it is then obvious that when we count the number of components failing in each fixed, nonrandom time interval $\tau_i < t \leq \tau_{i+1}$, the counts are all mutually independent. But by the discussion in Section 7.1, the no-memory property of the components means that this new preventive maintenance policy doesn't change the operation of the machine at all, probabilistically speaking. The automatically installed components have the same *total* lifetime probability as do the remaining lifetimes of the components they replace. So in the original model, the failure counts in each given time interval must be independent.

Example 7.4

Suppose the failure rate of the components in the replacement model is $\lambda = 3$ per year. Let $N(t)$ be the number of components that have failed up to time t ($t = 0$ corresponds to the moment that the machine was first put into use, and we will measure time in years). We can use the basic properties of shift-invariance and independent increments to answer the following questions:

a. What is the probability that at least two components will fail in a particular 6-month time period?

b. What is the probability that the elapsed time between the fourth and fifth component replacement will be at least 6 months?

Solution, Part (a) This probability is $P\{N(t + 0.5) - N(t) \geq 2\}$, if the 6-month period begins at time t. By the shift-invariance of the process, this probability is equal to $P\{N(0.5) \geq 2\}$. There are two ways to carry out the calculation:

(i) Use the Poisson distribution, with $\lambda t = 1.5$:

$$\begin{aligned} P\{N(0.5) \geq 2\} &= 1 - P\{N(0.5) = 0\} - P\{N(0.5) = 1\} \\ &= 1 - e^{-1.5} - 1.5e^{-1.5} \\ &= 1 - 2.5e^{-1.5} \\ &= 0.44 \end{aligned}$$

(ii) Since the event $\{N(0.5) \geq 2\}$ is the same as the event $\{S_2 \leq 0.5\}$, we can use the probability density of S_2, which is gamma with parameters $n = 2$ and $\lambda = 3$:

$$\begin{aligned} P\{S_2 \leq 0.5\} &= \int_0^{0.5} (\lambda t) \cdot \lambda e^{-\lambda t}\, dt \\ &= 9 \int_0^{0.5} t e^{-3t}\, dt \end{aligned}$$

Integrating by parts then gives the same answer as before. (Check this by carrying out the integration.)

Solution, Part (b) This event can be described by $\{T_5 \geq 0.5\}$. Since T_5 is an exponential random variable with parameter $\lambda = 3$, the probability is $e^{-3/2} = 0.22$. ■

The memoryless property of the component lifetimes is the reason underlying the completely random nature of the Poisson process illustrated in the following example.

Example 7.5

Suppose the failure rate for the electronic chip is λ per hour. In a particular 24-hour operating period, exactly one chip failed. What is the probability distribution of the clock time T $(0 \leq T \leq 24)$ when the chip failed?

Solution By the no-memory property of the chips, we may assume that the chip was first installed at time $t = 0$, so its lifetime is the random variable T_1. The conditional cumulative distribution function of T, given that $N(24) = 1$, is then given by

$$\begin{aligned} F_T(t) &= P\{T_1 \leq t \mid N(24) = 1\} \\ &= \frac{P\{N(t) = 1 \text{ and } N(24) = 1\}}{P\{N(24) = 1\}} \\ &= \frac{P\{N(24) - N(t) = 0 \text{ and } N(t) = 1\}}{P\{N(24) = 1\}} \\ &= \frac{P\{N(24) - N(t) = 0\} \cdot P\{N(t) = 1\}}{P\{N(24) = 1\}} && \text{(independent increments)} \\ &= \frac{P\{N(24 - t) = 0\} \cdot P\{N(t) = 1\}}{P\{N(24) = 1\}} && \text{(shift invariance)} \end{aligned}$$

We can evalutate this probability by using the fact that $N(\tau)$ is Poisson, with parameter $\lambda\tau$. Taking $\tau = t, 24$, and $24 - t$ in the formula for the Poisson probability mass function, substituting in the preceding formula, and simplifying, we get the result:

$$F_T(t) = \frac{t}{24} \quad \text{for } 0 \leq t \leq 24$$

(this is an easy algebraic exercise). This formula shows that T is **uniformly distributed** over the 24-hour period. Thus, if we know that *exactly one* component has failed in the past 24 hours, we can conclude that it was just as likely to have failed between midnight and 1 A.M. as between noon and 1 P.M. Furthermore, the failure rate λ doesn't appear in the answer at all! ■

With more detailed calculations along the same lines as those used in Example 7.5,we could show the following. Suppose that exactly n events of a Poisson process are known to have occurred in a time interval $(0, t)$, and that we calculate conditional probabilities relative to the event $\{N(t) = n\}$. Then the random instants $X_1 < X_2 < \cdots < X_n$ at which these events happened have the same joint conditional probability distribution as do the order statistics from a random sample of size n from a population that is uniformly distributed on $(0, t)$. (This means that we can get random variables with the same conditional joint distribution as $X_1, \ldots, X_n$ by taking n random variables $U_1, \ldots, U_n$ that are independent and uniform on $(0, t)$ and rearranging them in increasing order.) In particular, this joint conditional distribution doesn't depend on the rate λ of the process. In the next example we apply this property in describing how a random observer sees a Poisson process.

Example 7.6

The Poisson process often gives a good model for traffic flow of light to moderate intensity on a single-lane highway. Suppose that vehicles pass a particular point on the highway at a Poisson rate of 10 per hour. (This means that, if at time $t = 0$ we start counting the vehicles that pass this point, the number $N(t)$ of vehicles that have passed after t hours gives a Poisson process with rate $\lambda = 10$.)

For making traffic counts, an automatic recording mechanism is generally used. Suppose that this recording machine is turned off and on at various times. Take a specific time interval of length t, and assume that n vehicles passed the recording point during this interval. Let p be the proportion of time that the recorder was on (where $0 < p \leq 1$). What is the probability that a single vehicle chosen at random from those that passed was counted by the machine?

Solution Let T be the moment at which the randomly chosen vehicle passed by. From Example 7.5 and the remarks after it, we know that the *conditional* distribution of T, given that $\{N(t) = n\}$, is uniform on the interval $(0, t)$. Hence, the probability that the recording machine was on at the random moment T is p (the proportion of time that the machine was on). Thus

$$P\{\text{Random vehicle was recorded}\} = \text{Proportion of time the recorder was on} \quad (7.13)$$

This property of a Poisson process is summarized by the phrase

"Poisson arrivals see time averages"

(we shall call this the PASTA property). Notice that neither the rate λ of the process nor the length t of the time interval appear in equation (7.13). ■

7.5 Differential Equations for a Poisson Process

We continue our study of the Poisson process, which we have defined using a replacement model. We now examine more closely the probabilities for replacements in a short time interval of duration h. If λ is the rate of the process, then

$$\begin{aligned}
P\{\text{ No replacement in } (t, t+h]\} &= e^{-\lambda h} \\
&= 1 - \lambda h + \frac{(\lambda h)^2}{2} - \cdots \\
P\{\text{ Exactly one replacement in } (t, t+h]\} &= \lambda h\, e^{-\lambda h} \\
&= \lambda h - (\lambda h)^2 + \cdots
\end{aligned}$$

If we subtract both of these probabilities from 1, we get

$$P\{\text{ More than one replacement in } (t, t+h]\} = \frac{(\lambda h)^2}{2} + \cdots$$

In all these equations, the symbol $\cdots$ stands for a convergent infinite series of terms in higher powers of h than the terms displayed. A convenient notation for such calculations involving increasing powers of h is the symbol $o(h)$. By definition, $o(h)$ denotes *any* function of h such that

$$\lim_{h \to 0} \frac{o(h)}{h} = 0.$$

For example, $h^2 = o(h)$ and $o(h) + o(h) = o(h)$ are two valid equations in this notation.

We can rewrite the equations for replacement probabilities as

$$\begin{aligned} P\{\text{ No replacements in } (t, t+h]\,\} &= 1 - \lambda h + o(h) \\ P\{\text{ Exactly one replacement in } (t, t+h]\,\} &= \lambda h + o(h) \\ P\{\text{ More than one replacement in } (t, t+h]\,\} &= o(h) \end{aligned}$$

Of course, in each of these equations the symbol $o(h)$ stands for a different function. Note also that the first equation is a consequence of the other two equations.

Now suppose that we have a replacement model, as in Section 7.2, and let $N(t)$, for $t \geq 0$, be the total number of components replaced up to time t. Instead of making any explicit assumption about the distribution of the lifetimes of the components, we simply *assume* that the following properties hold for the family of random variables $\{N(t)\}$:

(i) $N(0) = 0$ (counting starts at $t = 0$).

(ii) The process has **independent increments** (events in the model that occur in disjoint time intervals are mutually independent).

(iii) $P\{\, N(t+h) - N(t) = 1 \,\} = \lambda h + o(h)$.

(iv) $P\{\, N(t+h) - N(t) > 1 \,\} = o(h)$.

In property (iii), λ is a constant that is naturally called the **replacement rate**. We verified in Section 7.4 and at the beginning of this section that properties (i) through (iv) are satisfied by a Poisson process. The converse is also true:

Theorem 7.2. *If the replacement process* $\{N(t)\}$ *satisfies properties* (i) *through* (iv), *then, for any* $s \geq 0$ *and* $t > 0$, *the number of replacements that occur in the time interval* $(s, s+t]$ *is a Poisson random variable with mean* λt.

Proof It suffices to consider the case $s = 0$. (If $s > 0$, we can replace $N(t)$ by $N(t+s) - N(s)$ and still have properties (i) through (iv) satisfied.) Set

$$p_n(t) = P\{\, N(t) = n \,\}$$

for n a nonnegative integer, and t a nonnegative real number. We shall calculate these functions of t by first finding a system of differential equations that they satisfy. This is done, using properties (i) through (iv), as follows.

Take a small number $h > 0$ and count the number of replacements in the interval $(0, t+h]$ by making separate counts in the long interval $(0, t]$ and the short interval

$(t, t+h]$. By the Rule of Total Causes, the probability of n replacements in the interval $(0, t+h]$ is equal to

$$\sum_{k=0}^{n} P\{\, k \text{ replacements in } (t, t+h] \text{ and } n-k \text{ replacements in } (0, t]\,\}$$

By property (ii) each term in this sum is a product of probabilities. Thus,

$$p_n(t+h) = \sum_{k=0}^{n} P\{\, k \text{ replacements in } (t, t+h]\,\} \cdot p_{n-k}(t)$$

Now observe that, by properties (iii) and (iv),

$$P\{\,\text{No replacements in } (t, t+h]\,\} = 1 - \lambda h + o(h)$$

By property (iv), the terms involving two or more replacements in $(t, t+h]$ are of the form $o(h)$. Hence, we get the equation

$$p_n(t+h) = (1 - \lambda h)\, p_n(t) + \lambda h\, p_{n-1}(t) + o(h)$$

Move the term involving p_n to the left side of this equation, divide by h, and let $h \to 0$. This gives the differential equation

$$p_n'(t) + \lambda p_n(t) = \lambda p_{n-1}(t) \tag{7.14}$$

where $'$ denotes derivative relative to t and n can be any positive integer. When $n = 0$, we get the equation

$$p_0'(t) + \lambda p_0(t) = 0 \tag{7.15}$$

It remains for us to solve equations (7.14) and (7.15). Notice that there is a shift from $n-1$ to n in equation (7.14). If we know $p_{n-1}(t)$, then we can solve for $p_n(t)$ using the standard formula for a first-order linear differential equation. To specify the solution completely, we need the *initial conditions*

$$p_n(0) = 0 \quad \text{for } n > 0 \tag{7.16}$$

and

$$p_0(0) = 1 \tag{7.17}$$

that follow from the definition of $p_n(t)$.

We start with $n = 0$. It is obvious that the function $p_0(t) = e^{-\lambda t}$ satisfies equations (7.15) and (7.17). Thus for $n = 1$, we can write equation (7.14) as

$$p_1'(t) + \lambda p_1(t) = \lambda e^{-\lambda t}$$

Multiply both sides of this equation by $e^{\lambda t}$:

$$p_1'(t) e^{\lambda t} + \lambda p_1(t) e^{\lambda t} = \lambda$$

Now the left side is the derivative of the function $p_1(t)\, e^{\lambda t}$, and the right side is the derivative of λt. Hence,

$$p_1(t)\, e^{\lambda t} = \lambda t + C$$

where C is a constant of integration. But by definition of a replacement process, we know that $N(0) = 0$, so $p_1(0) = P\{N(0) = 1\} = 0$. Hence $C = 0$, and we have found $p_1(t)$:

$$p_1(t) = \lambda t\, e^{-\lambda t}$$

The case for any value of $n > 1$ is just like the case $n = 1$. If we assume by induction that

$$p_{n-1}(t) = \frac{(\lambda t)^{n-1}}{(n-1)!}\, e^{-\lambda t}$$

then from equation (7.14) we get

$$e^{\lambda t}\, p_n'(t) + \lambda e^{\lambda t}\, p_n(t) = \lambda\, \frac{(\lambda t)^{n-1}}{(n-1)!}$$

The left side of this equation is the derivative of the function $e^{\lambda t}\, p_n(t)$, while the right side is the derivative of the function $(\lambda t)^n/n!$. Thus, the difference of these functions is a constant, since it has derivative zero. Since both of these functions take the value 0 at $t = 0$, they must be equal for all t. Multiplying through by $e^{-\lambda t}$, we see that $p_n(t)$ is the probability mass function for a Poisson random variable with mean λt. ■

Exercises

1. An electronic component has an exponentially distributed lifetime, with mean 1,000 hours.

   a. What is the probability that the component fails during the time period between 900 and 1,000 hours after it is put into service?

   b. If the component is functioning after 900 hours of service, what is the probability that its lifetime will be less than 1,000 hours?

2. A machine consists of components A and B connected in series; thus, the machine will only function if both of the components are in working order. Assume that the components have independent exponentially distributed random lifetimes, and that the mean lifetime of component A is 100 hours. What must the mean lifetime of component B be in order for

$$P\{\text{Machine is operating after 50 hours}\} = 0.5?$$

3. Suppose that a communications satellite consists of components A and B, and that the satellite will function only if *both* of the components are in working order. Assume that the components have mutually independent exponentially distributed random lifetimes, with failure rate $\lambda = 2$ per year for A and $\mu = 1$ per year for B.

   a. What is the probability that the satellite will fail within the first year after being launched?

   b. What is the probability that the satellite will fail in the first year *and* that component A will be the cause of its failure?

4. a. Suppose that X is a geometric random variable with parameter p. Show that $P\{X \geq k\} = q^k$ for $k = 1, 2, 3, \ldots$, where $q = 1 - p$, as usual.

   b. Suppose that X is a discrete random variable which takes on values $0, 1, 2, \ldots$. Assume that X has *no memory*, in the sense that

$$P\{X \geq k + m \mid X \geq m\} = P\{X \geq k\} \quad \text{for all integers } k, m \geq 0$$

   Prove that X is a geometric random variable.

   (HINT: Prove that $P\{X \geq k\} = q^k$ for $k = 1, 2, 3, \ldots$ and use part (a).)

5. Let $\{N(t)\}$ be a Poisson process, and let S_n, for $n = 1, 2, \ldots$, be the time of occurrence of the nth event of the process.

   a. Let $t > 0$. Prove that $N(t) = \max\{k : S_k \leq t\}$. (If this set of integers is empty, then $N(t) = 0$ by definition.)

   b. Let $n \geq 1$. Prove that $S_n = \min\{t : N(t) \geq n\}$.

   (HINT: See Figure 7.1.)

6. Assume that the number of telephone calls arriving at a switchboard by time t (minutes) is described by Poisson process $\{N(t)\}$. On average, one call comes in every 10 minutes.

   a. Calculate the probability that no calls will occur between $0 < t \leq 10$ and that exactly one call will occur in $10 < t \leq 15$.

   b. What is the expected time at which the fourth call will arrive?

   c. What is the probability that two or more calls will occur in $10 < t \leq 20$?

   d. Given that exactly one call arrived in $0 < t \leq 15$, calculate the (conditional) probability that it occurred in $10 < t \leq 15$.

   e. Given that exactly one call arrived in $0 < t \leq 15$, calculate the (conditional) probability that at least one call will arrive in $15 < t \leq 20$.

7. Sam is playing his favorite computer game during a thunderstorm. Lightning strikes the power lines in accordance with a Poisson process with rate $\lambda = 3$ per hour. If it takes Sam s minutes to play the game, what is the probability that he will be able to complete it? (Assume that each strike of lightning interrupts the power and causes the computer to cease functioning.) Do this calculation for $s = 2$, 10, and 20 minutes.

8. In Exercise 7, suppose that Sam installs a backup system so that a single power interruption doesn't cause his computer to go down (but two successive interruptions do knock out the system). Now what are his chances of finishing the game? Do this calculation for $s = 2$, 10, and 20 minutes.

9. Suppose that $\{N(t)\}$ is a Poisson process with rate λ, and that $0 \leq s < t$. Calculate $\text{Cov}(N(t), N(s))$.

   (HINT: Write $\text{Cov}(N(t), N(s)) = \text{Cov}(N(t) - N(s), N(s)) + \text{Cov}(N(s), N(s))$, and use general properties of the Poisson process to calculate each term.)

10. Suppose that $\{N(t)\}$ is a Poisson process with rate λ, and that $0 < s < t$. Show that the *conditional* distribution of the random variable $N(s)$, given that $N(t) = n$, is binomial, with parameters n and $p = s/t$.

11. People arrive at a bus stop to wait for the bus according to a Poisson process with rate λ per hour, and wait for the bus. Assume that the arrival time T (measured in hours) of the bus is a random variable that is *uniformly* distributed on $(0, 1)$, and that T is independent of the arrival times of the people. Let N be the number of people who are at the bus stop when the bus arrives.

    a. Suppose the bus arrives at time t, where $0 < t < 1$. What is the probability that there are n people at the bus stop, and what is the expected number of people at the bus stop, given this information?

    b. What is the probability distribution of the random variable $E[N \mid T]$?

    c. Calculate $E[N]$.

    d. Calculate the probability that no one is at the bus stop when the bus arrives.

Chapter 8

Continuous-Time Stochastic Processes

The Poisson process introduced in Chapter 7 furnishes an example of a continuous-time **stochastic process**, and it models a situation in which a *single* kind of random event is taking place, such as the failure of a specific circuit component in a microprocessor. A more complete analysis of a complex system such as a computer network must take into account many other random quantities, such as the time required to execute programs and store or retrieve data. The probabilistic analysis of such problems leads to the subject of **queueing theory**.

In this chapter, we study processes related to the Poisson process in greater detail, and we introduce the basic aspects of queueing processes and their simulation.

8.1 Stochastic Processes

An observed value of a single random variable X is a single number x. If we have a collection of random variables $X_1, X_2, X_3, \ldots$, then an observed value of the whole collection consists of a sequence of numbers $x_1, x_2, x_3, \ldots$.

Example 8.1

Suppose that an experiment consists of rolling a fair die repeatedly. Let $X(n)$ be the number of points showing on the nth roll. We can simulate 30 rolls by using the first thirty entries in the first column of Table 4 at the end of the book, as in Example 2.1. The entries in this column are five-digit random numbers. We divide the numbers from 0 to 99999 into six equal ranges to make each entry in the column correspond to a particular face of the die. Suppose we make the numbers in the range 0–16666 correspond to a 1 showing on the die, the numbers in the range 16667–33333 correspond to a 2 showing on the die, and so on. Then the first entry $10480 \to 1$, the second entry $22368 \to 2$, and so on. Let $x(n)$ be the number of points determined by the nth entry in the column.

We can display the results of this simulation by drawing the bar graph of the sample function $x(n)$, as in Figure 8.1. Notice that the sample function shows the complete history of the experiment— not just the relative frequencies that appear in Table 2.1. For example, we can see that there was exactly one run of four consecutive 6's.

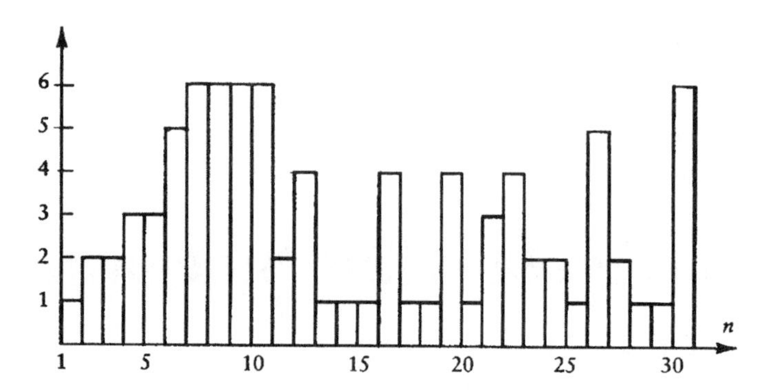

Figure 8.1: Sample Graph of Die Tossing

We can obtain another simulation of the experiment by starting at a different point in the random number table, or we can use the output of the random number key on a calculator. Each such simulation gives a particular graph. If we perform a large number of simulations, then the complete set of graphs summarizes the (simulated) experimental data. ■

In Example 8.1, the index n labeling the random variables has a natural interpretation as time measured in **discrete** units. (Imagine the die being rolled once a second.) The graphs of the sample functions have jumps at predictable times, but the jumps are of random height.

The Poisson process $\{N(t)\}$ in Chapter 7 is another example of a family of random variables. In this case the index t has a *continuous* range $0 \le t < \infty$. The sample functions have graphs with jumps that are always upwards by one unit, but the jumps occur at random times (see Fig. 7.1).

With these examples in mind, we define a **stochastic process** in general to be an *indexed family* of random variables. When the indexing parameter (call it n) takes only integer values and the family is $\{X(n)\}$, we refer to the process as a **discrete-time process** (but notice that the random variable $X(n)$ is allowed to be continuous or discrete).

Starting with independent Bernoulli trials with a fixed probability p of success in each trial (coin tosses), we can observe several discrete-time processes. The simplest process consists of taking $X(n)$ to be the indicator random variable for success on the nth trial. More complicated processes arise by taking $X(n)$ to be the number of successes in the first n trials or the length of the current run of successes (see Section 6.1). The Markov chains studied in Chapter 6 are an important general class of discrete-time processes.

When the indexing parameter (call it t) has a continuous range and the family is $\{X(t)\}$, we say that the process is a **continuous-time process** (as in the discrete-time case, the random variable $X(t)$ can be continuous or discrete). For both discrete and continuous time, a **sample path** of the process consists of the random function $x(n)$ or $x(t)$, for *all* values of the parameter n or t.

The simplest type of continuous-time stochastic process is a point process, such as the Poisson process $\{N(t)\}$, with discrete random variables and sample functions that are piecewise constant. In general, a **point process** (also called a **counting process**) describes events that occur at random times $t \geq 0$. For example, the failure of an electronic component was the event involved in the process $\{N(t)\}$ of Chapter 7. (This use of the word *event* is more specific than the use of the same word in the general sample-space terminology of earlier chapters.)

We can represent a point process graphically by a set of random marks on the positive t-axis, where each mark $\times$ corresponds to the occurrence of an event of the process. We define the random variable

$$N(t) = \{\text{ Number of marks in the interval } (0, t] \}$$

when $t > 0$, and we set $N(0) = 0$. Here $(a, b]$ denotes the set of numbers s in the range $a < s \leq b$. We *assume* that, in any interval of finite length, there are only a finite number of marks. Thus the random variable $N(t)$ has finite values $0, 1, 2, \ldots$. A sample path $n(t)$ of a point process therefore has a graph like the one shown in Figure 8.2.

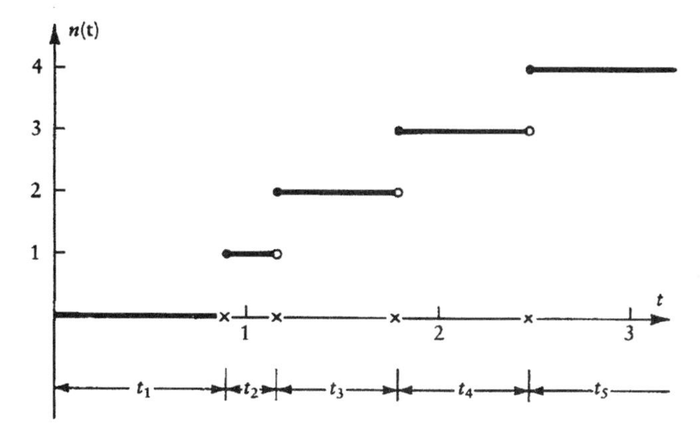

Figure 8.2: Sample Observation of a Point Process

Notice that the definition of $N(t)$ implies that $n(t)$ is right-continuous and constant on intervals, with upward jumps at the random moments marked by $\times$'s on the t-axis in Figure 8.2.

8.2 Characterization of the Poisson Process

In Chapter 7, we obtained the Poisson process as a consequence of the no-memory property, which we assumed for the lifetime of solid-state electronic components. The Poisson process in fact serves as a model for an enormous range of physical systems, including radioactive disintegration, chromosome interchanges in cells, connections to wrong telephone numbers, bacteria and blood counts, misprints in a book, and many others. In this section, we will try to understand why this is so.

We begin by recalling the two fundamental properties of the Poisson process $\{N(t)\}$ that we derived in Section 7.4:

(i) *Shift Invariance*: For any $s \geq 0$ and $h > 0$, the random variable

$$N(s+h) - N(s) = \begin{array}{l}\text{Number of marks in the} \\ \text{interval } s < t \leq s+h\end{array} \tag{8.1}$$

has a probability distribution that only depends on the length h of the interval—not on location s of the left endpoint of the interval.

In fact, we know that this probability distribution is Poisson, with parameter λh, when $N(t)$ is the Poisson process. The crucial point, however, is that the probabilities are the same for any interval of length h on the positive t-axis, no matter how we shift the interval around.

(ii) *Independent Increments*: Fix any set of successive times $0 = \tau_0 < \tau_1 < \tau_2 < \ldots < \tau_n$, and calculate

$$X_i = N(\tau_i) - N(\tau_{i-1}) = \begin{array}{l}\text{Number of marks in the} \\ \text{interval } (t_{i-1}, t_i]\end{array} \tag{8.2}$$

Then the random variables $X_1, X_2, \ldots, X_n$ are mutually independent.

Both of these assumptions are intuitively reasonable for the examples mentioned above.

Let's assume that we have a point process $\{N(t)\}$ that satisfies properties (i) and (ii). We do *not* assume that we know the exact form of the probability distribution of $N(t)$. We shall *prove* that it must be Poisson. As a first step in doing this, we construct the following **discrete-time approximation** of the process.

Fix $t > 0$, and subdivide the interval $(0, t]$ into n disjoint subintervals

$$I_j = (\tau_{j-1}, \tau_j] \qquad j = 1, \ldots, n$$

each of width t/n. Define

$$Y_j = \begin{cases} 1 & \text{if } I_j \text{ contains one or more marks} \\ 0 & \text{if } I_j \text{ contains no marks} \end{cases}$$

Then, by properties (i) and (ii), it is clear that $Y_1, Y_2, \ldots, Y_n$ are identically distributed independent Bernoulli random variables with parameter p (which depends on n and t). Thus by grouping all the marks in each subinterval into a single lump, we convert the process into a sequence of Bernoulli (coin-tossing) trials. But notice that the number n of subintervals was arbitrary—we can divide $(0, t]$ as finely as we wish and still observe this Bernoulli-trial behavior. (This also helps to explain why the various examples previously mentioned fit into this model.) One consequence of this property of unlimited divisibility of time intervals is Lemma 8.1

Lemma 8.1. *Given any $t > 0$, there is a positive probability that no event occurs in the time interval $(0, t]$. Thus $P\{N(t) = 0\} > 0$.*

Proof If $N(t) = 0$, then each of the random variables $Y_j = 0$; so, by independence,

$$P\{N(t) = 0\} = (1 - p)^n \tag{8.3}$$

Suppose, for the sake of contradiction, that the probability in equation (8.3) were zero. Then $p = 1$, and each of the n intervals would contain at least one mark. Thus the total number of marks $N(t) \geq n$ in this case. But n can be an arbitrarily large integer, so this contradicts the basic assumption about a point process: every interval of finite length has only a finite (random) number of marks in it. Hence we must have $P\{N(t) = 0\} > 0$. ∎

For any point process we can measure the random times between the events of the process—that is, the distances between the $\times$-marks on the time axis in Figure 8.2. We define

$$\begin{aligned}
T_1 &= \text{Time of occurrence of first event} \\
T_2 &= \text{Elapsed time between first and second event} \\
T_3 &= \text{Elapsed time between second and third event} \\
\ldots
\end{aligned}$$

just as in Section 7.2 (where T_j was the lifetime of the jth component). We shall now prove that properties (i) and (ii) above, together with a single parameter λ (the **rate** or **intensity** of the process) completely determine this stochastic process.

Theorem 8.1. *If the point process $\{N(t)\}$ is shift-invariant and has independent increments, then $\{T_n\}$ is a sequence of independent, identically distributed exponential random variables. Hence $N(t)$ is a Poisson random variable with mean λt. Here, the parameter λ is determined by the equation*

$$\lambda = E[N(1)] \tag{8.4}$$

It satisfies the equation $1/\lambda = E[T_n]$.

Proof We shall translate statements about the random variables T_n into statements about the random variables $N(t)$. As a first step, we observe that, for any $t > 0$, the following events are the same:

$$\{T_1 > t\} = \{N(t) = 0\} \tag{8.5}$$

By Lemma 8.1, this event has positive probability, so it makes sense to calculate the following conditional probabilities:

$$\begin{aligned}
P\{T_1 > t + s \mid T_1 > t\} &= P\{N(t+s) = 0 \mid N(t) = 0\} \\
&= P\{N(t+s) - N(t) = 0 \mid N(t) = 0\} \quad \text{(Use } N(t) = 0\text{)} \\
&= P\{N(t+s) - N(t) = 0\} \quad \text{(independent increments)} \\
&= P\{N(s) = 0\} \quad \text{(shift invariance)} \\
&= P\{T_1 > s\}
\end{aligned}$$

From this string of equalities, we conclude that the random variable T_1 has the no-memory property, and hence is an exponential random variable with some parameter λ, by Theorem 7.1. Hence, equation (8.5) implies that

$$P\{N(t) = 0\} = e^{-\lambda t}$$

Next, we check that T_1 and T_2 are mutually independent, by conditioning on the event $\{T_1 = t_1\}$, with $t_1 > 0$. Since T_1 is an exponential random variable, this event has a positive probability density and can also be described as

$$\{N(t_1) = 1 \text{ and } N(t) = 0 \text{ for } t < t_1\} \tag{8.6}$$

Let $t_2 > 0$. The event $\{T_1 = t_1 \text{ and } T_2 > t_2\}$ can also be described as

$$\{T_1 = t_1 \text{ and } N(t_1 + t_2) - N(t_1) = 0\}$$

Now we calculate conditional probabilities, using equation (8.6):

$$\begin{aligned} P\{T_2 > t_2 \quad &| \quad T_1 = t_1\} \\ &= P\{N(t_1 + t_2) - N(t_1) = 0 \mid T_1 = t_1\} \\ &= P\{N(t_1 + t_2) - N(t_1) = 0 \mid N(t_1) = 1, N(t) = 0 \text{ for } t < t_1\} \\ &= P\{N(t_1 + t_2) - N(t_1) = 0\} \end{aligned}$$

In the last line, we have used the independent increments property (ii). By the shift-invariance of the process, this probability is

$$P\{N(t_2) = 0\}$$

and doesn't depend on t_1 at all. Since this is true for any fixed time t_1, the random variable T_2 must be independent of the random variable T_1. From the previous calculation, we see that these two random variables have the same distribution function. We already know that T_1 is an exponential random variable with parameter λ; therefore, the same must be true for T_2.

This same argument now applies to T_3, if we condition on the values of T_1 and T_2. Proceeding inductively, we verify the stated property of $\{T_n\}$. Given this information about $\{T_n\}$, we conclude from the results of Section 7.2 that $N(t)$ is a Poisson random variable with mean λt, since $N(t)$ can be defined in terms of the random variables $\{T_n\}$. ■

8.3 Discrete-Time Approximation to the Poisson Process

The discrete approximation to the Poisson process that was used in the proof of Lemma 8.1 is worth examining more closely. Let $\{N(t)\}$ be a Poisson process with rate λ. Fix a discrete time step Δt. Then

$$P\{N(\Delta t) = 0\} = e^{-l\Delta t} \approx l - \lambda\Delta t$$

We now associate a success or failure with each time interval $(k\Delta t, (k+1)\Delta t]$ on the basis of the occurrence or the non-occurrence of at least one event of the Poisson process within that interval. The calculation just made shows that the probability of a success is the same for all of the intervals, and is approximately $p = \lambda\Delta t$.

Suppose that we replace the Poisson process $N(t)$ by a sequence of Bernoulli trials with probability $p = \lambda\Delta t$, where one trial is performed in each time interval $(k\Delta t, (k+1)\Delta t]$. We now measure time in discrete multiples of Δt; and if $t = n\Delta t$, then we let $B(t)$ be the total number of successes up to time t. Of course, $B(t)$ is a binomial random variable, with parameters n, p. It is clear that the family of random variables $\{B(t) : t = n\Delta t,\ n = 1, 2, \ldots\}$ satisfies a discrete-time version of the shift-invariance and independent increments properties (with the time step Δt fixed). We now prove that as $\Delta t \to 0$, the probability distribution of $B(t)$ converges toward the distribution of $N(t)$.

Proposition 1. *Fix $t > 0$. Given a positive integer n, set $\Delta t = t/n$ and let the binomial random variable $B(t)$ be defined relative to $N(t)$ and Δt. Then, for every integer $k \geq 0$,*

$$\lim_{n\to\infty} P\{B(t) = k\} = P\{N(t) = k\} \tag{8.7}$$

Proof Since $p = \lambda t/n$, we have $q = \big(1 - \lambda/n\big)$. Thus we can write the binomial probabilities $\binom{n}{k} p^k q^{n-k}$ as

$$P\{B(t) = k\} = \frac{(\lambda t)^k}{k!} \cdot \frac{n(n-1)\cdots(n-k+1)}{n^k} \cdot \left(1 - \frac{\lambda t}{n}\right)^{n-k} \tag{8.8}$$

When $n \to \infty$, we have

$$\frac{n(n-1)\cdots(n-k+1)}{n^k} \to 1$$

since there are k factors, each of which approaches 1 (k is fixed). Furthermore,

$$\left(1 - \frac{\lambda t}{n}\right)^{n-k} \to e^{-\lambda t}$$

by the fundamental limit formula for the exponential function. (Again k is fixed, so we can replace $n - k$ by n in the exponent when taking the limit as $n \to \infty$.) Starting with equation (8.8) and applying these limit results, we get equation (8.7). ■

Example 8.2 Quality Control

Suppose that in the manufacture of magnetic recording tape, flaws occur at a Poisson rate, with an average of one per 1000 inches. That is, if $N(t)$ is the total number of flaws in a section of tape t inches long, then $\{N(t)\}$ is a Poisson process with rate $\lambda = 0.001$.

For quality control purposes the tape is tested in segments of length $\Delta t = 0.1$ inch for flaws, with each flawless segment being classified as good and each segment with one or more flaws classified bad. By the information above, any single segment has probability

$$P\{N(0.1) = 0\} = e^{-l\cdot\Delta t} = e^{-0.0001} = 0.9999$$

of being good. Thus the probability that a particular 0.1 inch test segment is bad is essentially $\lambda \Delta t = 0.0001$. Let $B(t)$ denote the number of bad segments in a sample of length t. Then $B(t)$ is a binomial random variable with parameters $n = t/\Delta t = 10t$ and $p = 1 - e^{-\lambda \cdot \Delta t} \approx 0.0001$.

Suppose now that we are manufacturing tape cassettes containing 3,000 inches of tape (about enough for 30 minutes at 1 7/8 inches per second). The quality control specifications allow a cassette to have at most 5 bad segments. What proportion of the cassettes are acceptable?

Since $n = (10) \cdot (3000)$ is large and $p = 0.0001$ is small, we can use Proposition 8.7 to estimate the probability that a cassette meets specifications (notice that $\lambda t = (0.001)(3000) = 3$):

$$\begin{aligned} P\{B(3000) \leq 5\} &\approx P\{N(3000) \leq 5\} \\ &= e^{-3}\left(1 + 3 + \frac{3^2}{2!} + \frac{3^3}{3!} + \frac{3^4}{4!} + \frac{3^5}{5!}\right) \\ &= 0.916 \end{aligned}$$

Thus with this test procedure, about 8% of the cassettes will be rejected. ■

8.4 Compound Poisson Process

Several stochastic processes are closely related to the Poisson process. Following is one type that arises frequently in applications. Start with a Poisson process $\{N(t)\}$ with rate λ. Suppose that $\{Y_n\}$ is a family of identically distributed random variables, which we assume is independent of the family of random variables $\{N(t)\}$. Consider the continuous-time stochastic process

$$X(t) = \sum_{n=1}^{N(t)} Y_n \tag{8.9}$$

Here, we add up a random number of the random variables Y_n, with the number of terms in the sum determined by the counting process $N(t)$. Such a stochastic process $\{X(t) : t \geq 0\}$ is called a **compound Poisson process.**

The simplest example of a compound Poisson process occurs when $\{Y_n\}$ is a sequence of Bernoulli random variables with fixed parameter p. The sum in formula (8.9) is then the number of Y_n that equal 1, for $1 \leq n \leq N(t)$. We can describe the random variable $X(t)$ in a more intuitive way in this case as follows. Each time an event of the Poisson process $\{N(t)\}$ occurs, we flip a biased coin that has probability p of showing heads. When the coin shows heads, we count the event. When the coin shows tails, we ignore the event. Then $X(t)$ is the total number of events that we count in the time interval $(0, t]$.

We shall call this procedure a **random sampling** of the Poisson process $\{N(t)\}$. It turns out to have remarkably simple probability properties. To describe these, it is convenient to introduce the notation

$$\begin{aligned} N_1(t) &= X(t) \\ N_0(t) &= N(t) - X(t) \end{aligned}$$

We can visualize the single stream of events $N(t)$ as being randomly split into two streams $N_0(t)$ and $N_1(t)$ by a random gate (see Figure 8.3), where an event has probability p of going into one stream and probability q of going into the other (as usual, $q = 1 - p$).

Figure 8.3: Random Sampling of Poisson Process

The two output processes $\{N_0(t)\}$ and $\{N_1(t)\}$ are obviously counting processes. Under the independence assumptions above, we can completely describe them.

Theorem 8.2. *In random sampling from a Poisson process $\{N(t)\}$, the output stream $\{N_0(t)\}$ is a Poisson process with rate λq, and the output stream $\{N_1(t)\}$ is a Poisson process with rate λp. Furthermore, these two output processes are mutually independent.*

Proof Suppose we describe the events counted by $N_1(t)$ as Type 1 events. The classification of an event as Type 1 is independent of the occurrence of the event, and successive classifications are independent of each other. It is thus obvious that the counting process $\{N_1(t)\}$ inherits from $\{N(t)\}$ the properties of shift-invariance and independent increments. Hence, by Theorem 8.1, we know that $\{N_1(t)\}$ is a Poisson process.

We shall calculate the rate of the process $\{N_1(t)\}$ by conditioning. Observe that, if we know that $N(t) = n$, then $N_1(t)$ is a binomial random variable with parameters p, n. Hence

$$E[N_1(t) \mid N(t) = n] = np$$

This says that, as a random variable,

$$E[N_1(t) \mid N(t)] = pN(t)$$

Taking expectations on both sides gives us

$$E[N_1(t)] = p\lambda t \tag{8.10}$$

So we see from equation (8.10) and the preceding argument that the Type 1 events occur according to a Poisson process with rate λp.

Call the events counted by $N_0(t)$ the Type 0 events. The same argument shows that the counting of these events is a Poisson process with rate λq. It remains for us to prove that the Type 0 events and Type 1 events are independent, which is the only unexpected part of the Theorem. The joint probability mass function is

$$P\{\, N_0(t) = m \text{ and } N_1(t) = n \,\} =$$
$$P\{\, m + n \text{ events occur by time } t \,\} \cdot P\{\, n \text{ Type 1 events occur out of } m + n \,\}$$

(Here we have used the independence of the incoming Poisson process and the random gate.) The right side of the equation is the product of a Poisson and a binomial probability mass function. Using the equation $p + q = 1$, we can write it as:

$$e^{-\lambda t(p+q)} \frac{(\lambda t)^m (\lambda t)n}{(m+n)!} \cdot \frac{(m+n)!}{m!n!} p^n q^m$$

Rearranging the terms—and remembering that $N_0(t)$ is a Poisson random variable with mean λqt and that $N_1(t)$ is a Poisson random variable with mean λpt—we recognize this product as simply

$$P\{N_0(t) = m\} \cdot P\{N_1(t) = n\}$$

This factorization proves the independence of the two processes. ■

Example 8.3

Suppose that $N(t)$ counts the number of telephone calls arriving at a switchboard during a time interval of length t. It is often reasonable to assume that the occurrence of calls fits a Poisson process (recall the characterizing properties from Section 8.2), and that the classification of calls as local or long-distance is independent of the arrival of the calls at the switchboard. Define $Y_n = 1$ if the nth call is local; otherwise set $Y_n = 0$. Then $\{Y_n\}$ is a sequence of Bernoulli random variables that we assume to be mutually independent. The random splitting of the incoming stream of calls is defined by

$$\begin{aligned} N_1(t) &= \text{Number of local calls up to time } t \\ N_0(t) &= \text{Number of long-distance calls up to time } t \end{aligned}$$

As a numerical example, suppose that the average proportion of local telephone calls is $p = 0.6$, and suppose that calls arrive at the rate of $\lambda = 100$ per hour. In a particular 1-hour period, 30 local calls arrive. Given this information, what is the expected number of long-distance calls during the period?

Solution The number of long-distance calls during any 1-hour period is a Poisson random variable X, with $E[X] = q\lambda = 40$. If Y is the number of local calls during the given 1-hour period, then X and Y are independent, by Theorem 8.2. Hence,

$$E[X \mid Y = 30] = E[X] = 40$$

Thus we can ignore the additional information about the local calls. ■

Remark In this example, the ratio of the *observed* number of local calls to the *expected* number of long-distance calls in the 1-hour period was $30/40$. This ratio is a *sample value* of the random variable $Y/40$; it should not be confused with the ratio of *expected values*

$$\frac{E[\text{Number of local calls in 1 hour}]}{E[\text{Number of long-distance calls in 1 hour}]} = \frac{6}{4}$$

Example 8.4

Customers pass through a checkout counter at a food market at a Poisson rate λ. Let Y_n be the amount spent by the nth customer. Assume a homogeneous population of customers, so that the sequence $\{Y_n\}$ of random variables is independent and identically distributed. Then $X(t)$ defined in formula (8.9) is the total receipts at the counter, up to time t.

The exact probability distribution of $X(t)$ can be quite complicated. We can easily calculate the basic parameters for this process, however, by formulas (5.16) and (5.17) of Section 5.4. Let μ and σ be the mean and standard deviation of Y_n. Then

$$E[X(t)] = \mu \cdot E[N(t)] = \mu\lambda t$$

and

$$\mathrm{Var}[X(t)] = \lambda t\sigma^2 + \lambda t\mu^2$$

(Remember that a Poisson random variable has equal variance and mean.) ■

8.5 Nonhomogeneous Poisson Process

In stochastic models involving random arrivals of customers at a service facility, for example, it is often unrealistic to assume that the average arrival rate is constant. We need a modified version of the Poisson process that takes into account naturally occurring slack periods and busy periods.

To set up such a model, we recall from Section 7.5 that a Poisson process with constant rate λ satisfies the properties

$$\begin{aligned} P\{\text{ Exactly one event in } (t, t+h]\,\} &= \lambda h + o(h) \\ P\{\text{ More than one event in } (t, t+h]\,\} &= o(h) \end{aligned}$$

for all small values of h.

It is now fairly obvious how to describe a model in which the arrival rate λ fluctuates. Suppose that $\lambda(t) > 0$ is a given, non-random continuous function of t, defined for $0 \le t < T$ (with $T \le \infty$). We want to construct a counting process $\{N(t)\}_{0\le t<T}$ for which $\lambda(t)$ is the **intensity** of occurrence of events. With the preceding equations for the ordinary Poisson process in mind, we require that the following properties hold:

(i) $N(0) = 0$ (counting starts at $t = 0$).

(ii) The process has independent increments (events occurring in disjoint t-intervals are mutually independent).

(iii) $P\{\, N(t+h) - N(t) = 1\,\} = \lambda(t)h + o(h)$.

(iv) $P\{\, N(t+h) - N(t) > 1\,\} = o(h)$.

We can obtain such a process very simply by making a nonrandom *change of clock* in an ordinary Poisson process: Define τ as a function of t for $0 \le t < T$ by

$$\tau = \int_0^t \lambda(s)\, ds \tag{8.11}$$

Since $\lambda(s) > 0$ and is continuous, τ is a strictly monotone increasing function of t.

Theorem 8.3. *Let $\{N_0(t)\}$ be an ordinary Poisson process with rate $\lambda = 1$. Define $N(t) = N_0(\tau)$ for $0 \le t < T$, where τ is obtained from t by integral (8.11). Then $\{N(t)\}$ is a counting process that satisfies properties* (i) *through* (iv). *Furthermore, $N(t+s) - N(t)$ is a Poisson random variable, with mean $\tau(t+s) - \tau(t)$.*

Proof Since $\tau(0) = 0$ and τ is strictly monotone increasing, properties (i) and (ii) are satisfied by $\{N(t)\}$. Since $\lambda(t)$ is continuous, we have

$$\tau(t+h) = \tau(t) + h \cdot \lambda(t) + o(h)$$

Hence, properties (iii) and (iv) follow immediately from the corresponding properties of the process $\{N_0(t)\}$ recalled earlier. For any pair of times $\tau < \tau'$, the random variable $N_0(\tau') - N_0(\tau)$ is Poisson, with mean $\tau' - \tau$, since the process $\{N_0(t)\}$ has rate 1. ■

Remark The equation for the average number of events that occur in an interval $(t, t+r]$ can be written as

$$\begin{aligned}\tau(t+r) - \tau(t) &= \int_t^{t+r} \lambda(s)\, ds \\ &= r \cdot \big(\text{Average intensity over } (t, t+r]\big)\end{aligned}$$

where the **average intensity** over the interval is defined as

$$\frac{1}{r}\int_t^{t+r} \lambda(s)\, ds$$

In cases where $\lambda(s) = \lambda$ is constant, this formula gives λr as the average number of events, as before.

Example 8.5 Customers Arriving at a Bank

Suppose that a branch office of a bank is open between 10 A.M. and 2 P.M., and that the pattern of customer arrivals follows a nonhomogeneous Poisson process. Assume that, between 10 and 11 A.M. customers arrive at a constant Poisson rate of 10 per hour. Between 11 A.M. and 12 noon, the rate increases steadily to 20 per hour and remains constant until 1 P.M. From 1 P.M. to 2 P.M. the rate decreases steadily to 0.

a. What is the probability distribution of the number of customers in a day?

b. What is the probability that no customers arrive in the period between 1:45 P.M. and 2 P.M.?

Solution To answer these questions, we must calculate the function $\tau(t)$ for $0 \le t < 4$. (Here it is natural to measure time in hours, with $t = 0$ corresponding to 10 A.M.) The intensity function $\lambda(t)$, which is continuous and piecewise-linear, is given by the formula

$$\lambda(t) = \begin{cases} 10 & \text{for } 0 \le t < 1 \\ 10t & \text{for } 1 \le t < 2 \\ 20 & \text{for } 2 \le t < 3 \\ 20(4-t) & \text{for } 3 \le t \le 4 \end{cases}$$

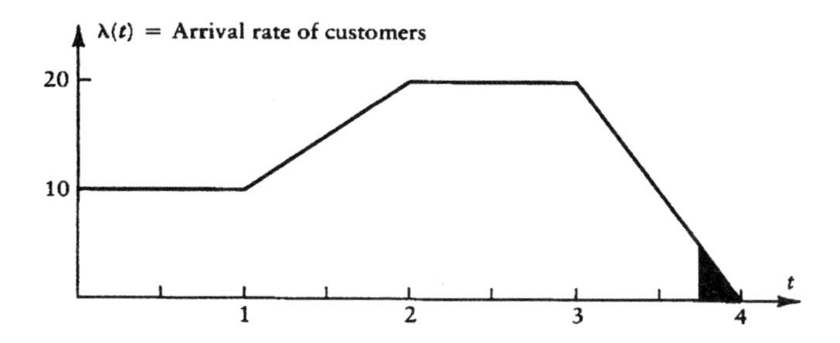

Figure 8.4: Variable Arrival Rate

The graph of $\lambda(t)$ is shown in Figure 8.4.

The function $\tau(t)$ is the area under the graph of $\lambda(s)$ between $s = 0$ and $s = t$. An explicit formula for $\tau(t)$ can be found by integrating the formulas for $\lambda(t)$ in each segment. From this, we get the graph in Figure 8.5.

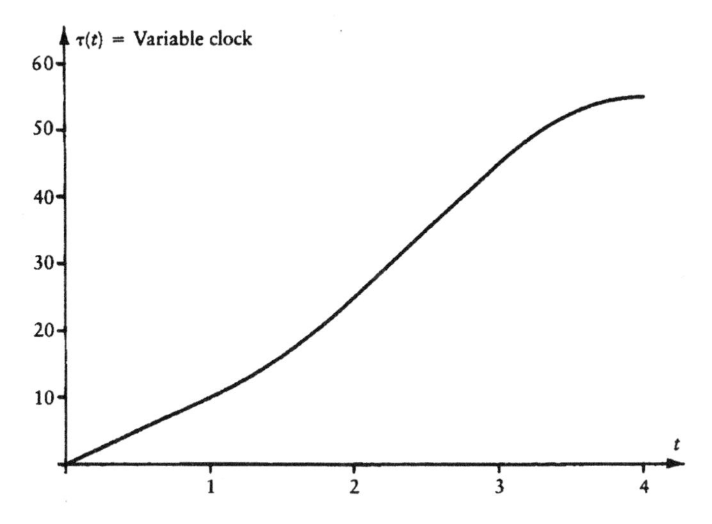

Figure 8.5: Change of Clock

To get the single value $\tau(4)$, we can simply use elementary geometry to calculate the area under the graph of $\lambda(t)$ between 0 and 4:

$$\tau(4) = 1 \cdot 10 + 1 \cdot 15 + 1 \cdot 20 + 1 \cdot 10 = 55$$

Thus, by Theorem 8.3, we know that $N(4)$, the number of customers in a day, is a Poisson random variable with mean 55, which answers question (a). To answer question (b), we calculate again by geometry that

$$\tau(4) - \tau(3.75) = \frac{1}{2} \cdot 5 \cdot \frac{1}{4} = 0.625$$

(the shaded area in the graph of $\lambda(t)$). Thus, the number of customers arriving between 1:45 P.M. and 2 P.M. is a Poisson random variable with mean 0.625. In particular, the probability that no customers will arrive during this period is $e^{-0.625} = 0.535$. ■

8.6 Queueing Processes

The stochastic theory of waiting lines, usually called *queues*, originated in connection with telephone networks in the early part of the twentieth century. It is a rich source of examples of continuous-time stochastic processes.

We start with some of the terminology of the subject. A **queue** is a waiting line of units demanding service at a service facility (called the **counter**), the unit demanding service is called the **customer**, and the service is provided by the **server**. These terms must be interpreted in a symbolic way: the server could be a person and the customer a computer, or vice versa.

Here are a few realistic examples of the customer-server mechanism:

a. Vehicles needing service arrive at a garage. Depending on the number of repairmen, one or more vehicles may be repaired at a time.

b. Patients arrive at a doctor's clinic for treatment. Even if some appointment system exists, there are always random elements involved because of emergency service rendered, and patients generally have to wait some time beyond the scheduled appointment time.

c. Computer programs submitted from timesharing terminals arrive at a central processor to be run. The programs wait in a storage buffer until the central processing unit is free.

You can easily add to the list above from your own daily experiences of waiting lines. Although these situations differ from each other in several details, they have some common basic characteristics, which we indicate schematically in Figure 8.6.

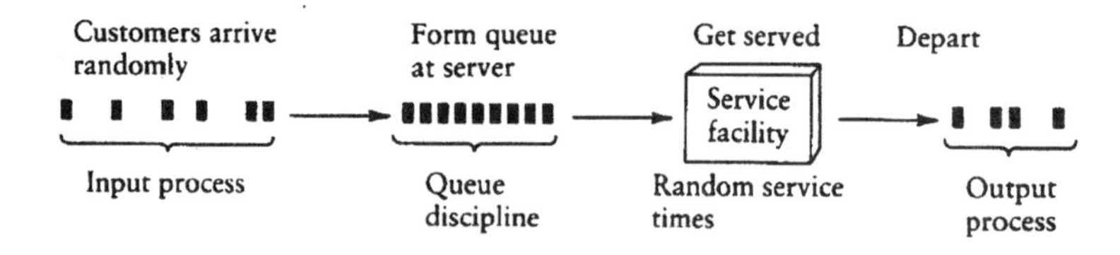

Figure 8.6: General Queueing System

When we observe the system over a period of time, we can measure several important random quantities, such as the current queue length (number of customers waiting at a particular time t), the waiting time for a newly arrived customer before service commences, and the busy periods (intervals of time during which the server is continuously busy).

The basic random variable associated with the queueing system is

$$X(t) = \text{Number of customers present at time } t$$

(we count both customers in the queue and customers being served). For example, if the service facility accomodates one customer at a time (the **single-server queue**), then

$$\begin{aligned} X(t) = 0 &\quad \leftrightarrow \quad \text{System is empty} \\ X(t) = 1 &\quad \leftrightarrow \quad \text{Queue is empty; one customer is in service} \\ X(t) = n &\quad \leftrightarrow \quad n-1 \text{ customers are in queue; one is in service} \end{aligned}$$

(if $n > 1$). If we start observing the system at time $t = 0$, then the family of random variables $\{X(t)\}_{t \geq 0}$ is a continuous-time stochastic process. We shall say that the system is in **state** n at time t if $X(t) = n$.

To specify a particular queueing model, we have to describe the input process, queue discipline, and service facility. We shall generally assume the following four characteristics:

a. The queue discipline is first come, first served. That is, arriving customers join the end of the queue and customers are served in order of arrival from the front of the queue.

b. The times $T_1, T_2, \ldots$ between successive arrivals of customers form a sequence of mutually independent, identically distributed random variables.

c. The service times $S_1, S_2, \ldots$ for successive customers form a sequence of mutually independent, identically distributed random variables.

d. The service times are independent of the arrival times.

When no particular restrictions are imposed on the distribution function of the T_i or the distribution function of the S_i, and there are n servers, then the queueing model is called the **G/G/n queue** (where G stands for *general*).

The most random queueing model occurs when the interarrival times are exponential random variables, with rate λ, and the service times are also exponential random variables, with rate μ. This model is called the **M/M/n queue**, where M stands for *Markov*. (By the results of Section 7.1, we can also think of M as standing for *memoryless*.) Thus, in the $M/M/n$ system, the input process is a Poisson process, with rate λ, as we saw in Section 7.2. During a busy period of a single server, the departures from this server will also form a Poisson process, with rate μ. However, the server is not always busy, and this makes the output process more complicated.

Simulation of a Queueing Process

Suppose that customers arrive at a single server according to a Poisson process with rate $\lambda = 4$ per hour ($= \frac{1}{15}$ per minute). Assume that the service time of a customer (in minutes) is uniformly distributed on the interval $(5, 15)$. To simulate the operation of this system, we can proceed as follows:

Step 1 Generate a sequence of uniformly distributed random numbers $u_1, u_2, \ldots$ in $(0, 1)$. Convert these random numbers into sample values for the exponential interarrival times $t_1, t_2, \ldots$ of a Poisson process with rate $\lambda = \frac{1}{15}$ by the formula

$$t_i = -\frac{1}{\lambda} \log u_i = -15 \log u_i \tag{8.12}$$

(see Sections 3.6 and 7.2).

For example, suppose that we read down from the top of column 10 in Table 4 at the back of the book and get the sequence $u_i = 0.362, 0.341, 0.321, 0.570, \ldots$ (rounded to three decimal places). Then the corresponding sequence t_i of interarrival times obtained from equation (8.12) is $15, 16, 17, 8, 7 \ldots$ (rounded to the nearest minute). Thus, in this simulation, the first four customers arrive at $t = 15, 31, 48, 56$, and 63 minutes after observations begin.

Step 2 Generate another sequence of uniformly distributed random numbers $v_1, v_2, \ldots$ in $(0, 1)$, independent of the u_i sequence. Convert these random numbers into sample values for the service times s_i of the customers, by setting

$$s_i = 5 + 10 v_i. \tag{8.13}$$

(see Section 3.6).

For example, suppose that we start at the top of column 6 in Table 4. Then the sequence v_i is $0.916, 0.892, 0.648, 0.164, \ldots$. The corresponding sequence s_i of service times for these customers obtained from equation (8.13) is $14, 14, 11, 7, \ldots$ minutes.

Step 3 Combine the information about arrival times and service time to describe the simulated operation of the queueing system as follows:

CLOCK TIME	CHANGE OF STATE DESCRIPTION
15	Customer 1 arrives and enters service ($s_1 = 14$)
29	Customer 1 finishes service and departs, server idle
31	Customer 2 arrives and enters service ($s_2 = 14$), queue empty
45	Customer 2 finishes service and departs, server idle
48	Customer 3 arrives and enters service ($s_3 = 11$), queue empty
56	Customer 4 arrives and queues (customer 3 still being served)
59	Customer 3 departs, customer 4 begins service ($s_4 = 7$)

Since customer 5 arrives after $t = 60$, we can construct the graph of the sample function $x(t)$ in the range $0 \leq t \leq 60$ from the description just given (see Figure 8.7).

8.7 M/M/1 Queue and Exponential Models

We concentrate now on the $M/M/1$ queueing model. From the simulation example in Section 8.6, it is evident that quite a bit of bookkeeping is needed to go from the sequences of interarrival times T_n and service times S_n to the state $X(t)$ of a queueing process at time t.

For the $M/M/1$ model, where T_n and S_n are exponentially distributed, we can use the no-memory and independence properties of these random variables to obtain a simpler way of generating the changes of state of the system. As we already saw in the simulation example in Section 8.6, these changes of state arise as follows:

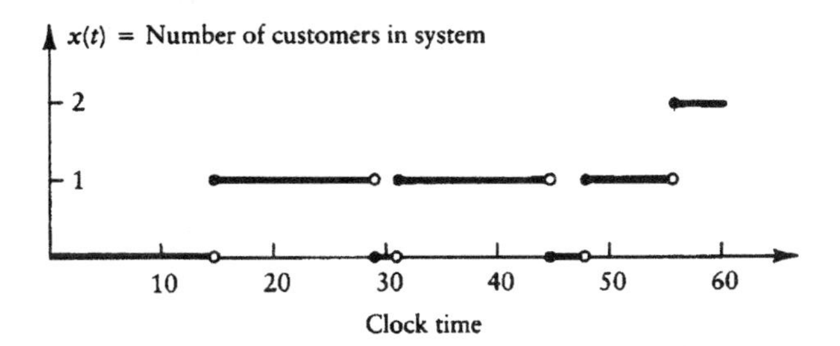

Figure 8.7: Sample Function for Queueing Simulation

STATE TRANSITION	DESCRIPTION
$n \to n+1$	New customer arrives before customer in service departs
$n \to n-1$	Customer in service departs before new customer arrives

(Since the probability is zero that arrivals and departures occur at exactly the same time, these are the only observable changes of state.)

Theorem 8.4. *For an M/M/1 queue with arrival rate λ and service rate μ, let $R_0, R_1, R_2, \ldots$ be the lengths of the successive time intervals between changes of state of the system. Then*

(i) *R_i is an exponential random variable with rate $\lambda + \mu$, if the system is in state $n \geq 1$ during this interval, and with rate λ if $n = 0$.*

(ii) *When there is a change of state of the system, the transition $n \to n+1$ occurs with probability $p = \lambda/(\lambda + \mu)$, and the transition $n \to n-1$ occurs with complementary probability $q = \mu/(\lambda + \mu)$, if $n \geq 1$. If $n = 0$, then the transition $0 \to 1$ occurs with probability 1.*

Furthermore, the random variables in part (i) *and the jump events in part* (ii) *are all mutually independent.*

Proof Suppose that the system jumped to state n at time t. If $n = 0$, the next customer will arrive at time $t + T$, where T is an exponential random variable with rate λ, and the system will jump to state 1. If $n \geq 1$, then the customer in service at time t will depart at time $t + S$, where S is an exponential random variable with rate μ. The next customer will arrive at time $t + T$, where T is an exponential random variable with rate λ. (By the lack of memory of the interarrival and service times, we don't need to know when the previous arrival occurred, or when the service time of the current customer started.) The state of the system goes from $n \to n+1$ if $T < S$, and from $n \to n-1$ if $S < T$, as noted earlier. Thus, starting at t, the time R spent in state n is the minimum of S and T.

Recall from Lemma 7.1 that, if T and S are independent exponential random variables, with rates λ and μ respectively, then $R = \min\{T, S\}$ is an exponential random variable, with rate $\lambda+\mu$. Also $P\{\, R = T\,\} = \lambda/(\lambda+\mu)$ and $P\{\, R = S\,\} = \mu/(\lambda+\mu)$. These properties imply parts (i) and (ii) of Theorem 8.4. The mutual independence of the random variables $\{R_i\}$ and the random jump also follows from Lemma 7.1, by the same no-memory argument and by the independence of arrival and service times. ■

Exponential Model Simulation of $M/M/1$ Queue

To generate sample functions $x(t)$ for the M/M/1 queue, we can treat this system as an exponential model, as follows:

1. Specify an initial state for the system—say, $x(0) = n$. Generate a sample value r of an exponentially-distributed random variable with rate $\lambda + \mu$ (if $n \geq 1$) or rate λ (if $n = 0$). This value determines the length of the first horizontal segment of the graph of $x(t)$.

2. If $n = 0$, then $x(t)$ jumps up to 1 at the end of the interval generated in step (1). If $n \geq 1$, generate a sample value b of a Bernoulli random variable, with mean $p = \lambda/(\lambda + \mu)$. (This sample should be independent of the sample value r in step (1).) If $b = 1$, then $x(t)$ jumps to $n + 1$ at the end of the interval generated in step (1). Otherwise, it jumps to $n - 1$.

3. Repeat steps (1) and (2), with the successive random sample values of r and b being chosen independently.

Remark By the results of Section 3.6, we can generate sample values of an exponential random variable R with rate α from sample values of U (uniform on $(0, 1)$), by setting

$$R = -\frac{1}{\alpha} \log U$$

To get sample values of a Bernoulli random variable B with mean p from U, set

$$B = \begin{cases} 1 & \text{if } U \leq p \\ 0 & \text{if } U > p \end{cases}$$

Then $P\{ B = 1 \} = p$, as required.

Example 8.6

Suppose that $\lambda = 0.5$ and $\mu = 0.6$. We shall follow the preceding algorithm to generate a sample function $x(t)$ with x(0) = 0.

Using the random-number key on a calculator, we generate a sample

$$u_0 = 0.71, \quad u_1 = 0.44, \quad u_2 = 0.17, \quad u_3 = 0.74$$

of a uniformly distributed random variable to use in generating the values r_i. To determine the direction of the jumps, we generate another independent random number sample

$$u_1' = 0.67, \quad u_2' = 0.13$$

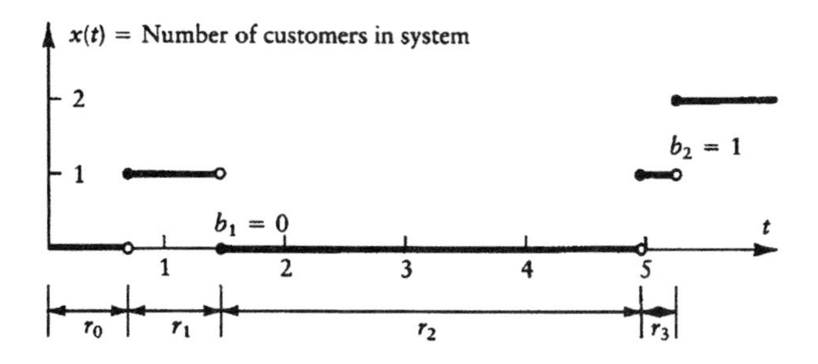

Figure 8.8: Exponential Model Simulation

Since $p = \frac{5}{11} \approx 0.45$, this gives us the sample

$$b_1 = 0, \quad b_2 = 1$$

of a Bernoulli random variable, by the remark above.

We now apply the algorithm, using these random data. Since we are starting in state 0, we have

$$r_0 = -\left(\frac{1}{0.5}\right) \log 0.71 = 0.68$$

Thus, at time $t = 0.68$, the system jumps to state 1 and stays there for an interval of length

$$r_1 = -\left(\frac{1}{1.1}\right) \log 0.44 = 0.75$$

We now take the sample value $b_1 = 0$, which generates a jump back down to state 0. The system stays there for an interval of length

$$r_2 = -\left(\frac{1}{0.5}\right) \log 0.17 = 3.54$$

The next jump is automatically back up to state 1 and doesn't require any random number data. This second visit to state 1 lasts

$$r_3 = -\left(\frac{1}{1.1}\right) \log 0.74 = 0.27$$

time units. Now we take the sample value $b_2 = 1$, which generates a jump up to state 2, and so on.

The graph of the sample function $x(t)$ generated by this exponential model simulation is given in Figure 8.8. ■

Exercises

1. Suppose that $\{N_1(t)\}$ and $\{N_2(t)\}$ are independent Poisson processes with rates λ_1 and λ_2, respectively. Show that $N(t) = N_1(t) + N_2(t)$ is a Poisson process with rate $\lambda_1 + \lambda_2$.

   (HINT: No calculation is necessary if you use the Theorem 8.1.)

2. Suppose coauthors Mark and Twain each spend 1 hour proofreading different chapters of the manuscript of their next book. Assume that the number of typographic errors that Mark finds is a Poisson random variable X with mean 10 per hour. The number of typographic errors that Twain finds is a Poisson random variable Y with mean 15 per hour. Mark and Twain work independently of each other.

   a. What is the probability that Mark and Twain together find a total of 20 errors?

   (HINT: See Exercise 1.)

   b. Suppose that they jointly find a total of 20 errors. Calculate the conditional probability distribution of X, given this information, and the conditional expectation of X.

3. In Example 8.2, suppose that we divide the tape into test segments of length $\Delta t = 1$ inch. Find the probability that a 100-inch length of tape has no more than two defective segments.

   a. Calculate this probability using a binomial distribution.

   b. Calculate this probability using the Poisson approximation to the binomial distribution.

4. Suppose that $\{B(n) \ : \ n = 1, 2, 3, \ldots\}$ is a discrete-time counting process; that is, at times $t = 1, 2, 3, \ldots$ a random event either occurs or does not occur, and $B(n)$ counts the total number of occurrences up to (and including) $t = n$. Suppose that the process is shift-invariant and has independent increments. Prove that $B(n)$ is a binomial random variable with parameters n, p, where $p = P\{\, B(1) = 1 \,\}$.

5. Vehicles pass a point on a road at a Poisson rate of 1 per minute. Suppose that 20% of the vehicles using the road are trucks.

   a. What is the probability that at least two trucks pass in a 5-minute period?

   b. Given that two trucks have passed in a 5-minute period, what is the expected number of vehicles to have passed in that time?

   c. If ten vehicles have passed in a 5-minute period, what is the probability that two of them were trucks?

6. Suppose that the number of jobs arriving at a computer center can be described by a Poisson process with rate $\lambda = 3$ per minute. Assume that an arriving job either goes immediately to the printer, with probability 0.2, or else goes to the central processing unit (each job randomly choosing the printer independently of the other jobs).

   a. What is the probability that, during a 5-minute period, no jobs will go to the printer?

   b. Suppose that four arriving jobs went to the printer during a particular 5-minute period. What is the conditional expectation (given this information) of the total number of jobs that arrived during this period?

7. Incoming orders arrive at a mail-order business according to a Poisson process with rate $\lambda = 10$ per day. Clerks A and B process orders, with order 1 going to A, order 2 going to B, and so on, in a *nonrandom* alternation. Let $N_A(t)$ be the total number of orders received by clerk A up to time t, and let $T_1, T_2, \ldots$ be the (random) interarrival times for the process $\{N_A(t)\}$.

   a. What is the probability distribution of T_j?

   (HINT: Relate the sequence T_j to the interarrival times for the process $\{N(t)\}$.)

   b. Is $\{N_A(t)\}$ a Poisson process?

8. An insurance company pays out claims on its life insurance policies in accordance with a Poisson process having rate $\lambda = 5$ per week. If the amount of money paid on each policy is exponentially distributed with mean \$ 2000, what is the mean and variance of the amount of money paid by the company over a 4-week period?

9. Men and women enter a supermarket according to independent Poisson processes having respective rates α and β per minute.

   a. Starting at an arbitrary time t, let X be the number of men who enter before the next woman enters. Show that X is a geometric random variable with parameter $p = \beta/(\alpha + \beta)$.

   (HINT: Use Lemma 7.1 and the no-memory property of the interarrival times to calculate $P\{X = k\}$.)

   b. Starting at an arbitrary time t, let S be the number of men who enter before the next r women enter, where r is a positive integer. Show that S is a negative-binomial random variable with parameters $p = \beta/(\alpha + \beta)$ and r.

   (HINT: Use part (a) and the no-memory property of the interarrival times.)

10. In Example 8.5 (nonhomogeneous Poisson process), what is the probability that two or more customers will arrive

    a. between 1:30 P.M. and 1:45 P.M.?

    b. between 1:45 P.M. and 2:00 P.M.?

11. In Example 8.5, find an explicit formula for the variable clock function $\tau(t)$, in the range $0 \leq t \leq 4$.

12. Let $\{N(t)\}$ be a non-homogeneous Poisson process with variable rate $\lambda(t) = 3t^2$.

    a. Calculate the expected number of events of the process in $(0, 1]$.

    b. Calculate the expected number of events of the process in $(1, 2]$.

    c. Let T be the time of occurence of the first event of the process. Show that

    $$P\{T > t\} = \exp(-t^3)$$

    What is the probability density function of T?

13. Consider a single-server queueing system. Assume that the customers arrive according to a Poisson process with rate $\lambda = 0.2$ customers per second. Assume that the service times (in seconds) are uniformly distributed on the interval $(5, 10)$. Use the general method of queue simulation in Section 8.6 in the following calculations:

    a. Generate the arrival times of the first four customers by applying a suitable probability transform to the uniform random number sequence 0.843, 0.157, 0.900, 0.912.

    b. Generate the service times of the first four customers by applying a suitable probability transform to the uniform random number sequence 0.774, 0.244, 0.817, 0.799.

    c. Plot the graph of the number of customers $x(t)$ in the system at time t, for $0 \leq t \leq 20$, assuming that the fifth customer arrives after $t = 20$. Calculate the average queue length and the average queue time per customer during $0 \leq t \leq 20$.

14. Consider an $M/M/1$ queueing system, with arrival rate $\lambda = 0.2$ customers per second and service rate $\mu = 0.1$ customers per second. Assume that the system starts in state 0 at time $t = 0$. Use the exponential model method of queue simulation in Section 8.7 to perform the following calculations:

    a. Determine $x(2)$, $x(3)$, and $x(4)$ by applying a suitable probability transform to the uniform random number sequence 0.482, 0.973, 0.118 (by assumption, $x(0) = 0$ and $x(1) = 1$).

    b. Generate the first four time intervals during which $x(t)$ is constant, by applying a suitable probability transform to the uniform random number sequence 0.751, 0.172, 0.406, 0.308 (the choice of transform uses the information from part (a)).

    c. Plot the graph of this simulation of $x(t)$ for $0 \leq t \leq 15$, assuming that the fifth jump occurs after $t = 15$. How many customers arrived during this period? What percentage of time was the server idle?

Chapter 9

Birth and Death Processes

In this chapter we set up a more general class of exponential models called birth–death processes; this class includes the Poisson process and the $M/M/1$ queue as special cases. We show how to simulate these models by a simple generalization of the method used for the $M/M/1$ queue in Chapter 8.

Finding exact analytic formulas for the probabilities in these models is more difficult and involves solving a system of differential equations. In this chapter, we obtain these differential equations; they involve the parameters (birth and death rates) of the models. We get explicit solutions for special case that is similar to the differential equations for the Poisson process from Chapter 7. In Chapter 10, we will use the differential equations to determine the behavior of the models over long time intervals.

9.1 Linear Growth Model

We introduce the notion of a **birth–death** process with the following exponential waiting time model. Consider a biological population (for example, a colony of algae) whose members can give birth to new members and can die. Assume that, starting at any time t, each individual in the population will either give birth to a new member at time $t + B$, or else die at time $t + D$, whichever event occurs first. Here, we assume that B and D are mutually independent exponential random variables with parameters λ and μ, respectively. We also suppose that the birth and death of each individual are independent of the other members of the population.

Define the **state** $X(t)$ of the system at time t to be the total number of individuals in the population at that moment. Of course, $X(t)$ is a random variable, and we would like to determine its properties, either by simulation or by analytic methods. For example, what is the expected population size at time t? What is the long-term population behavior as $t \to \infty$?

The probabilistic analysis of this example is similar to the exponential model version of the $M/M/1$ Queue in Section 8.7 (where the arrival of a new customer corresponds to a birth, and the departure of a customer after service is completed corresponds to a death.) Suppose that, at time t, we observe $X(t) = n$. We want to determine the elapsed time R, starting at t, until the population changes because of a birth or a death.

Label the individuals in the population at time t as $1, \ldots, n$ and consider the jth individual. By our assumptions, this individual will either die or else give birth to a new individual at time $t + R_j$, where $R_j = \min\{B_j, D_j\}$. Here B_j and D_j are independent exponential random variables with parameters λ and μ, respectively. By Lemma 7.1, R_j is an exponential random variable with parameter $\lambda + \mu$; also,

$$P\{\text{birth}\} = \frac{\lambda}{\lambda + \mu} = p \qquad P\{\text{death}\} = \frac{\mu}{\lambda + \mu} = q$$

Now consider the whole population of n individuals at time t. By the analysis just made, we find that $R = \min\{R_1, R_2, \ldots, R_n\}$. Thus R is exponentially distributed with parameter (rate) $n\lambda + n\mu$, since by assumption the random variables R_j are mutually independent. The jump to the next state ($n + 1$ or $n - 1$) is decided by whichever event occurs first: a birth ($n \to n+1$) with probability $p = \lambda/(\lambda+\mu)$, or a death ($n \to n - 1$) with probability $q = \mu/(\lambda + \mu)$; see again Lemma 7.1. Repeating this process starting in the new state, we eventually arrive at time s with the system in some state $X(s)$. (If the population drops to zero at any time, then it remains at zero from that time on, for this particular model.)

Note that the birth rate is $n\lambda$ and the death rate is $n\mu$, when the system is in state n. Since these birth and death rates depend on the population size, the system is more complicated than the $M/M/1$ queue model, in which the transition rates are the same for all states (except state 0).

9.2 Birth-Death Processes

We now introduce a general class of continuous-time stochastic processes that includes the Poisson process, the $M/M/1$ queue and the linear growth model as special cases. Consider a stochastic model for a population. Here we use the term *population* in a figurative way; it could refer to such varied examples as customers in line for service, fish in a lake, or misprints in a book. We assume that the population size $X(t)$ at time $t \geq 0$ is a random variable. We shall refer to the model as a *system*, and we say the system is in state n at time t, when $X(t) = n$. If the maximum population size is restricted so that $X(t) \leq N < \infty$ for some fixed integer N and all $t \geq 0$, then we say that the system has **capacity** N. If there is no restriction on population size, we say that the system has **unlimited capacity**.

Definition *The continuous-time stochastic process* $\{X(t) : t \geq 0\}$ *is a* **birth-death process** *if there are sequences of non-negative birth rates* λ_n, *for* $n \geq 0$, *and death rates* μ_n, *for* $n \geq 1$, *such that the system evolves in time as follows:*

(i) *The elapsed time (***sojourn time***) between entering state n and leaving state n is an exponential random variable with rate* $\lambda_n + \mu_n$.

(ii) *The possible changes of state are from state n to states* $n \pm 1$. *The choice is determined by a Bernoulli random variable, with probability* $p_n = \lambda_n/(\lambda_n+\mu_n)$ *of jumping to state* $n+1$, *and the complementary probability* $q_n = \mu_n/(\lambda_n+\mu_n)$ *for jumping to state* $n - 1$.

Furthermore, all of the sojourn times in property (i) *and the Bernoulli random variables in property* (ii) *are assumed to be mutually independent.*

Remark In this definition, we always take $\mu_0 = 0$, corresponding to the restriction $X(t) \geq 0$. We assume that $\lambda_n + \mu_n > 0$ for all $n \geq 0$ if the system has unlimited capacity. When the system has finite capacity N, we assume that $\lambda_n + \mu_n > 0$ for $0 \leq n \leq N$. Furthermore, $\lambda_n = 0$ for $n \geq N$ and $\mu_n = 0$ for $n > N$ in the case of capacity N. Given these assumptions, the rates and probabilities in properties (i) and (ii) make sense.

Example 9.1 Pure Birth Processes

When $\mu_n = 0$ for all $n \geq 1$, the population can never decrease in size, and we have a *pure birth* process. For example, the definition we gave for the Poisson process (see Section 7.2) shows that it is a pure birth process for which $\lambda_n = \lambda$ is independent of n. ■

Example 9.2

The $M/M/1$ queue is a birth-death process, by Theorem 8.4. ■

Notice that, by the no-memory property of the exponential random variable, the knowledge that a birth-death process is in state n at a particular moment gives us sufficient information to calculate the probabilities for all possible future states of the system; we don't need to know how long it has been since the latest birth or death. (This is often called the *Markov* property of the system.) We shall continue to use the biological birth-death language for its picturesque quality. In the case of a Poisson process associated with a replacement model, a birth in this new terminology corresponds to the *failure* of a component.

Simulation of a Birth-Death Process

From the exponential model definition of a birth-death process, we can easily generate a simulation of the process, using an obvious modification of the procedure followed for the $M/M/1$ queue in Section 8.7.

An interval of time during which the system remains in a particular state is called a **sojourn time**. To simulate a sojourn time r after the system jumps into state n, we generate a random number u uniformly distributed on $(0, 1)$ and then transform it into an exponentially distributed random number (with rate $\lambda_n + \mu_n$), using the transformation

$$r = -\frac{\log u}{\lambda_n + \mu_n}$$

To determine the state transitions at the end of a sojourn time in state n, we generate another (independent) random number v uniformly distributed on $(0, 1)$. Either the system jumps to state $n + 1$ if $v \leq p_n$ or else it jumps to state $n - 1$ if $v > p_n$. Here,

$$p_n = \frac{\lambda_n}{\lambda_n + \mu_n}$$

If $n = 0$, the jump is always to state 1 and we don't need to generate v.

By starting the system in any given state at time $t = 0$ and following this procedure, we can generate sample functions for the process whose graphs will look like those generated in Sections 8.6 and 8.7. ■

9.3 State-Transition Diagrams

Setting up a birth-death process involves specifying the sequences $\{\lambda_n\}$ and $\{\mu_n\}$ of birth and death rates. A convenient schematic way of displaying these parameters is the **state-transition diagram** for the process, as illustrated in Figure 9.1.

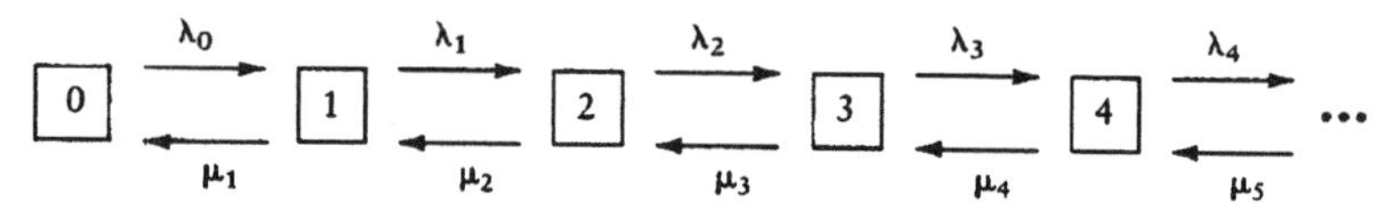

Figure 9.1: State-Transition Diagram

The numbered boxes in the figure correspond to the states of the system (the population size). The arrows pointing to the right describe births, with the label on the arrow giving the birth rate for the state. The arrows pointing to the left describe deaths and are similarly labeled with the death rates. (This diagram is not the same as the transition diagrams for Markov chains introduced in Chapter 6. Here, the labels λ_n and μ_n are *rates* of change, not probabilities. The sum of the labels on the arrows leaving a state need not add up to 1 in Figure 9.1.)

We have already described the Poisson process and the $M/M/1$ queueing system as birth-death processes; the state-transition diagrams are obvious for these examples. Following are some other important examples.

Example 9.3 $M/M/1$ Queue with Finite Capacity N

We assume that a single server working at an exponential rate μ. Customers arrive according to a Poisson process at rate λ, but there is room for at most N customers in the system ($N - 1$ is the maximum queue length allowed). For example, customers might be computer programs arriving at a central processing unit for execution, where the storage buffer can only hold a maximum of $N - 1$ programs.

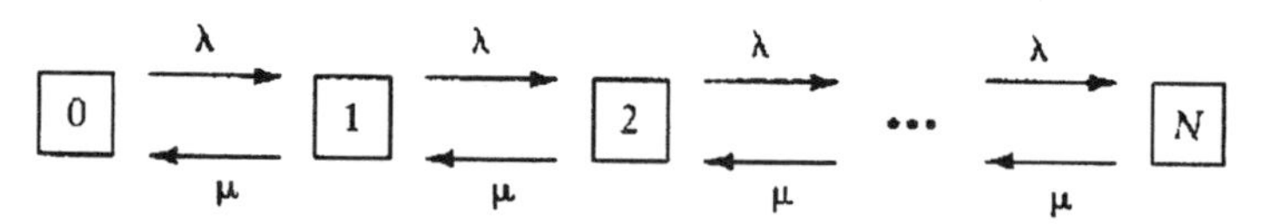

Figure 9.2: $M/M/1$ Queue, Capacity N

The birth rates are

$$\lambda_n = \begin{cases} \lambda & \text{if } n < N \\ 0 & \text{if } n \geq N \end{cases}$$

The possible states of the system are $0, 1, \ldots, N$. The state-transition diagram is shown in Figure 9.2. ∎

Example 9.4 $M/M/s$ Queueing System

Assume the Poisson-process arrivals of customers at rate λ, an unlimited system capacity, and s servers ($1 \leq s < \infty$), each working independently at an exponential rate μ (the same rate for each server). The customers form a single queue, and the customer at the head of the queue goes to the first available server. (This is a standard multiple-server queueing setup, used at banks and airline ticket counters, for example.) The queue discipline is FIFO: first in–first out. Figure 9.3 shows a schematic description of the system.

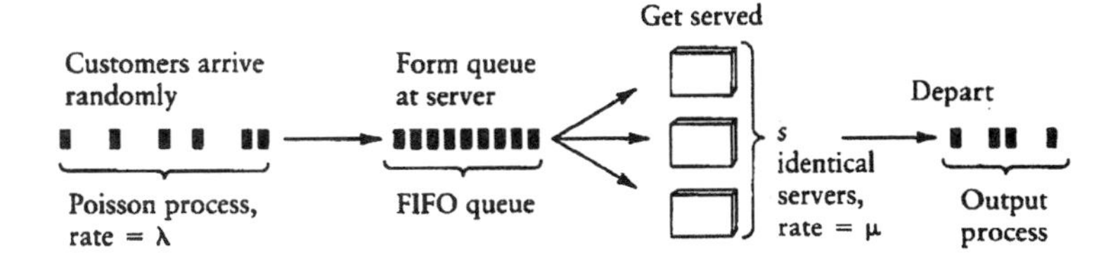

Figure 9.3: $M/M/s$ Queueing System

Clearly the birth rates are $\lambda_n = \lambda$. To determine the death rates, notice that, if there are n servers working at a particular moment, then the elapsed time until the next death (departure of a customer from the system) is

$$T = \min\{T_1, \ldots, T_n\}$$

where the random variables T_j are the remaining service times of the customers currently at the servers. By the no-memory property of the exponential random variable, each of these random variables is exponential, with rate μ. Since the servers are assumed to work independently of each other, the random variables T_j are mutually independent. Hence T is an exponential random variable with rate $n\mu$ (see Lemma 7.1). It follows that

$$\mu_n = \begin{cases} n\mu & \text{if } 0 < n \leq s \\ s\mu & \text{if } n > s \text{ (all servers busy)} \end{cases}$$

Thus we obtain the state-transition diagram shown in Figure 9.4. ■

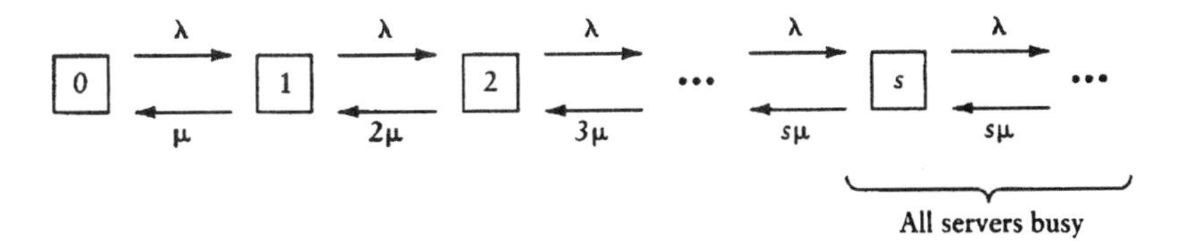

Figure 9.4: $M/M/s$ Transition Diagram

Example 9.5 Machine Repair Model

Assume that a factory has M machines and s repairmen, with $s < M$. Suppose that

a. Each machine operates for an exponentially distributed length of time (mean time $= 1/\lambda$) before breaking down.

b. When a machine breaks down, the first available repairman fixes it (if all the repairmen are busy, then the machine waits in the repair queue). The repair time is exponentially distributed (mean $= 1/\mu$).

c. The breakdown times and repair times are mutually independent random variables.

We let the state $X(t)$ of the system be the number of broken machines at time t. Thus, a *birth* occurs when a machine breaks down; a *death* occurs when a broken machine is restored to working order.

By the same argument as in example 9.4, we see that, when the system is in state $n \leq M$, there are $M - n$ machines still running; therefore, the birth rate (= breakdown rate) in this state is

$$\lambda_n = (M - n)\lambda$$

When the system is in state n, there are $\min\{n, s\}$ repairmen working, so the death rate (= repair rate) is

$$\mu_n = \begin{cases} n\mu & \text{if } n \leq s \\ s\mu & \text{if } n > s \end{cases}$$

The state-transition diagram for the case $s = 2$ and $M = 4$ is shown in Figure 9.5. ■

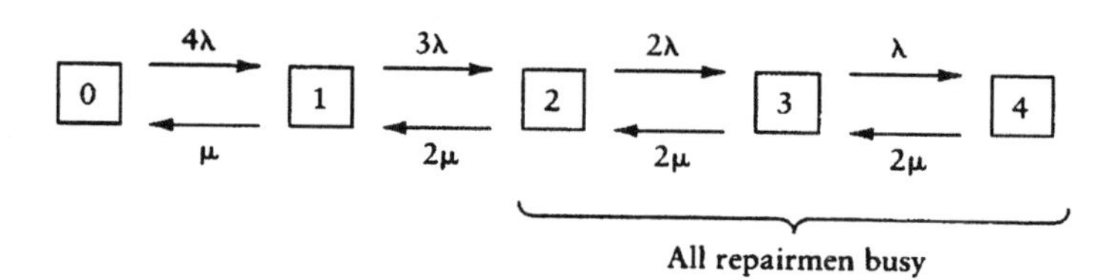

Figure 9.5: Machine Repair System

9.4 Transition Probabilities

We return now to the study of a general birth–death process. Recall that $X(t)$ denotes the population size (= number of customers in the system) at time t; when there are n customers present, the birth rate is λ_n and the death rate is μ_n. We say the system is in state n when there are n customers present.

Starting from the definition of a birth–death process as an exponential model, we can construct sample functions $x(t)$ for the process using pseudo-random numbers, as described at the end of Section 9.2. Thus from the point of view of numerical simulation, the random variable $X(t)$ is well-defined once we specify the probability distribution of $X(0)$.

Following the same procedure we used in Sections 2.1 and 6.1, we can generate a large number of independent numerical simulations $x(t)$—all with the same starting value $x(0) = i$. From these, we can obtain an experimental estimate for $P\{X(t) = j \mid X(0) = i\}$ by taking the observed proportion of sample functions that satisfy $x(t) = j$. By the Law of Large Numbers, these relative frequencies will converge to the exact probabilities as the number of simulation trials increases. We shall describe the results of some such simulations in Section 11.8.

This simulation approach, while immediately applicable, has the drawback of requiring extensive calculations. Furthermore, the statistical properties of simulation runs can be quite complicated, so that it is often hard to assess the reliability of simulation results. For these and other reasons it is important to try to obtain an *analytic* definition of the random variable $X(t)$—that is, a formula for computing $P\{X(t) = j\}$.

For any real number $t \geq 0$, we define the **transition probabilities** from state i to state j as

$$P_{ij}(t) = P\{X(t) = j \mid X(0) = i\} \quad \text{for } i, j = 0, 1, 2, \ldots$$

Our notational convention for conditional probabilities requires a switch in order between i and j on the two sides of this equation; state i is the **initial state**, and state j is the **final state**. Since the sojourn times in each state have the no-memory property, these conditional probabilities only depend on the *relative* clock time, not the *absolute* clock time. Thus, if we observe that there are i customers in the system at some particular moment s, then $P_{ij}(t)$ is the conditional probability that there will be j customers in the system t units of time later:

$$P_{ij}(t) = P\{X(t+s) = j \mid X(s) = i\} \tag{9.1}$$

This shift-invariance of the time variable is usually called the **stationary** property of the process.

By their very definition, the transition probabilities satisfy the *initial conditions*

$$P_{ij}(0) = \begin{cases} 1 & \text{if } i = j \\ 0 & \text{if } i \neq j \end{cases} \tag{9.2}$$

We have argued, using the exponential-model simulation approach, that $\{X(t)\}$ is uniquely defined as a stochastic process, once the birth and death rates are specified. (For finite-state systems this is always true; for infinite-state systems analytic subtleties arise whose discussion is beyond the level of this book.) Assuming this to be the case, it follows that, for each fixed time $t \geq 0$ and initial state i, $P_{ij}(t)$ is a probability mass function in the variable j (the final state):

$$\sum_{j=0}^{\infty} P_{ij}(t) = 1 \tag{9.3}$$

Example 9.6 Poisson Process

Suppose that all the death rates $\mu_n = 0$, and the birth rates $\lambda_n = \lambda > 0$ are all equal. Then the process is an ordinary Poisson process, and the stationary property expressed by equation (9.1) was called *shift-invariance* in Section 7.4. If the system is in state i a particular time, then at any later time it will be in some state j with $j \geq i$. Since $P_{ij}(t)$ is the probability of $j - i$ events of the process occurring in a time interval of length t, we know by the formulas for a Poisson process given in Section 7.2 that

$$P_{ij}(t) = \begin{cases} \frac{(\lambda t)^{j-i}}{(j-i)!} e^{-\lambda t} & \text{when } i \leq j \\ 0 & \text{when } i > j \end{cases}$$

Notice that the transition probability from state i to state j only depends on the difference $i - j$. This reflects the fact that the birth and the death rates in this model are the same in all states of the system. (For a matrix interpretation of this property, see the end of this section.) ■

So far, we have been considering only conditional probabilities. At every moment s, the random variable $X(s)$ has a probability mass function

$$q_i(s) = P\{X(s) = i\}$$

If at time s the system is in state i, then the probability $q_j(s+t)$ that the system be in state j at time $t + s$ is given by

$$q_j(s+t) = \sum_{i=0}^{\infty} q_i(s) \cdot P_{ij}(t) \tag{9.4}$$

This follows from the Rule of Total Causes (formula (2.24)), by conditioning on the possible states of the system at time s.

We shall now relate the transitions occurring $t + h$ time units in the future to the transitions occurring t time units in the future, where h is a positive number. We observe that, in order for the system to go from state i at time s to state j at time $s + t + h$, it first must go to some intermediate state k by time $s + t$, and then from state k to state j in the remaining time h. The probability of such a sequence of events is

$$P\{\, X(s+t+h) = j \mid X(s+t) = k \,\} \cdot P\{\, X(s+t) = k \mid X(s) = i \,\}$$

By the stationary property (9.1) of the transition probabilities, we may replace s with 0 in this calculation. Thus, the sequence of transitions $i \to k \to j$ just described has probability

$$P_{ik}(t) \cdot P_{kj}(h)$$

Summing these probabilities over all the possible intermediate states k gives us the **Chapman–Kolmogorov equation**

$$P_{ij}(t+h) = \sum_{k=0}^{\infty} P_{ik}(t) \cdot P_{kj}(h) \tag{9.5}$$

Equation (5) is the continuous-time analog of equation (6.14). We can visualize this equation as shown in Figure 9.6.

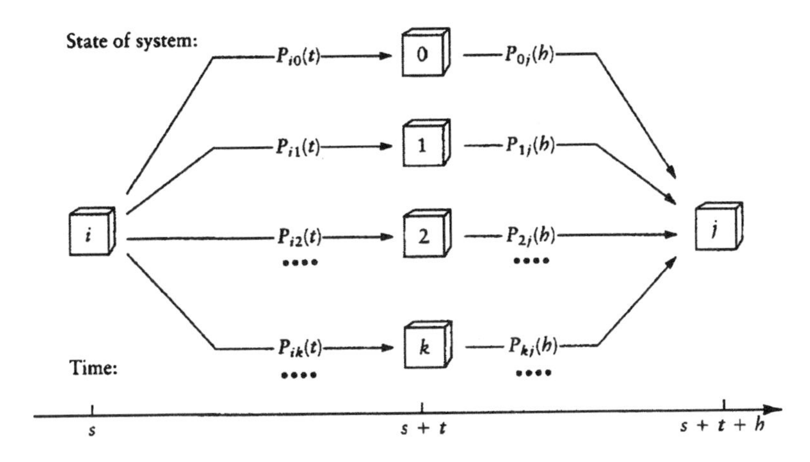

Figure 9.6: Transitions from State i to State j in Time $t + h$

If we are analyzing a **finite-capacity system**, in which there can be at most N customers at any one time, then $P_{ij}(t) = 0$ if either $i > N$ or $j > N$. In this case, we can form the $N + 1 \times N + 1$ matrix

$$\mathbf{P}(t) = \left[P_{ij}(t)\right]_{0 \leq i, j \leq N}$$

The Chapman-Kolmogorov equation then can be written as the matrix equation

$$\mathbf{P}(t + h) = \mathbf{P}(t) \cdot \mathbf{P}(h) \tag{9.6}$$

where the right side is the product of the matrices. If we let

$$\mathbf{q}(s) = \left[\begin{array}{cccc} q_0(s) & q_1(s) & \ldots & q_N(s) \end{array}\right]$$

be the row vector of probabilities that the system is in state i at time s, then equation (9.4) can be written as the vector-matrix product

$$\mathbf{q}(s + t) = \mathbf{q}(s) \cdot \mathbf{P}(t) \tag{9.7}$$

When the system has unlimited capacity, we can still form the *infinite* matrix $\mathbf{P}(t)$, and equation (9.6) still holds, with the obvious definition of multiplication of infinite matrices in terms of rows and columns. We can also form the infinitely long row vector $\mathbf{q}(s)$, and equation (9.7) holds. (All under our standing assumption that the stochastic process $\{X(t)\}$ exists.)

Example 9.7 Transition Matrix for Poisson Process

For the Poisson process the matrix $\mathbf{P}(t)$ can be given explicitly:

$$\mathbf{P}(t) = e^{-\lambda t} \begin{bmatrix} 1 & \lambda t & (\lambda t)^2/2! & (\lambda t)^3/3! & \cdots \\ 0 & 1 & \lambda t & (\lambda t)^2/2! & \cdots \\ 0 & 0 & 1 & \lambda t & \cdots \\ 0 & 0 & 0 & 1 & \cdots \\ \vdots & \vdots & \vdots & \vdots & \ddots \end{bmatrix}$$

Notice that, along any diagonal of this matrix, the entries are all the same. When $t \approx 0$, all the entries located one position above the main diagonal are proportional to t (up to an error of size $o(t)$), while those entries located more than one diagonal away from the main diagonal are of size $o(t)$. Here we are using the notation introduced in Section 7.5. We shall see in the next section that this is a general property of the transition matrix for a birth-death process. ■

9.5 Differential Equations for Transition Probabilities

The Chapman–Kolmogorov equations will be our starting point in deriving a system of differential equations for the transition probabilities. We first observe that the matrix form $\mathbf{P}(t+h) = \mathbf{P}(t)\,\mathbf{P}(h)$ of the Chapman–Kolmogorov equations is similar to the formula for multiplying an exponential function

$$e^{\alpha(t+h)} = e^{\alpha t}e^{\alpha h}$$

where α is a constant. To obtain α from the function $f(t) = e^{\alpha t}$, we calculate the tangent-line approximation to $f(h)$ for small values of h:

$$f(h) = 1 + h\alpha + o(h)$$

Conversely, once we know α, we can obtain the function $f(t)$ by solving the differential equation

$$f'(t) = \alpha f(t)$$

with the initial condition $f(0) = 1$.

The situation for the *matrix-valued function* $\mathbf{P}(t)$ is quite similar: if we can find a matrix $\mathbf{A}$ such that

$$\mathbf{P}(h) = \mathbf{I} + h\,\mathbf{A} + o(h)$$

for small values of h (where $\mathbf{I}$ is the identity matrix), then we can obtain $\mathbf{P}(t)$ by solving the matrix differential equation

$$\mathbf{P}'(t) = \mathbf{A}\,\mathbf{P}(t)$$

with the initial condition $\mathbf{P}(0) = \mathbf{I}$. (This equation has a unique solution when the system has finite capacity. For many systems of infinite capacity, such as the $M/M/s$ queue, there is also a unique solution, under suitable conditions on the birth and death rates. It is beyond the scope of this book, however, to describe the subtle points of analysis that arise with infinite-capacity systems.)

We now calculate the matrix $\mathbf{A}$ in terms of the given parameters of the process (the birth and death rates). To do this, we must determine the behavior of the individual transition probabilities over a short time interval. The calculation follows the same pattern as in Section 7.5 for the Poisson process, although the final form of the equations is considerably more complicated.

We will use the exponential model realization of a birth-death process to estimate the probabilities $P_{ij}(h)$ when h is a small positive number, as illustrated in the following example.

Example 9.8 Machine Repair Model

Consider the machine-repair model (Example 9.5). Suppose that there are four machines and two repairmen. The breakdown rate is $\lambda = 1$ per hour and the service rate is $\mu = 2$ per hour. At time 10 A.M. you observe that both repairmen are idle. Estimate the probability that *both* repairmen will be busy at some time within the next minute. (It is obvious from the breakdown and service rates that the probability of this event is small.)

Solution The system is in state 0 at 10 A.M. and will remain in this state for time T_0, where T_0 is an exponential random variable with rate 4 per hour. Let B be the event that both repairmen are simultaneously busy at some moment between 10:00 and 10:01. For B to occur, both state transitions $0 \to 1$ and $1 \to 2$ must occur within a time interval of $h = 1/60$ hour. But a sojourn time T_1 in state 1 is an exponential random variable with rate $\lambda_1 + \mu_1 = 5$ (see the state-transition diagram in Example 9.5). Hence, if B occurs, then both $T_0 \leq h$ and $T_1 \leq h$ (this is a very crude estimate, but it will suffice since h is so small). Thus we can estimate

$$P\{B\} \leq P\{\, T_0 \leq h \text{ and } T_1 \leq h \,\}$$

But the random variables T_0 and T_1 are mutually independent, so the probability on the right side of this inequality is a product:

$$P\{B\} \leq \left(1 - e^{-4h}\right) \cdot \left(1 - e^{-5h}\right)$$

Finally, we can use the inequality $1 - e^{-x} \leq x$ for $x \geq 0$ to get the estimate

$$P\{B\} \leq (4h) \cdot (5h) = 20h^2 = \frac{20}{3600} = 0.0055.$$

■

We return to a general birth-death system. As usual, the random variable $X(t)$ is the state of the system (population size) at time t. In Example 9.8 we found that events involving more than one change of state during a short time interval of length h have probability $o(h)$. This is true in general.

Lemma 9.1. *The probability that two or more jumps of $X(t)$ will occur during a time interval of length h is $o(h)$.*

Proof We follow the same method as in Example 9.8. Suppose that the time interval is $(s, s+h]$, and $X(s) = i$. If there are at least two jumps of $X(t)$ between time s and time $s+h$, let T_1 and T_2 be the first two sojourn times of the process in this interval. Then obviously $T_1 \leq h$ and $T_2 \leq h$. Thus

$$P\{\text{ Two or more jumps in } (s, s+h]\} \leq P\{T_1 \leq h \text{ and } T_2 \leq h\} \tag{9.8}$$

Since T_1 and T_2 are independent, we can calculate the probability on the right side of equation (9.8) by multiplying the separate probabilities. This gives the upper bound

$$P\{\text{ Two or more jumps in } (s, s+h]\} \leq P\{T_1 \leq h\} \cdot P\{T_2 \leq h\}$$

Now recall from the definition of a birth-death process (Section 9.2) that T_j is an exponential random variable with rate α_j. Here,

$$\alpha_1 = \lambda_i + \mu_i \quad \text{and} \quad \alpha_2 = \lambda_{i\pm 1} + \mu_{i\pm 1}$$

since the system is in state i during the first sojourn time, and then is in either state $i+1$ or state $i-1$ during the second sojourn time. Thus, if we set $\alpha = \max\{\alpha_1, \alpha_2\}$, then

$$P\{T_j \le h\} = 1 - e^{-h\alpha_j} \le 1 - e^{-h\alpha}$$

Since $1 - e^{-x} \le x$ for $x \ge 0$, we finally conclude that the right side of equation (9.8) is bounded by $(\alpha h)^2$. This quantity is of size $o(h)$, if the initial state i is fixed. ■

Remark The proof of Lemma 9.1 shows that the $o(h)$ estimate for the probability does not depend on the state of the system if the birth-death rates are *uniformly bounded*:

$$\lambda_i + \mu_i \le M \quad \text{for all } i$$

where M is some fixed constant. This condition is satisfied by all finite-capacity systems and also by systems such as the $M/M/s$ queue.

From Lemma 9.1 we obtain the behavior of the transition probabilities over a short time interval:

Theorem 9.1. *For small positive values of h, the transition probabilities of a birth-death process are related to the birth and death rates as follows. For each fixed state i,*

$$\begin{aligned}
P_{i,i+1}(h) &= \lambda_i h + o(h) && \text{(birth in state } i\text{)}\\
P_{i,i-1}(h) &= \mu_i h + o(h) && \text{(death in state } i\text{)}\\
P_{ii}(h) &= 1 - \lambda_i h - \mu_i h + o(h) && \text{(no change in state } i\text{)}
\end{aligned}$$

Furthermore, for fixed state i,

$$\sum_{j:|i-j|>1} P_{ij}(h) = o(h)$$

Proof A transition from state i to state j when $|i - j| > 1$ involves at least two changes of state. Thus, Lemma 9.1 gives the last statement of the theorem. For the same reason, in analyzing the transitions $i \to i \pm 1$ we only need to calculate the probability of exactly one jump occuring between time t and time $t + h$. We calculate the probability of such an event as follows:

$i \to i+1$: This means one birth and no deaths in $(t, t+h]$. By independence of births and deaths, the probability is

$$(\lambda_i h + o(h)) \cdot (1 - \mu_i h + o(h)) = \lambda_i h + o(h)$$

(see the calculations in Section 7.5).

$i \to i-1$: This means one death and no birth in $(t, t+h]$. In this case, again by independence of births and deaths, the probability is

$$(\mu_i h + o(h)) \cdot (1 - \lambda_i h + o(h)) = \mu_i h + o(h)$$

Subtracting the probabilities in these two case from 1 gives the probability that no change of state will occur between t and $t+h$. ∎

We can combine Theorem 9.1 and the Chapman–Kolmogorov equation (9.5) to obtain a system of differential equations satisfied by the transition probabilities. Indeed, we now know that, when h is small, the only significant terms on the right side of equation (9.5) are those with $k = j-1$, j, and $j+1$. From the formulas for these terms in Theorem 9.1, we get

$$\begin{aligned} P_{ij}(t+h) &= P_{i,j-1}(t) \cdot \lambda_{j-1}h + P_{i,j+1}(t) \cdot \mu_{j+1}h \\ &\quad + P_{ij}(t) \cdot (1 - h\lambda_j - h\mu_j) + o(h) \end{aligned}$$

If we move the term $P_{ij}(t)$ to the left side of this equation, divide by h, and let $h \to 0$, we obtain **Kolmogorov's forward equations** for a birth-death process:

$$\frac{d}{dt}P_{ij}(t) = \lambda_{j-1}P_{i,j-1}(t) - (\lambda_j + \mu_j)P_{ij}(t) + \mu_{j+1}P_{i,j+1}(t) \qquad (9.9)$$

These equations are called *forward equations* because the probabilities on the right side all involve transitions from a fixed state i at time 0 to states $j-1$, j, and $j+1$ at a future time t. This system of equations holds for any initial state $i = 0, 1, 2, \ldots$ and final state $j = 1, 2, \ldots$. The equations are also valid for $j = 0$, provided that we define $\lambda_{-1} = 0$ and $\mu_0 = 0$. That is, for $j = 0$, equation (9.9) reads:

$$\frac{d}{dt}P_{i0}(t) = -\lambda_0 P_{i0}(t) + \mu_1 P_{i1}(t)$$

Suppose that we return to equation (9.4), taking the initial time $s = 0$ (for example), and differentiate the right side term-by-term with respect to t (assuming that the infinite series of the derivatives is the derivative of the infinite series). If we use Kolmogorov's forward equation (9.9), we obtain the set of differential equations

$$\begin{aligned} \frac{d}{dt}q_j(t) =& \lambda_{j-1}\sum_{i=0}^{\infty} q_i(0)P_{i,j-1}(t) - (\lambda_j + \mu_j)\sum_{i=0}^{\infty} q_i(0)P_{ij}(t) \\ &+ \mu_{j+1}\sum_{i=0}^{\infty} q_i(0)P_{i,j+1}(t) \end{aligned} \qquad (9.10)$$

If we use equation (9.4) again, for $j-1$, j, and $j+1$, we can write the set of equations (9.10)

$$\frac{d}{dt}q_j(t) = \lambda_{j-1}q_{j-1}(t) - (\lambda_j + \mu_j)q_j(t) + \mu_{j+1}q_{j+1}(t) \qquad (9.11)$$

In this derivation, the equation for $j = 0$ reads

$$\frac{d}{dt}q_0(t) = -\lambda_0 q_0(t) + \mu_1 q_1(t)$$

9.6 Pure Birth Process

Suppose we have a pure birth process, such as the Poisson process. Since all the death rates $\mu_i = 0$, equation (9.11) simplifies to

$$\frac{d}{dt}q_j(t) + \lambda_j q_j(t) = \lambda_{j-1}q_{j-1}(t) \quad \text{for } j \geq 1 \qquad (9.12)$$

and the equation for $j = 0$ reads

$$\frac{d}{dt}q_0(t) = -\lambda_0 q_0(t) \tag{9.13}$$

Equation (9.13) obviously has the solution

$$q_0(t) = q_0(0)e^{-\lambda_0 t}$$

Equation (9.12) can be written as

$$\frac{d}{dt}\left(e^{\lambda_j t}\, q_j(t)\right) = \lambda_{j-1}\, e^{\lambda_j t}\, q_{j-1}(t) \tag{9.14}$$

for $j = 1, 2, \ldots$. We can successively calculate $q_j(t)$ from $q_{j-1}(t)$ by integrating the right side of this equation from 0 to t. We use the initial probabilities $q_j(0)$ to determine the constants of integration, just as we did for the Poisson process in Section 7.5.

Example 9.9 Yule Process

Assume that the birth rate is proportional to the population size (this model is often called the **Yule process**; it is the limiting case of the linear growth Model in Section 9.1, when the death rate $\mu \to 0$). Thus, we have

$$\lambda_j = j\lambda$$

for $j = 0, 1, 2, \ldots$ and some fixed constant $\lambda > 0$. Suppose that the initial population at time $t = 0$ is $n \geq 1$. Since there are no deaths, the population size at any time $t > 0$ is at least n. Hence,

$$q_j(t) = 0 \quad \text{for } j < n$$

To calculate the probabilities $q_j(t)$ for $j \geq n$, it is convenient to introduce the new variable $z = e^{\lambda t}$ and the functions $f_j(z) = z^j\, q_j(t)$. Then equation (9.14) can be written as

$$\frac{d}{dt} f_j(z) = (j-1)\lambda z f_{j-1}(z) \tag{9.15}$$

By the chain rule,

$$\frac{d}{dz} f_j(z) = \left(\frac{dt}{dz}\right) \cdot \left(\frac{df_j(z)}{dt}\right)$$

Since $t = (\log z)/\lambda$, we have $dt/dz = 1/(\lambda z)$. Hence,

$$\frac{df_j(z)}{dt} = (\lambda z) \cdot \left(\frac{df_j(z)}{dz}\right)$$

Thus, equation (9.15) can be simplified as follows:

$$\frac{d}{dz} f_j(z) = (j-1) f_{j-1}(z) \tag{9.16}$$

Since $t = 0$ corresponds to $z = 1$, we also have the initial conditions

$$f_j(1) = \begin{cases} 1 & \text{if } j = n \\ 0 & \text{otherwise} \end{cases} \tag{9.17}$$

Notice that

$$f_j(z) = 0 \quad \text{for } j < n \tag{9.18}$$

by the corresponding property of $q_j(t)$.

Now we can integrate equation (9.16) for each value of $j \geq n$, taking into account equations (9.17) and (9.18). For $j = n$, $f_n(z)$ has zero derivative (by equations (9.16) and (9.18); hence, it is constant. By equation (9.17), this constant is 1. Thus, we have

$$f_n(z) = 1.$$

Now take $j = n + 1$ in equation (9.16) and use the value of $f_n(z)$ we just found:

$$\frac{d}{dz} f_{n+1}(z) = n$$

Hence, $f_{n+1}(z) = nz + C$, where the constant of integration C has to be chosen to satisfy equation (9.17). Thus $C = -n$, so

$$f_{n+1}(z) = n(z - 1)$$

Next, take $j = n + 2$ in equation (9.16), and use the value of $f_{n+1}(z)$ we just found:

$$\frac{d}{dz} f_{n+2}(z) = n(n + 1)(z - 1)$$

Hence, $f_{n+2}(z) = [n(n + 1)/2](z - 1)^2 + C$, where the constant of integration C has to be chosen to satisfy (9.17). But obviously $C = 0$, so that

$$f_{n+2}(z) = \frac{n(n + 1)}{1 \cdot 2}(z - 1)^2$$

By this argument and induction on k, we prove that

$$f_{n+k}(z) = \frac{n(n + 1) \cdots (n + k - 1)}{k!}(z - 1)^k \quad \text{for } k = 1, 2, \ldots$$

Going back to the original variable t and the probabilities $q_j(t) = e^{-j\lambda t} f_j(e^{\lambda t})$, where $j = n + k$, we have the following results:

$$q_n(t) = e^{-n\lambda t}$$

and for $j = n + 1, n + 2, \ldots$

$$q_j(t) = \frac{n(n + 1) \cdots (j - 1)}{1 \cdot 2 \cdots (j - n)} e^{-j\lambda t} \left(e^{\lambda t} - 1\right)^{j-n} \tag{9.19}$$

We can reduce formula (9.19) to a more recognizable form as follows. First, multiply numerator and denominator by $(n-1)!$ so that the numerical factor is obviously a binomial coefficient. Define

$$a(t) = e^{-\lambda t} \quad \text{and} \quad b(t) = 1 - e^{-\lambda t}$$

Then we can write $e^{-j\lambda t} = a(t)^n \cdot a(t)^{j-n}$. Making obvious algebraic manipulations, we get

$$q_j(t) = \binom{j-1}{n-1} a(t)^n \, b(t)^{j-n} \tag{9.20}$$

for $j = n, n+1, \ldots$. We observe that formula (9.20) is a *negative-binomial* distribution with parameters $r = n$ and $p = e^{-\lambda t}$ (see Section 3.2). That is, the probability of attaining population size j by time t starting with population n at time $t = 0$ is the same as the probability of winning n times in j Bernoulli trials, where the win probability in each trial is $e^{-\lambda t}$. (Notice that the win probability decreases to zero as $t \to \infty$.)

Now that we have found the probability distribution of $X(t)$, we see that there is another probability model hidden in the Yule process. Recall that the number of Bernoulli trials needed to get n wins is the sum of the number of trials to get the first win, plus the number of trials after the first win to get the second win, and so on. Thus, if $X(0) = n$, then the population at time t can be written as

$$X(t) = Y_1(t) + Y_2(t) + \cdots + Y_n(t) \tag{9.21}$$

where $Y_i(t)$ is a *geometric* (discrete waiting-time) random variable with parameter $p = e^{-\lambda t}$. Here the random variables $Y_1(t), \ldots, Y_n(t)$ are mutually independent.

We may associate the random variable $Y_i(t)$ with the ith member of the original population at time 0. Then, by formula (9.21), the population size at time t is the sum of the number of independent Bernoulli trials that each of the original n members would have to play to get a single win (where the win probability is $e^{-\lambda t}$). ■

Exercises

1. (Logistic process) Modify the linear growth model of Section 9.1 to take into account the effect of increasing population on individual birth and death rates, as follows. Assume that the total population size cannot exceed a fixed number N. Let the state of the system $X(t)$ at time t be the number of individuals in the population that moment. Suppose that when the population is n, then the individual birthrate is $\alpha(N-n)$, and the individual deathrate is βn, where α and β are fixed positive numbers. (Thus in times of low population, the individual birthrate is high and the individual deathrate is low; when the population approaches its limit N, this relation is reversed.) The independence assumptions about births and deaths are as expressed in Section 9.1.

   a. Set up this system as a birth–death system, and draw a state-transition diagram.

   b. Suppose $\alpha = \frac{1}{4}$, $\beta = \frac{1}{5}$, and $N = 5$. Let B = elapsed time, starting at $t = 0$, until the population changes (birth or death). Calculate $P\{B \geq 1 \mid X(0) = n\}$ for $n = 2$ and $n = 4$.

2. (Linear Growth Model, with Immigration) Modify the linear growth model in Section 9.1 by assuming that there is an external source of population growth due to immigration. Specifically, assume that the waiting time until the next immigrant arrives is an exponential random variable T with rate θ. Here, we assume that T is independent of the times until the next birth or death in the existing population, and that the times between the arrivals of immigrants are mutually independent. Once an immigrant joins the population, it is assumed to have the same birth and death rates as the native population. However, immigration is not allowed when the population size is N or larger. Set up this system as a birth–death system, and draw a state-transition diagram.

3. Consider the machine-repair model involving two repairmen and four machines. Assume that the breakdown rate is $\lambda = 1$ machine per hour, and that the service rate is $\mu = 2$ machines per hour per serviceman. Let $X(t)$ be the number of machines broken at time t.

   a. Using suitable probability transforms applied to the same sequences of uniform random numbers as in Chapter 8, Exercise 14, generate the first four sojourn times and jumps of the system, and plot the graph of the sample function $x(t)$ in the range $0 \leq t \leq 1$. Assume that $x(0) = 0$ and that the fifth jump occurs after $t = 1$.

   b. How many machines broke down and how many were repaired during this one-hour simulation?

4. Consider the same machine-repair model as in Exercise 3. Suppose that you observe at 11 A.M. that two of the machines are operating and two of the machines are being repaired.

   a. What is the probability that, in the next $\frac{1}{4}$ hour there will be no change in the state of the system?

   b. What is the probability that both of the machines being repaired at 11 A.M. will be fixed before any more breakdowns occur?

5. Suppose that a birth–death system has finite capacity N (thus $\lambda_j = 0$ for $j \geq N$). Let $\mathbf{q}(t)$ be the row vector with components $\{q_j(t)\}$. Show that the system of differential equations (9.11) can be written in vector-matrix form as

$$\mathbf{q}'(t) = \mathbf{q}(t)\,\mathbf{A}$$

   where $\mathbf{A}$ is the matrix

$$\begin{bmatrix} -\lambda_0 & \lambda_0 & 0 & \cdots & 0 & 0 \\ \mu_1 & -(\lambda_1 + \mu_1) & \lambda_1 & \cdots & 0 & 0 \\ 0 & \mu_2 & -(\lambda_2 + \mu_2) & \cdots & 0 & 0 \\ \vdots & \vdots & \vdots & \ddots & \vdots & \vdots \\ 0 & 0 & 0 & \cdots & -(\lambda_{N-1} + \mu_{N-1}) & \lambda_{N-1} \\ 0 & 0 & 0 & \cdots & \mu_N & -\mu_N \end{bmatrix}$$

6. Consider a pure birth process with linear birth rates $\lambda_j = j\lambda$ and death rates $\mu_j = 0$ (see Section 9.6). Let $X(t)$ be the population size at time t.

   a. Suppose that, at time $t = 0$, the population consists of n individuals. Calculate $E[X(t)]$.

   b. Suppose that, because of experimental errors, the population size at time $t = 0$ can only be estimated with certain probabilities (for example, the population might be a bacteria culture whose exact size is impossible to count). Calculate $E[X(t)]$ in terms of λ and $E[X(0)]$.

   (HINT: Condition on $X(0)$).

   c. Let $\phi(t) = E[X(t)]$. Show that $\phi(t)$ satisfies the differential equation

$$\phi'(t) = \lambda\phi(t)$$

7. (Continuation of Exercise 6) Consider a pure birth process with linear birth rates $\lambda_j = j\lambda$ and death rates $\mu_j = 0$. Assume that the population size is N at time $t = 0$, where N is a random variable.

   a. Show that the population size at time $t > 0$ can be written as a sum of a random number of random variables; namely,

$$X(t) = Y_1(t) + Y_2(t) + \cdots + Y_N(t)$$

   where $Y_i(t)$, for $i = 1, 2, \ldots$, are mutually independent geometric random variables with parameter $p = e^{-\lambda t}$, and $Y_i(t)$ is independent of N.

   b. Use the result in part (a) to calculate $\mathrm{Var}(X(t))$.

8. Consider the Chapman–Kolmogorov equation (9.5) in the form

$$P_{ij}(h+t) = \sum_{k=0}^{\infty} P_{ik}(h) \cdot P_{kj}(t)$$

   where the first time interval is of length h and the second is of length t.

   a. Use this equation and Theorem 9.1 to show that

$$P_{ij}(h+t) = P_{i+1,j}(t) \cdot \lambda_i h + P_{i-1,j}(t) \cdot \mu_i h + P_{ij}(t) \cdot (1 - \lambda_i - \mu_i)h + o(h)$$

   b. Use the result of part (a) to derive the **Kolmogorov backward equations** for a birth–death process:

$$\frac{d}{dt} P_{ij}(t) = \lambda_i P_{i+1,j}(t) - (\lambda_i + \mu_i) \cdot P_{ij}(t) + \mu_i P_{i-1,j}(t)$$

   These equations are called the *backward equations* because the probabilities on the right side all involve transitions into state j from states $i-1$, i, and $i+1$ over a time interval of length t in the past.

   c. Suppose that a birth–death system has finite capacity N. Let $\mathbf{P}(t)$ be the $(N+1) \times (N+1)$ matrix of transition probabilities, and let $\mathbf{A}$ be the matrix in Exercise 5. Use the Kolmogorov forward and backward equations to show that $\mathbf{P}'(t) = \mathbf{A}\,\mathbf{P}(t) = \mathbf{P}(t)\,\mathbf{A}$, where $\mathbf{P}'(t)$ denotes the matrix whose elements are the t-derivatives of the elements of $\mathbf{P}(t)$.

Chapter 10

Steady-State Probabilities

We continue studying the birth–death processes introduced in Chapter 9. We show that, when the birth and death rates of such a system are compatible, we can find the long-term steady-state behavior of the system by solving a set of *balance equations*. These linear algebraic equations are the time-independent version of the Kolmogorov differential equations for the system. We illustrate this analysis for the $M/M/1$ queueing model and the machine-repair model.

10.1 Approach to Equilibrium

Consider a birth–death process with *birth rates* λ_j and *death rates* μ_j. Recall that this means that, when the population size (the *state* of the system) is j, then the waiting time until the next *event* (either a birth or a death) is exponentially-distributed with rate $\nu_j = \lambda_j + \mu_j$. The chance that this event will be a birth is λ_j/ν_j, and the chance for a death is μ_j/ν_j. The system is completely determined by this description, together with the initial probability distribution of the population at time $t = 0$. We have already seen that it is quite simple to simulate sample paths of the system, using a random number generator. In this chapter we consider the behavior of a birth–death system over a long time interval.

Let $X(t)$ be the total population at time t. Because most of our examples of birth–death processes will be queueing models, we shall generally refer to the members of the population as *customers*. In our analysis of such a probability model, the basic quantities of interest to us are the probabilities

$$q_j(t) = P\{X(t) = j\}$$

that we will find j customers in the system if we observe it at time t. In general, these probabilities will depend both on the time of observation and on the initial state of the system at time $t = 0$ (see equation (9.4)).

If we recall some real-life examples of queues—for example, queues at turnpike toll booths or store check-out counters—we would expect to observe the following sort of behavior of the system over the course of time:

a. There is an initial start-up period (such as when the store opens or when another checkout counter or toll booth is opened to relieve congestion). During this period the probability of finding j customers present at a particular time depends strongly on how many were initially present.

b. If the service facilities are adequate for the system, the system gradually settles down to a steady-state behavior, in the sense that the probability of finding j customers in the system is essentially constant, no matter when we observe the system. For example, when an extra tollbooth is opened on a turnpike in response to heavy traffic, then the congestion that had built up gradually dissipates. After some time all we observe is a generally heavier traffic load, but not the size of the traffic jam before the extra server was added to the system. (This assumes that the system with the added server now has adequate capacity to handle the traffic load.)

c. If the service facilities are not adequate for the system, the customer departures (*deaths*) do not occur rapidly enough to balance the arrivals (*births*), and the number of customers keeps growing without limit as time goes on. There is no steady-state behavior.

We shall now try to see if the mathematical model of a birth–death process that we have been developing follows the pattern just described. For the moment, we ignore the unstable situations (a) and (c), and focus on the steady-state situation (b).

10.2 Steady-State Equations

We continue studying a birth–death process. Let $q_j(t) = P\{X(t) = j\}$, as in Section 10.1. In Section 9.5 we showed that the functions $q_j(t)$ satisfy the following system of linear differential equations (with coefficients determined by the birth–death rates):

$$\left(\frac{d}{dt}\right) q_j(t) = \lambda_{j-1} q_{j-1}(t) - (\lambda_j + \mu_j) q_j(t) + \mu_{j+1} q_{j+1}(t) \qquad (10.1)$$

Recall that, in order to have these equations also be valid for $j = 0$ we *defined* $\lambda_{-1} = 0$ and $\mu_0 = 0$. That is, equation (10.1) for $= 0$ reads

$$\left(\frac{d}{dt}\right) q_0(t) = -\lambda_0 q_0(t) + \mu_1 q_1(t) \qquad (10.2)$$

Suppose that, as $t \to \infty$, the birth–death system approaches steady-state behavior, in the sense described in Section 10.1, situation (b). That is, we assume that, when t is large, the probability mass function $\{q_j(t)\}$ doesn't vary appreciably with t. This means that

$$p_j = \lim_{t \to \infty} q_j(t) \qquad (10.3)$$

exists for $j = 0, 1, 2, \ldots$. Although we don't know the constants p_j at the moment, the existence of limit (10.3), together with equations (10.1) and (10.2), guarantees that

$$\lim_{t \to \infty} \left(\frac{d}{dt}\right) q_j(t)$$

also exists. Call the limiting value of this derivative C_j. We claim that all the constants C_j must be zero. To see this, recall that by the mean-value theorem of calculus,

$$q_j(t+1) - q_j(t) = q_j'(\tau) \tag{10.4}$$

where τ is a number between t and $t+1$. If we take t sufficiently large, the right side of equation (10.4) is essentially C_j, no matter what τ happens to be, and the left side is essentially zero. Since C_j is a constant not depending on t, this forces $C_j = 0$. Hence we can conclude that if the limit (10.3) exists, the limiting values p_j satisfy the time-independent equations that we obtain from equations (10.1) and (10.2) by letting $t \to \infty$ and replacing the derivative of $q_j(t)$ by zero:

$$0 = \lambda_{j-1} p_{j-1} - (\lambda_j + \mu_j) p_j + \mu_{j+1} p_{j+1} \tag{10.5}$$

These equations must hold for $j = 0, 1, \ldots$ (with $\lambda_{-1} = \mu_0 = 0$).

What other properties do the limiting values p_j have? Clearly $p_j \geq 0$. Furthermore, since $\{q_j(t)\}$ is a probability distribution for each fixed value of t, we should expect that

$$\sum_{j=0}^{\infty} p_j = 1 \tag{10.6}$$

Notice that equation (10.6) is obtained formally by interchanging the two limiting operations $t \to \infty$ and $j \to \infty$. If this interchange is legitimate and if the limits in expression (10.3) exist, then equation (10.6) holds. It is possible in some cases, however, for the limiting probabilities to exist but to be *defective*; that is, the sum in equation (10.6) can be strictly less than 1. This means that there is a positive probability for the population to become infinite as $t \to \infty$. Indeed, for a pure birth process such as the Poisson process, the limiting probabilities are all zero.

10.3 Limiting Probabilities

In deriving equation (10.5), we have **assumed** the existence of the limiting probabilities $\{p_j\}$. In this section we will explicitly solve the Kolmogorov system of differential equations for the time-dependent probabilities $q_j(t)$ in some special cases. From the general form of the solution it will be obvious that there is a unique limiting probability distribution when $t \to \infty$.

Example 10.1 Queue with Capacity 1

Consider the $M/M/1$ queue with capacity $N = 1$. This is the system in which arriving customers leave immediately without queuing if the server is busy (for example, a telephone operator with no hold button). The possible states are 0 and 1. If λ is the arrival rate of customers, and μ is the service rate, then equation (10.2) is

$$\left(\frac{d}{dt}\right) q_0(t) = -\lambda q_0(t) + \mu q_1(t)$$

Since $q_0(t) + q_1(t) = 1$, we can eliminate $q_1(t)$ from this equation and obtain the inhomogeneous first-order differential equation

$$\left(\frac{d}{dt}\right) q_0(t) = -(\lambda + \mu) q_0(t) + \mu$$

A particular solution to this equation is the *constant* solution $\mu/(\lambda + \mu)$. The general solution is then the sum of this particular solution and the general solution $Ce^{-(\lambda+\mu)t}$ to the homogeneous equation $dy/dt + (\lambda + \mu)y = 0$. Thus we have

$$q_0(t) = \frac{\mu}{\lambda + \mu} + C_0 e^{-(\lambda+\mu)t}$$

where the constant C_0 depends on the initial probabilities at time 0. Using the relation $q_0(t) + q_1(t) = 1$, we find that

$$q_1(t) = \frac{\lambda}{\lambda + \mu} - C_0 e^{-(\lambda+\mu)t}$$

From these formulas it is clear that the general pattern described in situations (a) and (b) of Section 10.1 does hold. The negative exponential term is the transient part, depending on the initial configuration of the system. As $t \to \infty$, the system has the limiting probabilities

$$p_0 = \frac{\mu}{\lambda + \mu} \qquad p_1 = \frac{\lambda}{\lambda + \mu}$$

which are the solutions of equations (10.5), in this case. These limiting probabilities are always the same, no matter what the state of the system was at time t = 0. ■

In the next example, we prove that the situation in Example 10.1 also holds true for many finite-capacity birth–death processes. Specifically, we shall call a birth–death system of capacity N **ergodic** if the birth and death rates are **strictly positive** for all possible transitions:

$$\lambda_i > 0 \quad \text{for } 0 \le i \le N - 1 \quad \text{and} \quad \mu_i > 0 \quad \text{for } 1 \le i \le N .$$

In terms of the state-transition diagram for the system, the ergodic condition means that, for $1 \le i \le N$, each pair of adjacent states $i - 1$ and i is connected by a left-pointing and a right-pointing arrow (for examples of such systems, see Section 9.3). In the language of Markov chains (see Section 6.5), all states of the system *communicate* and form a single class.

Example 10.2 Finite Ergodic Process

Assume we have a finite-state, ergodic birth and death process. Then *the limiting probabilities exist and are always the same, whatever the initial probability distribution may be at time* $t = 0$. The proof of this result requires some results from linear algebra and differential equations and will occupy the rest of this section. It may be omitted on first reading without loss of continuity.

The investigation of the time-dependent probabilities for the birth–death system is simplified by using vector-matrix notation. If $\mathbf{q}(t)$ is the row vector with components

$q_j(t) = P\{X(t) = j\}$, then the system of differential equations (9.10) can be written in vector-matrix form as

$$\mathbf{q}'(t) = \mathbf{q}(t)\,\mathbf{A} \tag{10.7}$$

where $\mathbf{A}$ is the matrix

$$\begin{bmatrix} -\lambda_0 & \lambda_0 & 0 & \cdots & 0 & 0 \\ \mu_1 & -(\lambda_1+\mu_1) & \lambda_1 & \cdots & 0 & 0 \\ 0 & \mu_2 & -(\lambda_2+\mu_2) & \cdots & 0 & 0 \\ \vdots & \vdots & \vdots & \ddots & \vdots & \vdots \\ 0 & 0 & 0 & \cdots & -(\lambda_{N-1}+\mu_{N-1}) & \lambda_{N-1} \\ 0 & 0 & 0 & \cdots & \mu_N & -\mu_N \end{bmatrix}$$

(see Exercise 5 in Chapter 9). Recall from linear algebra that the **eigenvalues** (or **characteristic values**) of a square matrix $\mathbf{A}$ are the roots of the **characteristic polynomial**

$$M(x) = \det(x\mathbf{I} - \mathbf{A})$$

where $\mathbf{I}$ denotes the $(N+1)\times(N+1)$ identity matrix. Because the matrix $\mathbf{A}$ in this case is of such a special type, we can say quite a lot about its eigenvalues:

$$\begin{gathered}\text{The matrix } \mathbf{A} \text{ has } N+1 \textit{ distinct} \text{ eigenvalues} \\ 0 = \alpha_0 > \alpha_1 > \alpha_2 > \cdots > \alpha_N\end{gathered} \tag{10.8}$$

We shall prove fact (10.8) at the end of the example. Assuming this result, we then find $N+1$ (left) eigenvectors

$$\mathbf{v}_0, \mathbf{v}_1, \ldots, \mathbf{v}_N$$

one for each eigenvalue α_i. Here, $\mathbf{v}_i$ is a non-zero row vector with $N+1$ components that satisfies the homogeneous equation

$$\mathbf{v}_i\,\mathbf{A} = \alpha_i \mathbf{v}_i$$

Since $\alpha_0 = 0$, the equation for $\mathbf{v}_0$ is the vector form of equation (10.5) for steady-state probabilities (see Exercise 1 at the end of this chapter). Thus we can choose the vector $\mathbf{v}_0$ to be the vector $\mathbf{p}$ of steady-state probabilities.

Because $\mathbf{A}$ has $N+1$ distinct eigenvalues, the set of eigenvectors is a basis for the vector space of row vectors with $N+1$ components. Thus we can write the initial probability distribution $\mathbf{q}(0)$ in terms of this basis:

$$\mathbf{q}(0) = c_0\mathbf{p} + c_1\mathbf{v}_1 + \cdots + c_N\mathbf{v}_N \tag{10.9}$$

In this expansion, the coefficient $c_0 = 1$. This follows by noting that the $(N+1)\times 1$ column vector $\mathbf{u} = \begin{bmatrix} 1 & 1 & \ldots & 1 \end{bmatrix}^{\mathrm{T}}$ is in the right null space of $\mathbf{A}$, and hence it is orthogonal to the left eigenvectors $\mathbf{v}_i$ for $i \leq 1$, which are in the row space of $\mathbf{A}$. Multiplying both sides of equation (10.9) by $\mathbf{u}$, we get $1 = c_0$, since $\mathbf{p}\cdot\mathbf{u} = \mathbf{q}(0)\cdot\mathbf{u} = 1$.

From the elementary theory of systems of differential equations with constant coefficients, we now get the solution $\mathbf{q}(t)$ immediately from fact (10.8) and equation (10.9):

$$\mathbf{q}(t) = \mathbf{p} + c_1 e^{\alpha_1 t}\,\mathbf{v}_1 + \cdots + c_N e^{\alpha_N t}\,\mathbf{v}_N \tag{10.10}$$

Notice that the coefficient of $\mathbf{p}$ in this formula is constant in time, because the corresponding eigenvalue is $\alpha_0 = 0$. All the other coefficients have a negative-exponential time dependence, since $\alpha_i < 0$ for $i \geq 1$. Hence as $t \to \infty$, these coefficients rapidly approach zero. This proves that

$$\lim_{t\to\infty} \mathbf{q}(t) = \mathbf{p} \tag{10.11}$$

for any choice of the initial condition $\mathbf{q}(0)$, with the rate of convergence controlled by the negative eigenvalue α_1 nearest zero.

We now go back and prove the key fact (10.8), which really contains two assertions:

(i) The characteristic polynomial $M(x) = \det(x\mathbf{I} - \mathbf{A})$ has $N + 1$ distinct real roots.

(ii) If α is a nonzero root of $M(x)$, then $\alpha < 0$.

This second assertion is easy to prove. If α is an eigenvalue for $\mathbf{A}$, then we can find a **right** eigenvector

$$\mathbf{w} = \begin{bmatrix} w_0 & w_1 & \dots & w_N \end{bmatrix}^{\mathrm{T}}$$

corresponding to α. The eigenvector equation $\mathbf{A}\mathbf{w} = \alpha\mathbf{w}$ is the same as the system of equations

$$\mu_j w_{j-1} + \lambda_j w_{j+1} = (\lambda_j + \mu_j + \alpha) w_j \quad \text{for } j = 0, \dots, N \tag{10.12}$$

Pick k so that $|w_k| \geq |w_j|$ for all values of j. Since $\mathbf{w}$ is a nonzero vector, we know that $w_k \neq 0$. Set $v_j = w_j / w_k$. With this choice of k, we have $|v_j| \leq 1$ for all j. Taking equation (10.12) with $j = k$ and dividing by w_k, we get

$$\mu_k v_{k-1} + \lambda_k v_{k+1} = \lambda_k + \mu_k + \alpha \tag{10.13}$$

The left side of equation (10.13) has an absolute value of at most

$$\mu_k |v_{k-1}| + \lambda_k |v_{k+1}| \leq \mu_k + \lambda_k$$

(recall that the birth and death rates are all positive). Comparing this bound with the right side of equation (10.13), we see that $\alpha \leq 0$. If we assume that $\alpha \neq 0$, it follows that α must be negative. This proves property (ii).

It remains to prove property (i). We shall assume that $N \geq 2$, since the case $N = 1$ was treated in Example 10.1. Because of the special tri-diagonal form of the matrix $\mathbf{A}$, we can calculate its characteristic polynomial $M(x)$ by means of the following recursive scheme.

Define a sequence of polynomials by $M_0(x) = 1$, $M_1(x) = x + \lambda_0$, and in general

$$M_k(x) = \text{Determinant of the } k \times k \text{ matrix formed by removing all but the first } k \text{ rows and } k \text{ columns from the matrix } x\mathbf{I} - \mathbf{A}$$

($M_k(x)$ is the kth *principal minor* of $x\mathbf{I} - \mathbf{A}$). By expanding the determinant $M_{k+1}(x)$ according to the elements of its last row, we find that, for $1 \leq k \leq N$, these polynomials satisfy the recurrence relation

$$M_{k+1}(x) = (x + \delta_k) M_k(x) - \lambda_{k-1}\mu_k M_{k-1}(x) \tag{10.14}$$

Here, we have written the diagonal elements of $\mathbf{A}$ as $-\delta_k$ (so $\delta_0 = \lambda_0, \delta_1 = \lambda_1 + \mu_1$, and so on). Notice that $M_{N+1}(x) = M(x)$, the characteristic polynomial of $\mathbf{A}$; the polynomial $M_k(x)$ is of degree k with leading coefficient 1.

The idea of the proof of property (i) is to show that the zeros of the polynomials $M_k(x)$ and $M_{k+1}(x)$ *interlace* on the x-axis (just as the zeros of the functions $\sin(x)$ and $\cos(x)$ do). Since $M_{k+1}(x) \approx x^{k+1}$ for large $|x|$, the interlacing property implies that $M_{k+1}(x)$ has one more real zero than $M_k(x)$. Since $M_0(x)$ has no zeros, we conclude that $M_1(x)$ has one real zero, ..., and $M_{N+1}(x)$ has $N + 1$ real zeros.

To see how we can prove this interlacing property, we look at the example $k = 1$. In this case, equation (10.14) reads

$$M_2(x) = (x + \delta_1)(x + \lambda_0) - \lambda_0\mu_1 \tag{10.15}$$

The first-degree polynomial $M_1(x)$ has a single zero at $x = -\lambda_0$. The second-degree polynomial $M_2(x)$ has two zeros. We could calculate them, of course, by the quadratic formula, but it is much more instructive to argue directly from formula (10.15), as follows. When $x = -\lambda_0$, then the first term on the right side of formula (10.15) vanishes; hence, $M_2(-\lambda_0)$ has the same sign as the second term. Thus, we see that

$$M_2(-\lambda_0) < 0$$

But since the coefficient of x^2 in $M_2(x)$ is 1, we know that $M_2(x)$ is positive for large $|x|$. Hence $M_2(x)$ has one real root x_1 to the left of $-\lambda_0$ and another real root x_2 to the right of $-\lambda_0$. Since $M_2(x)$ is of degree two, $\{x_1, x_2\}$ is the complete set of roots of $M_2(x)$. Furthermore,

$$M_1(x_1) < 0 \quad \text{and} \quad M_1(x_2) > 0$$

We shall now prove by induction on k that the pattern just observed for $k = 2$ is true in general. For $1 \le k \le N + 1$,

a. $M_k(x)$ has k real zeros.

b. At the k zeros of $M_k(x)$, the values of $M_{k-1}(x)$ are nonzero and alternate in sign as one moves from the smallest to the largest zero of $M_k(x)$.

In particular, property (i) follows from assertion (a) for the case $k = N + 1$.

Proof of Assertions (a) and (b) We have already verified that these assertions are true when $k = 0, 1, 2$. Assume that assertions (a) and (b) are true for k. We first observe from the recurrence relation (10.14) that $M_{k+1}(x)$ and $M_{k-1}(x)$ have opposite signs at each of the k zeros of $M_k(x)$ (the key fact being that the coefficient of $M_{k-1}(x)$ in equation (10.14) is positive). Since the signs of $M_{k-1}(x)$ alternate at these zeros by the induction hypothesis (b), it follows that there is a zero of $M_{k+1}(x)$ between every pair of zeros of $M_k(x)$. This gives $k - 1$ zeros for $M_{k+1}(x)$.

Let x_1 be the smallest zero and let x_k be the largest zero of $M_k(x)$. Since $M_{k-1}(x)$ has only $k - 1$ zeros, it must remain nonzero to the left of x_1 and to the right of x_k. But $M_{k+1}(x)$ has the opposite sign to $M_{k-1}(x)$ at x_1 and x_k. Since both $M_{k-1}(x)$ and $M_{k+1}(x)$ have either even or odd degree, it follows that $M_{k+1}(x)$ must have one change of sign to the left of x_1 and one change of sign to the right of x_k. This gives two more real zeros for $M_{k+1}(x)$, making a total of $k+1$. This verifies property (a) for the case $k+1$.

The argument just given also shows that between each pair of zeros of $M_{k+1}(x)$ there is exactly one zero of $M_k(x)$; in other words, the zeros of $M_{k+1}(x)$ and the zeros of $M_k(x)$ interlace. This implies the truth of assertion (b) for the case $k+1$ and completes the induction. ■

10.4 Balance Equations

We continue analyzing a birth-death system under the assumption that the **limiting probabilities** $p_j = \lim_{t\to\infty} P\{X(t) = j\}$ exist. In this section we calculate $\{p_j\}$ in terms of the birth and death rates of the system.

With Example 10.2 in mind, we shall assume that the system is **ergodic**: every pair of states can be connected by a path made up of arrows in the transition diagram (remember that, in the transition diagram, each arrow corresponds to a strictly positive birth or death rate). In the terminology of Section 6.5, all states of the system *communicate*.

We start by writing equation (10.5) in the form

$$\mu_{j+1} p_{j+1} = (\lambda_j + \mu_j) p_j - \lambda_{j-1} p_{j-1} \tag{10.16}$$

The equations obtained by taking $j = 1, 2, \ldots$ in (10.16) are called the **balance equations** for the system, for reasons that will be explained in more detail in the next section. For the present, let's try to solve this set of equations recursively, starting with $j = 0$ (remember that $\mu_0 = 0$ and $\lambda_{-1} = 0$):

$$\begin{aligned} \mu_1 p_1 &= \lambda_0 p_0 \\ \mu_2 p_2 &= (\lambda_1 + \mu_1) p_1 - \lambda_0 p_0 = \lambda_1 p_1 \\ \mu_3 p_3 &= (\lambda_2 + \mu_2) p_2 - \lambda_1 p_1 = \lambda_2 p_2. \end{aligned}$$

Continuing by induction, we see that the general relation is

$$\mu_n p_n = \lambda_{n-1} p_{n-1} \tag{10.17}$$

This equation has an obvious interpretation: the left side is the effective rate at which the system is leaving state n because of deaths, and the right side is the effective rate at which the system is entering state n because of births.

If we multiply both sides of equation (10.17) by $\mu_1 \mu_2 \cdots \mu_{n-1}$, we get

$$\mu_1 \mu_2 \ldots \mu_n p_n = \mu_1 \mu_2 \ldots \mu_{n-1} \lambda_{n-1} p_{n-1}$$

Since the factor $\mu_{n-1} p_{n-1}$ on the right side equals $\lambda_{n-2} p_{n-2}$, we have

$$\mu_1 \mu_2 \ldots \mu_n p_n = \mu_1 \mu_2 \ldots \mu_{n-2} \lambda_{n-2} \lambda_{n-1} p_{n-2}$$

Iterating this replacement process, we arrive at the formula

$$(\mu_1 \mu_2 \cdots \mu_n) \cdot p_n = (\lambda_0 \lambda_1 \cdots \lambda_{n-1}) \cdot p_0 \tag{10.18}$$

Now we can solve equation (10.18) for p_n in terms of p_0:

$$p_n = \frac{\lambda_0 \lambda_1 \cdots \lambda_{n-1}}{\mu_1 \mu_2 \cdots \mu_n} \cdot p_0 \tag{10.19}$$

In obtaining equation (10.19), we used the ergodic assumption: the death rates $\mu_j \neq 0$ for $1 \leq j \leq n$ if the system capacity is n or greater. From equation (10.19) we see that p_n is completely determined by the value of p_0; and if $p_0 \geq 0$ then $p_n \geq 0$.

It only remains to choose p_0 so that the probabilities $\{p_j\}$ sum to 1. It is precisely at this point that we must reconsider situation (c) of Section 10.1, where the births overtake the deaths. Namely, the products of birth–death ratios

$$\rho_n = \frac{\lambda_0 \lambda_1 \cdots \lambda_{n-1}}{\mu_1 \mu_2 \cdots \mu_n} \tag{10.20}$$

must get small sufficiently rapidly as $n \to \infty$ to satisfy

$$\sum_{n=1}^{\infty} \rho_n < \infty \tag{10.21}$$

Indeed, suppose that the series with terms $\{\rho_n\}$ converges. Then we can *define* p_0 by the equation

$$p_0 = \left(1 + \sum_{n=1}^{\infty} \rho_n\right)^{-1} \tag{10.22}$$

Since $p_n = \rho_n p_0$ for $n \geq 1$, we have

$$\sum_{n=0}^{\infty} p_n = \left(1 + \sum_{n=1}^{\infty} \rho_n\right) \cdot p_0 = 1$$

Thus, the sequence $\{p_n\}$ defined by equations (10.19) and (10.22) is the *unique* normalized solution to equation (10.16). This shows that *if the series in formula* (10.21) *converges and if there is a limiting normalized probability distribution* $\{p_n\}$, *then this distribution doesn't depend on the initial state of the system at* $t = 0$.

Suppose that these assumptions are satisfied. If the system starts in state i at $t = 0$, then $q_j(t) = P_{ij}(t)$ by definition. Hence, we conclude that the transition probabilities have limits as $t \to \infty$:

$$p_j = \lim_{t \to \infty} P_{ij}(t) \tag{10.23}$$

Notice that the limit (10.23) is the same for every choice of the initial state i.

For a *finite* ergodic birth-death system, with maximum population size N, the equations above still apply. In this case the ratios $\rho_n = 0$ for $n > N$, since the birth rate becomes zero once the population size reaches N. The series in (10.21) only has a finite number of terms now, so there is no question of convergence.

Remark In deriving equation (10.23), we assumed that the limits on the right side exist and give a normalized probability distribution. By Example 10.2, we know that this is the case for a finite-state ergodic birth–death system. When the theoretical capacity of the system is unlimited, then the sequences $\{\lambda_j\}$ and $\{\mu_j\}$ must satisfy additional conditions as $j \to \infty$ (in addition to the convergence condition (10.21), so that the Kolmogorov equations have a unique solution. (These conditions are satisfied by the models we shall study.) The proof that the limits (10.23) exist and sum to 1 when these additional conditions are satisfied requires advanced techniques that are beyond the scope of this book.

10.5 Time Averages

Suppose that we have an ergodic birth–death system, with birth rates λ_j and death rates μ_j. We are interested in calculating the proportion of time during which there are n customers in the system, over a long time interval.

Example 10.3 Queue with Capacity 1

Consider the $M/M/1$ queue system, with capacity 1, arrival rate λ, and service rate μ. By solving the Kolmogorov differential equations in Example 10.1, we found that the limiting probabilities for this system are $p_0 = \mu/(\lambda + \mu)$ and $p_1 = \lambda/(\lambda + \mu)$. Let us relate p_0 and p_1 to time averages.

Suppose that we start observing the system at the clock time $t = 0$ and find the server idle at this moment. The first customer arrives X_1 minutes later, and this customer's service lasts Y_1 minutes. Then the server is idle for X_2 minutes until the next customer arrives, and this customers's service lasts Y_2 minutes. Continuing in this way, we let $\{X_i\}$ be the sequence of successive sojourn times in state 0, and we let $\{Y_i\}$ be the sequence of successive sojourn times in state 1. The Nth customer leaves at the (random) clock time

$$T = X_1 + Y_1 + \cdots + X_N + Y_N$$

Recall that X_i is an exponential random variable with rate λ, and that Y_i is an exponential random variable with rate μ. Let $Z_i = X_i + Y_i$. Then

$$E[Z_i] = \frac{1}{\lambda} + \frac{1}{\mu} = \frac{\lambda + \mu}{\lambda\mu}$$

Because T is the sum of the N mutually independent random variables $Z_i = X_i + Y_i$, the strong form of the Law of Large Numbers (Section 4.6) asserts that

$$\lim_{N\to\infty} \frac{T}{N} = \frac{\lambda + \mu}{\lambda\mu} \tag{10.24}$$

In the interval $0 \le t \le T$, the proportion of time during which the server is idle is

$$\frac{X_1 + \cdots + X_N}{T}$$

Applying the Law of Large Numbers again and equation (10.24), we find that

$$\begin{aligned}\lim_{N\to\infty} \frac{X_1 + \cdots + X_N}{T} &= \left(\lim_{N\to\infty} \frac{N}{T}\right)\left(\lim_{N\to\infty} \frac{X_1 + \cdots + X_N}{N}\right) \\ &= \frac{\lambda\mu}{\lambda + \mu} \cdot \frac{1}{\lambda} \\ &= \frac{\mu}{\lambda + \mu} \end{aligned} \tag{10.25}$$

Since this limiting value is just the steady-state probability p_0 for the state 0 (see Example 10.1), we conclude that

$$p_0 = \text{Long-term proportion of time that the server is idle}$$

The same limiting argument, using Y_i in place of X_i in equation (10.25), shows that

$$p_1 = \text{Long-term proportion of time that the server is busy}$$

■

Consider now a general birth–death system, and let $X(t)$ as usual be the population size at time t. In Example 10.3, $X(t)$ takes only the values 0 and 1. In general, $X(t)$ can have any nonnegative integral value. To keep track of a particular state n, we define the following indicator random variable:

$$I_n(t) = \begin{cases} 1 & \text{if } X(t) = n \\ 0 & \text{otherwise} \end{cases}$$

If we generate a sample function $X(t)$ for the process by the simulation method described in Section 9.2, then $X(t)$ will have a graph like the one shown in Figure 10.1.

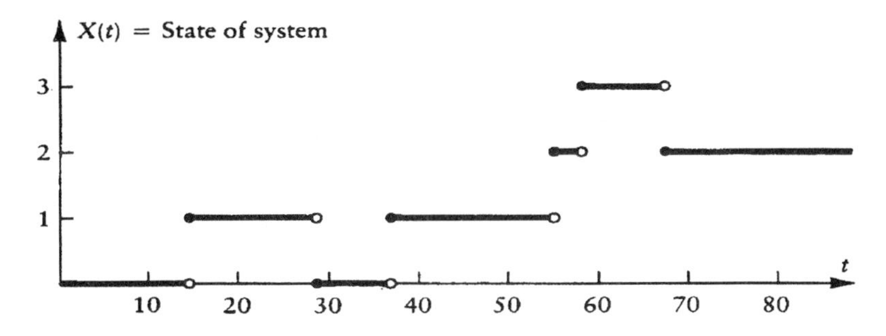

Figure 10.1: State of System at Time t

The corresponding sample functions $I_n(t)$ will have graphs with alternating segments on the t-axis and one unit above the t-axis. For $X(t)$ defined as in Figure 10.1, the indicator function for state 1 is shown in Figure 10.2.

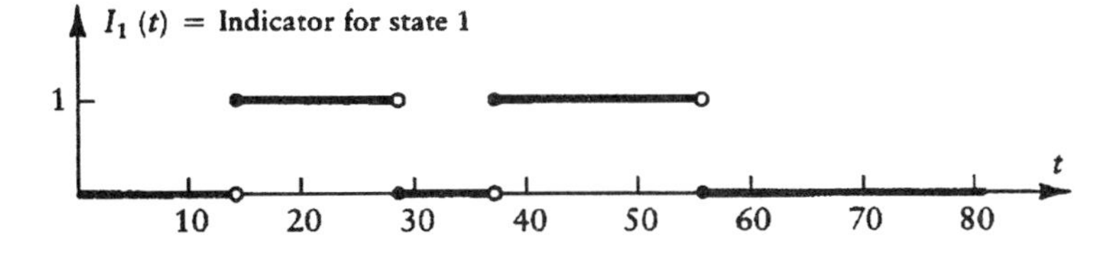

Figure 10.2: Indicator Random Variable for State 1

We observe that the area under the graph of $I_n(t)$ between 0 and T equals the total time that the system is in state n during this period. We can write this as the following **time average**:

$$\text{Average time in state } n = \frac{1}{T}\int_0^T I_n(t)\,dt \tag{10.26}$$

Of course, this average time depends both on the length T of the interval of observation of the system and on the particular sample function we have generated.

A basic property of ergodic systems is that the limit

$$\lim_{T\to\infty}\frac{1}{T}\int_0^T I_n(t)\,dt \tag{10.27}$$

exists and doesn't depend on the particular sample function generated (see Section 6.9 for the discrete-time version of limit (10.27)). But if limit (10.27) is a constant, we can determine its value by taking expectations:

$$\lim_{T\to\infty}\frac{1}{T}\int_0^T I_n(t)\,dt = \lim_{T\to\infty}\frac{1}{T}\int_0^T E[I_n(t)]\,dt \tag{10.28}$$

Notice that we interchanged the operations of taking expectations and taking the limit as $T \to \infty$ in the right side of equation (10.28).

We now relate the limiting time average in equation (10.28) to the limiting probabilities

$$p_n = \lim_{t\to\infty} P\{\, X(t) = n \,\}$$

that we calculated in Section 10.4. Since $I_n(t)$ is an indicator random variable, we have $E[I_n(t)] = P\{\, X(t) = n \,\}$. Hence,

$$p_n = \lim_{t\to\infty} E[I_n(t)] \tag{10.29}$$

If $\phi(t)$ is any bounded function that approaches a limit as $t \to \infty$, then it is easy to prove that

$$\lim_{t\to\infty}\phi(t) = \lim_{T\to\infty}\frac{1}{T}\int_0^T \phi(t)\,dt \tag{10.30}$$

Taking $\phi(t) = E[I_n(t)]$ in equation (10.30) and using equations (10.28) and (10.29), we finally obtain the main result of this section:

$$\begin{aligned} p_n &= \lim_{T\to\infty}\frac{1}{T}\int_0^T I_n(t)\,dt \\ &= \text{Long-term proportion of time spent in state } n \end{aligned} \tag{10.31}$$

Balance Equations

Recall from Section 10.4 the equations for the steady-state probabilities:

$$(\lambda_j + \mu_j)p_j = \lambda_{j-1}p_{j-1} + \mu_{j+1}p_{j+1} \tag{10.32}$$

We can now understand why these are called the *balance equations* for the system. On the left, $\lambda_n + \mu_n$ is the rate of leaving state n, while p_n is the proportion of time (under steady-state conditions) during which the system is in state n. Thus the product on the left is the long-term *average* rate of leaving state n. By the same analysis, the term $\lambda_{n-1}p_{n-1}$ on the right in equation (10.32) is the long-term average rate of entering state n from state $n-1$ due to a birth, while the term $\mu_{n+1}p_{n+1}$ is the long-term average rate of entering state n from state $n+1$ due to a death. Thus, equation (10.32) is the statement that

$$\text{Average rate of leaving state } n = \text{Average rate of entering state } n \tag{10.33}$$

under the assumption that the system has reached a steady-state mode of operation.

Remark In studying the steady-state behavior of ergodic birth–death systems, it is possible to start with equation (10.33) as the basic heuristic principle, and then to derive the balance equations directly from this principle. This approach bypasses the difficulties of Sections 9.4–9.5 and 10.2–10.3. For the rest of the book, we shall, in effect, follow this procedure. The main point of the discussion at the beginning of this section was to connect time averages with probabilities. (See the discussion of relative frequencies versus probabilities in Section 2.1 and the treatment of discrete time averages in Section 6.9.)

10.6 Steady-State Analysis of the $M/M/1$ Queue

We now examine in detail the steady-state behavior of the $M/M/1$ queue model introduced in Section 8.7. In this case, the birth rate $\lambda_n = \lambda$ is the customer arrival rate, and the death rate $\mu_n = \mu$ is the service rate. By formula (10.19),

$$p_n = \rho^n p_0$$

where $\rho = \lambda/\mu$ is the arrival rate/service rate ratio, which is often called the **traffic intensity** of the system. Notice that $1/\mu$ is the average service time, and that $1/\lambda$ is the average time between arrivals. Thus, we can also express the parameter ρ as follows:

$$\rho = \frac{\text{Average service time}}{\text{Average time between arrivals}}$$

For the existence of steady-state probabilities, the geometric series

$$\sum_{n=0}^{\infty} \rho^n$$

must converge. This will only happen if $\rho < 1$; in other words, the arrival rate must be strictly less than the service rate. When this is so,

$$p_0 = \left(1 + \sum_{n=1}^{\infty} \rho^n\right)^{-1} = 1 - \rho$$

by the formula for the sum of a geometric series. The steady-state probabilities are then

$$p_n = \rho^n(1 - \rho)$$

Thus *the proportion of time the system spends in each state has a modified geometric distribution*, under steady-state conditions (see Example 3.4).

$M/M/1$ Queue with capacity N

Consider the single-server exponential queueing model in which the queue length is restricted. Whenever there are N customers in the system, then the queue is full and any new arrivals are turned away.

Let the arrival rate be λ when there are fewer than N customers in the system, and let the service rate be μ. Just as for the $M/M/1$ queue we have

$$p_n = \rho^n p_0$$

when $n \leq N$, where $\rho = \lambda/\mu$. But now $p_n = 0$ if $n > N$, so the formula for p_0 becomes

$$p_0 = \left(1 + \sum_{n=1}^{N} \rho^n\right)^{-1} = \frac{1 - \rho}{1 - \rho^{N+1}}$$

by the formula for the sum of a finite geometric series. Thus

$$p_n = \rho^n \left(\frac{1 - \rho}{1 - \rho^{N+1}}\right)$$

for $0 \leq n \leq N$. This is a **truncated geometric distribution**.

Notice that, by the description of steady-state probabilities in terms of time averages,

$$\begin{aligned} p_0 &= \text{Proportion of time the server is idle} \\ p_N &= \text{Proportion of time the system is full} \end{aligned}$$

From the previous formula, we see that, if $r \approx 0$ (service rate much faster than arrival rate), then $p_0 \approx 1 - \rho$, so the system is empty most of the time. On the other hand, if ρ is very large (service rate much slower than arrival rate), then

$$p_N = \frac{1 - \rho^{-1}}{1 - \rho^{-N-1}} \approx 1 - \frac{1}{\rho}$$

so the system is full most of the time. This is in accordance with the way we would expect such a system to behave.

The Arriving Customer's View

For any queueing system with Poisson arrivals, if we know that k customers arrived in a particular time interval, then the customer arrival times are uniformly distributed over this interval (see Example 7.5). Thus, we can also describe the steady-state probabilities from the point of view of the arriving customer:

$$p_n = \text{ Proportion of arriving customers who find system in state } n$$

Thus, p_n can also be interpreted as the probability that (under steady-state conditions) an arriving customer will find n customers already in the system. This property is often phrased as "Poisson arrivals see time averages"—the PASTA principle (see Example 7.6). In particular, this property leads to another interpretation of p_N for the $M/M/1$ queue with capacity N that is very important in practical applications:

$$p_N = \text{ Proportion of arriving customers who are turned away}$$

10.7 Machine-Repair Model

Assume that a factory contains M machines and s repairmen (or repair teams), with $s < M$; the machines break down at an exponential rate λ, and repairs are made at an exponential rate μ, as in Example 9.5. Let $\rho = \lambda/\mu$ be the ratio of breakdown rate to repair rate. Recall that we have defined the *state* of the system to be the number of machines that are *not* in working order.

From the formulas for the birth and death rates derived in Section 9.3 and formula (10.19), we find that the birth–death ratios for this system are

$$\rho_n = \binom{M}{n} \rho^n \quad \text{if } n \leq s \tag{10.34}$$

and

$$\rho_n = \binom{M}{n} \frac{n!s^s}{s!} \left(\frac{\rho}{s}\right)^n \quad \text{if } s < n \leq M \tag{10.35}$$

Once the values of the parameters ρ, s, and M are specified, we can calculate the steady-state probabilities by the formulas

$$\begin{aligned} p_0 &= \left(1 + \rho_1 + \rho_2 + \cdots + \rho_M\right)^{-1} \\ p_n &= \rho_n \, p_0 \end{aligned}$$

As an application of this analysis, we shall determine how efficiently the repair system functions. Assuming that breakdowns and repairs are unavoidable, we observe that inefficiency in the system arises in two ways:

a. Broken machines sit idle waiting to be repaired because all of the repairmen are busy, and as a result some of the productive capacity of these machines is lost.

b. During the time periods when there are more repairmen than broken machines, some of the repairmen are idle and the cost of their labor is lost.

To measure losses of type (a), we define the *coefficient of loss for machines* to be

$$C_{\text{Machines}} = \frac{\text{Average number of machines waiting for repairs}}{\text{Total number of machines}}$$

We calculate the average, using the steady-state probabilities, as

$$C_{\text{Machines}} = \frac{1}{M} \sum_{n=s+1}^{M} (n-s)p_n$$

(Notice that, if there are n machines broken but only s repairmen, with $s < n$, then there are $n - s$ machines in the repair queue.)

To measure losses of type (b), we define the *coefficient of loss for repairmen* to be

$$C_{\text{Repairmen}} = \frac{\text{Average number of idle repairmen}}{\text{Total number of repairmen}}$$

Once again, the average can be calculated by using the steady-state probabilities, and the formula here is given by

$$C_{\text{Repairmen}} = \frac{1}{s} \sum_{n=0}^{s-1} (s-n)p_n$$

(Notice that, when there are n machines broken, then there are $s - n$ repairmen idle, provided that $n < s$.)

The actual calculation of these coefficients of loss is elementary and easily programmed. Extensive tables (for ranges of values of ρ, s, and M) have been published from which the operating characteristics of different working conditions can determined.

As a representative numerical example, if $\rho = 0.8$, $M = 3$, and $s = 1$, we calculate that

$$C_{\text{Machines}} = 0.32 \qquad C_{\text{Repairmen}} = 0.10$$

If the breakdown and repair times remain the same, but the number of machines and repairmen is doubled to $M = 6$ and $s = 2$, then the calculation yields

$$C_{\text{Machines}} = 0.29 \qquad C_{\text{Repairmen}} = 0.05$$

This is obviously a more efficient arrangement. The loss for machines has decreased slightly, and the loss for repairmen is only half what it was before.

Exercises

1. Consider a birth–death process with finite capacity N. Write the normalized solution to the balance equations as a row vector

$$\mathbf{p} = [\; p_0 \quad p_1 \quad p_2 \quad \cdots \quad p_N \;]$$

   a. Show that the system of balance equations (10.16) is equivalent to the single vector-matrix equation $\mathbf{p}\,\mathbf{A} = 0$, where $\mathbf{A}$ is the matrix in Example 10.2.

b. Suppose that the system is started up at $t = 0$, with the initial probability distribution $\mathbf{q}(0) = \mathbf{p}$. Use part (a) and the Kolmogorov backward equations (Chapter 9, Exercise 8) to show that $\mathbf{q}(t) = \mathbf{p}$ for all $t \geq 0$ in this case.

2. Consider an $M/M/1$ queueing system with capacity $N = 1$, arrival rate $\lambda = 0.2$ customers per minute, and service rate $\mu = 0.3$ customers per minute. Let $X(t)$ be the number of customers present at time t.

   a. Suppose that $X(0) = 0$. Calculate $P\{X(1) = 0\}$ and $P\{X(10) = 0\}$.

   b. Suppose that $X(0) = 1$. Calculate $P\{X(1) = 0\}$ and $P\{X(10) = 0\}$.

   c. Could you have predicted the difference between the answers in parts (a) and (b) by the general theory of limiting probabilities?

3. (Continuation of Exercise 2) Suppose that $X(0) = 0$. How large must t be so that the probability of finding no customer in the system at time t is within 10% of the steady-state value?

4. (Continuation of Exercise 2) Suppose that $X(0) = 0$. How large must T be so that the time-average

$$\frac{1}{T}\int_0^T P\{X(t) = 0\}\,dt$$

   is within 10% of the steady-state probability p_0?

5. Consider an $M/M/1$ queueing system with finite capacity $N = 2$. Let the matrix $\mathbf{A}$ be as in Example 10.2, with $N = 2$, $\lambda_0 = \lambda_1 = \lambda > 0$, and $\mu_1 = \mu_2 = \mu > 0$.

   a. Find the characteristic polynomial of $\mathbf{A}$.

   b. Show that the eigenvalues of $\mathbf{A}$ are $0, -\lambda-\mu-(\lambda\mu)^{1/2}$, and $-\lambda-\mu+(\lambda\mu)^{1/2}$. Prove directly that this last eigenvalue is negative.

6. Consider the linear growth model with immigration (Chapter 9, Exercise 2). Show that series (10.21) converges, and hence that steady-state probabilities exist, if the birth rate λ is less than the death rate μ.

   (HINT: Use the ratio test.)

7. Consider a birth–death process with finite capacity N. Assume that all of the birth rates $\lambda_j \neq 0$, for $0 \leq j < N$ (this means that state N can be reached from state j for every $j < N$).

   a. Express the steady-state probabilities p_j in terms of p_N.

   b. Find the normalization equation for p_N.

8. Consider a birth–death process with finite capacity $N = 2$. Assume that the birth rate is λ and that the death rate is μ when the population size is 1. When the population size is 0 or 2, then the birth and death rates are both zero.

   a. Draw the transition diagram, and write down the balance equations for this system.

b. Let $0 \leq p \leq 1$, and set $q = 1 - p$. Show that *every* probability distribution of the form $[p \;\; 0 \;\; q]$ satisfies the balance equations (so, in this case, the steady-state probabilities for the system are *not* unique).

9. Derive formulas (10.34 and (10.35).

Assume that steady-state conditions apply in Exercises 10 through 15.

10. A gasoline station has one pump. Potential customers go by at a Poisson rate of 20 cars per hour. If there are 2 or fewer cars already at the station, the next arriving customer stops for gas; otherwise, the customer does not stop at the station. Assume that the amount of time required to service a car is exponentially distributed with a mean of 5 minutes.

    a. During what proportion of the time is the attendant busy servicing cars?

    b. What proportion of potential customers enter the station for service?

    c. If the service times were twice as fast, how many more customers per hour would enter the station, on average?

11. A printing shop has four presses and two repairmen. The amount of time a printing press works before needing service is exponentially distributed, with a mean of 10 hours. Suppose that the amount of time it takes a single repairman to fix a machine is exponentially distributed, with a mean of 8 hours.

    a. What is the average number of machines not in use, and what is the coefficient of loss for machines?

    b. During what proportion of time are both repairmen busy, and what is the coefficient of loss for repairmen?

12. Suppose that the work assignments of the repairmen in Exercise 11 are changed so that each repairman has exclusive responsibility for two machines. Calculate the coefficient of loss for machines and the proportion of time each repairman is busy under these conditions. Compare these results with those of Exercise 11.

13. Consider an $M/M/1$ queueing system with total capacity $N = 2$. Suppose that customers arrive at the rate of λ per hour and are served at the rate of 8 per hour.

    a. What should the arrival rate be so that an arriving potential customer has a 50% chance of joining the queue?

    b. With λ chosen to satisfy the requirement of part (a), what percentage of the customers who actually enter the system get served immediately?

14. A bottling plant has one repairman and three machines. Each machine randomly breaks down at an exponential rate of two times every hour, and the repairman fixes them at an exponential rate of three machines per hour. Whenever all three machines are out of order, however, the plant manager assists the repairman, and the service rate becomes four machines per hour.

    a. Let the state of the system be the number of machines that are broken. Give the transition diagram for the system, and calculate the steady-state probabilities for the system.

b. Calculate the average number of machines that are broken and the average rate at which machine breakdowns occur under steady-state conditions.

15. Consider a singe-server queueing system in which the service times are exponential at rate 1 per minute. Suppose that the potential customers arrive at a Poisson rate of λ per minute. However, if a potential customer finds, upon arriving, that there are already n customers in the system, the probability is $1/(n+1)$ that the customer leaves immediately. (This is a model for **discouraged arrivals**. Notice that, the longer the queue, the greater the probability that an arriving customer won't wait.)

    a. Give the state-transition diagram, and calculate the steady-state probabilities for this system (notice that the queue length is unlimited).

    b. What is the average number of customers in the system, under steady-state conditions?

Chapter 11

General Queueing Systems

We return to considering a general queueing system, as described in Section 8.6. Without making any specific assumptions about the randomness of the arrivals of customers or the random distribution of service times, we introduce several random variables that measure the overall performance of the system. We derive Little's formula, relating average customer waiting time to average number of customers. This serves as a general tool for comparing the cost of customer waiting time against the cost of providing the service.

In Chapter 10, we obtained steady-state probabilities for birth–death processes via balance equations, and we interpreted these probabilities as time averages. In this chapter, we apply the same method to analyze more complicated exponential queueing models involving service in stages.

At the end of the chapter, we apply the theoretical queueing results to the problem of adjusting the parameters in a queueing system for *optimal* performance. Of course, there will usually be constraints (limitation on number of servers, arrival rates, service rates, queue capacity, and so on), and the criterion for optimality must be specified. We conclude with some examples of queue simulations, including models whose analytic solution we will obtain in Chapter 12.

11.1 Long-Term Average Queue Characteristics

Recall from Section 8.6 that a general queueing system with one service facility can be described in a schematic way by the diagram shown in Fig. 11.1. We shall look for ways to measure the overall performance of such a system. There are two conflicting points of view in a queueing system: the customer's and the server's. We shall analyze the operation of the system from both points of view.

Suppose we have a queueing system that settles down to a steady-state behavior after the initial start-up period, as described in Section 10.1. So far, we have concentrated on the random variable

$$L(t) = \text{Number of customers in system at time } t$$

(For consistency with the queueing literature, we now use $L(t)$ to replace the earlier notation $X(t)$.) For a more detailed understanding of the functioning of the system,

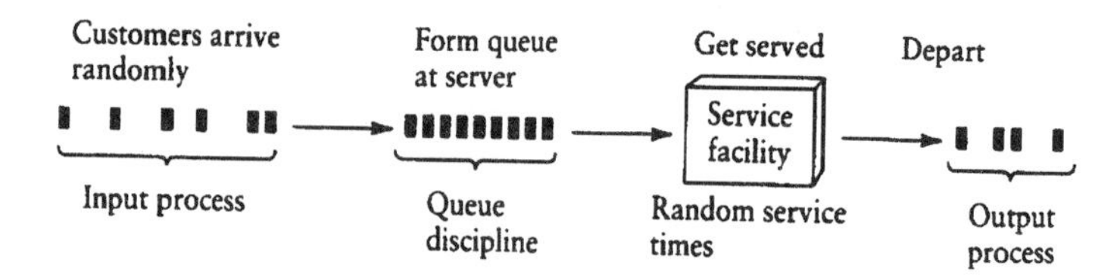

Figure 11.1: General Queueing System

we also need to know the following random variables:

$$\begin{aligned}
A(t) &= \text{Number of arrivals in interval } (0,t] \\
D(t) &= \text{Number of departures in interval } (0,t] \\
S(t) &= \text{Total time spent in system by } \textit{all} \text{ customers during } (0,t] \\
W(t) &= \text{Average time in system per customer during } (0,t] \\
L_Q(t) &= \text{Number of customers in queue at time } t \\
W_Q(t) &= \text{Average waiting time in queue per customer during } (0,t]
\end{aligned}$$

Notice that, by definition,

$$W(t) = \frac{S(t)}{A(t)}$$

Just as in Section 10.5, we can calculate the long-term time averages of these random variables:

Long-term average number of customers in system:

$$L = \lim_{T\to\infty} \frac{1}{T} \int_0^T L(t)\,dt \tag{11.1}$$

Long-term average arrival rate:

$$\lambda_a = \lim_{T\to\infty} \frac{A(T)}{T}$$

Long-term average time in system per customer:

$$W = \lim_{T\to\infty} W(T)$$

There are corresponding long-term averages L_Q and W_Q for the queue statistics.

We shall assume that the system has a steady-state behavior as $t \to \infty$ such that the limits just described exist. We also assume that the effects of the particular start-up conditions of the system at $t = 0$ disappear after a long time (see Section 10.3).

It is important to notice that, in calculating a long-term time average of some random variable such as $L(t)$, we could start sampling $L(t)$ at any fixed time $s \geq 0$ and still get the same limiting value:

$$L = \lim_{T\to\infty} \frac{1}{T} \int_s^{s+T} L(t)\,dt \tag{11.2}$$

We can see heuristically why the limits in equations (11.1) and (11.2) should be the same by the following argument. The difference between the integral in equation (11.1) and the integral in equation (11.2) is the sum of two integrals—one over $(0, s)$ and the other over $(T, T + s)$. The length of each interval of integration is the fixed number s; when we divide by T and let $T \to \infty$, the difference between the two integrals goes to zero. (This heuristic argument assumes that the integrals of $L(t)$ over a time interval of fixed length are bounded, which we anticipate is true for a system with steady-state behavior.) Thus L should only depend on the long-term behavior of the system.

We now assume that *any random quantity that only depends on the limiting behavior of the system as $t \to \infty$ is constant.* We have already used this principle—the ergodic hypothesis—in the derivation of equation (10.25) in Section 10.5. (The discrete-time version for regular Markov chains was proved in Section 6.9.) If this assumption is valid, then all the long-term averages such as L and W that have appeared previously are constants, rather than random variables. By running a long simulation of the system, we can estimate these quantities, as we shall illustrate in Section 11.8.

When the steady-state probabilities $\{p_n\}$ for having n customers in the system can be found by analytical methods, as we did in Section 10.4 for birth–death processes, then L can be calculated from the formula

$$L = \sum_{n=0}^{\infty} n \, p_n \tag{11.3}$$

This follows by taking the expectation of the right side of equation (11.1) and applying equation (10.31). (This argument assumes that, as $T \to \infty$, the limit of the expectation is the expectation of the limit.) We shall carry out the calculation of formula (11.3) for several examples later in the chapter.

We already know what the long-term average arrival rate is in many cases. For example, when the customers arrive at the system according to a Poisson process with rate λ, then

$$\lambda = \lambda_a$$

since $E[A(t)] = \lambda t$ in this case. For a general birth–death process, in which the arrival rate is λ_n when the system is in state n, the long-term average arrival rate is

$$\lambda_a = \sum_{n=0}^{\infty} \lambda_n \, p_n \tag{11.4}$$

This follows from the interpretation of p_n as the long-term proportion of time during which the system is in state n (see Section 10.5).

11.2 Little's Formula

A basic result in queueing theory is that, once we know two of the three quantities L, λ_a, and W, we can calculate the third, by Little's Formula:

$$L = \lambda_a W \tag{11.5}$$

To prove Little's Formula, we concentrate on the number of arrivals $A(t)$ and the number of departures $D(t)$ up to time t. Observe that $D(t) \leq A(t)$, and $L(t) = A(t) - D(t)$. Suppose that we simulate the system over some time interval and generate sample values for the random variables. Instead of drawing the sample graph of $L(t)$, as we did in Section 8.6, let's plot the sample graphs of both $A(t)$ and $D(t)$ on the same axes, as shown in Figure 11.2. At any time t, $L(t)$ is the difference of the ordinates of these two graphs.

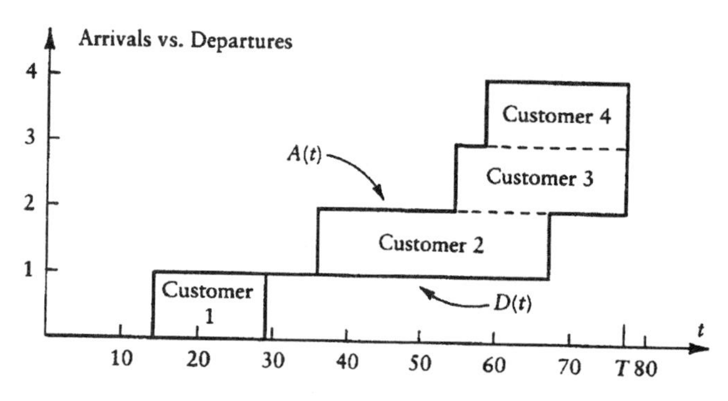

Figure 11.2: Total Time in System

In Figure 11.2, we have divided the area between the graphs up to time T into rectangles of height 1, with one rectangle for each customer that has arrived. Thus there are a total of $A(T)$ rectangles. Next, we observe that

$$\text{Area of each rectangle} = \text{Time spent in system by a customer}$$

Thus the total time spent in the system by all customers up to time T, which we denote by $S(T)$, is the area between the arrival and departure graphs, up to time T.

We have another way of calculating this area: integrate the sample function $L(t)$ between 0 and T. Since total customer time divided by the number of customers is the average time per customer, we obtain the equation

$$W(T) = \frac{S(T)}{A(T)} = \frac{1}{A(T)} \int_0^T L(t)\, dt$$

Thus, the long-term average system time per customer can be calculated by the formula

$$W = \lim_{T \to \infty} \frac{1}{A(T)} \int_0^T L(t)\, dt$$

But $A(T)/T \to \lambda_a$ as $T \to \infty$, by definition of the average arrival rate. Using this in the preceding formula, we get

$$W = \frac{1}{\lambda_a} \left(\lim_{T \to \infty} \frac{1}{T} \int_0^T L(t)\, dt \right) = \frac{L}{\lambda_a}$$

This is Little's formula.

If we apply the same arrival-departure argument to the queue, we obtain the analogous relation

$$L_Q = \lambda_a W_Q \tag{11.6}$$

for the queue statistics.

11.3 Steady-State Behavior of $M/M/1$ Queue

In this section we use Little's formulas (11.5) and (11.6) to analyze the $M/M/1$ queue from two points of view: the customer's and the server's.

Recall that

$$\begin{aligned} \lambda &= \text{Arrival rate} \\ \mu &= \text{Service rate} \\ \rho &= \lambda/\mu = \text{Utilization factor} \end{aligned}$$

We assume that $\rho < 1$, so that steady-state conditions will occur. We already calculated, in Section 10.6, that the steady-state probabilities are

$$p_n = \rho^n(1-\rho) \quad \text{for } n = 0, 1, \ldots$$

Notice that

$$1 - \rho = p_0 = \text{Proportion of time the system is empty}$$

Thus,

$$\rho = \text{Proportion of time the server is busy}$$

which explains why ρ is called the **utilization factor**.

The steady state probabilities p_n are the same as the probability mass function for a modified geometric random variable with parameter $p = 1 - \rho$. Hence, as we calculated in Section 3.2,

$$\begin{aligned} L &= \sum_{n=0}^{\infty} n p_n = \frac{\rho}{1-\rho} \\ &= \frac{\lambda}{\mu - \lambda} \end{aligned} \tag{11.7}$$

Since the arrivals are Poisson, with rate $\lambda = \lambda_a$, we have, by Little's formula,

$$W = \frac{\lambda}{L} = \frac{1}{\mu - \lambda} \tag{11.8}$$

Since the average service time is $1/\mu$, it follows that the average time per customer in the queue is

$$W_Q = W - \frac{1}{\mu} = \frac{\rho}{(1-\rho)\mu} \tag{11.9}$$

Thus, by Little's formula again, the average number of customers in the queue is

$$L_Q = \lambda W_Q = \frac{\rho^2}{1-\rho} \tag{11.10}$$

Notice that L and L_Q depend only on ρ, and they become infinitely large as $\rho \nearrow 1$.

How does the behavior of this system change if there is a constraint on the length of the queue? Suppose that *at most* N customers can be present in the system at any time ($N-1$ in line, and 1 at the server). Recall from Section 10.6 that the steady state probabilities now are

$$p_n = \frac{\rho^n(1-\rho)}{1-\rho^{N+1}} \qquad \text{for } 0 \le n \le N$$

So for this system we have

$$L = \frac{1-\rho}{1-\rho^{N+1}} \sum_{n=0}^{N} n\rho^n$$

We can evaluate this last sum by the same trick that we used in Section 3.2 for the geometric distribution:

$$\sum_{n=0}^{N} n\rho^n = \rho \frac{d}{d\rho} \sum_{n=0}^{N} \rho^n = \rho \frac{d}{d\rho} \left(\frac{1-\rho^{N+1}}{1-\rho} \right)$$

Calculating the derivative and substituting in the preceding equation, we find that

$$L = \left(\frac{\rho}{1-\rho} \right) \left(\frac{N\rho^{N+1} - (N+1)\rho^N + 1}{1-\rho^{N+1}} \right) \tag{11.11}$$

Formula (11.11) is valid for any $\rho \neq 1$. When $\rho < 1$, then the first factor in the formula is the value of L for the queue with unlimited capacity. The second factor approaches the limit 1 as $N \to \infty$. Thus the finite-capacity system converges toward the infinite-capacity system as the constraint is removed, provided that the infinite-capacity system has a steady-state limit.

At the other extreme, if N is fixed and ρ gets large (very slow service relative to arrival rate), then

$$L \approx N \qquad \text{when } r \to \infty$$

This simply says the system is essentially full, under steady-state conditions, when ρ is large.

To apply Little's formula, we must determine the average arrival rate λ_a of the system. For this, we argue as follows:

$$1 - p_N = \text{Proportion of time the system is not full}$$

Over a time interval $(0, T)$, the conditional distributions of the arrival times of the customers are uniform (see Example 7.5). Hence,

$$1 - p_N = \text{Proportion of Poisson arrivals who join queue}$$

(this is an application of the PASTA principle). It follows that

$$\lambda_a = (1 - p_N)\lambda$$

This result could also be obtained by formula (11.4), since we know that the arrival rates are $\lambda_n = \lambda$ for $n < N$ and $\lambda_n = 0$ for $n \geq N$. Thus, by Little's formula,

$$W = \frac{L}{(1 - p_N)\lambda} = \frac{1 - \rho^{N+1}}{1 - \rho^N} \cdot \frac{L}{\lambda} \tag{11.12}$$

To calculate the average time in the queue and average length of the queue, we proceed as before:

$$\begin{aligned} W_Q &= W - \frac{1}{\mu} \\ L_Q &= \lambda_a W_Q = L - \frac{\lambda_a}{\mu} \end{aligned}$$

In particular, the formula for L_Q shows that the average number of customers at the server is λ_a/μ, since this is the difference between L and L_Q. This quantity is the same as the utilization of the server, which (for the case of unlimited queue capacity) equals ρ. Taking into account the formula above for λ_a, we find that

$$\text{Utilization of server} = \rho \cdot (1 - p_N)$$

in the case of queue capacity N.

11.4 Service in Stages

In this section, we examine two systems in which customers have to visit more than one server before departing. This is a very common situation. For example, at a warehouse retail outlet, customers queue at the cashier's desk to pay for purchases, then queue at the loading dock to pick up purchases. At a time-sharing computer center, programs to be run queue up at the central processing unit and then queue again at the printer or disk storage device. We can indicate such a system schematically as shown in Figure 11.3. We shall assume that these systems can be treated as *exponential models*, in the sense of Section 8.7.

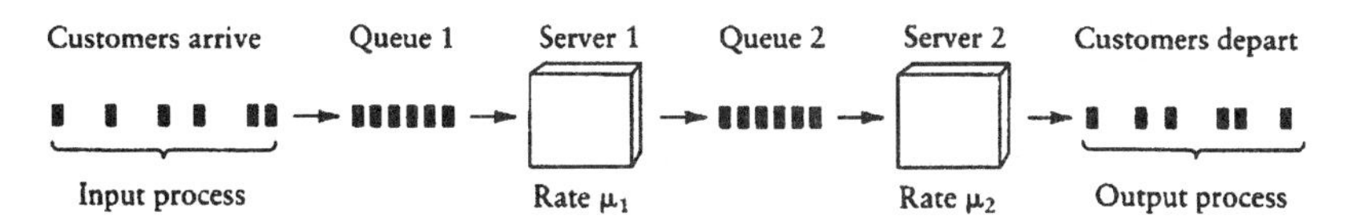

Figure 11.3: Service in Stages

Example 11.1 Car Wash and Vacuum Model

Consider a do-it-yourself car wash and vacuum facility. Assume that the time to wash a car is exponentially distributed, with rate μ_1 per hour, and that the time to vacuum the interior of a car is also exponentially distributed, with rate μ_2 per hour. There is only space for one car at a time at the facility (no space for a queue). Potential customers arrive according to a Poisson process with rate λ per hour. If the facility is empty, the customer enters, washes the car, vacuums it, and leaves. If the facility is busy, the potential customer's business is lost. In terms of the diagram above, the capacity of the queues is zero. Furthermore, entrance to server 1 is closed while server 2 is busy.

We can describe this system as an exponential model with three states:

$$\begin{aligned} \text{State 0} &= \text{System is empty} \\ \text{State 1} &= \text{Car is being washed} \\ \text{State 2} &= \text{Car is being vacuumed} \end{aligned}$$

The system stays in each state for an exponentially distributed length of time, and then jumps to another state. From the description of the system we obtain the transition diagram shown in Figure 11.4.

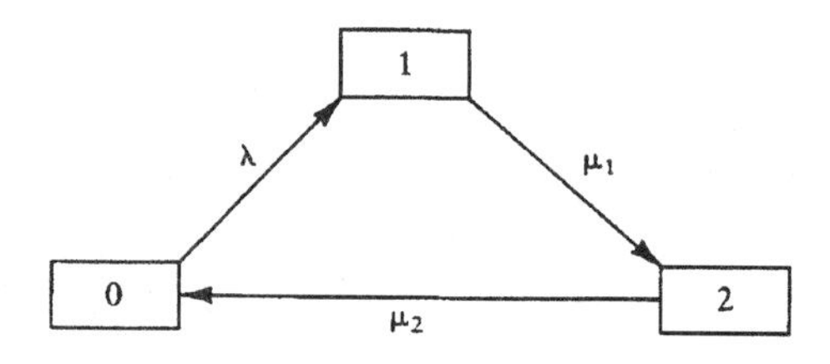

Figure 11.4: Wash and Vacuum Model

Any such exponential model has the *Markov* (memoryless) property: we can predict (probabilistically) the future states of the system from our knowledge of the current state. By the same method used in Sections 9.4 and 9.5, we can derive a system of differential equations (Kolmogorov equations) for the time-dependent transition probabilities. Assuming that the system settles down to steady-state behavior as $t \to \infty$, we then obtain a set of balance equations for the steady-state probabilities by equating the derivatives to zero in the Kolmogorov equations. The resulting equations are of the form

$$\text{Rate of leaving each state} = \text{Rate of entering each state} \tag{11.13}$$

Since the derivation of these results follows the same pattern as the derivation for the birth-death processes (which we treated in detail in Chapter 9 and 10), we shall not repeat it here.

Given the general principle of equation (11.13), we can use the transition diagram to write down the balance equations for the steady-state probabilities p_0, p_1, p_2 in this example:

STATE	RATE OF LEAVING	=	RATE OF ENTERING
0	λp_0	=	$\mu_2 p_2$
1	$\mu_1 p_1$	=	λp_0
2	$\mu_2 p_2$	=	$\mu_1 p_1$

From these equations, we have $p_1 = (\lambda/\mu_1)p_0$ and $p_2 = (\lambda/\mu_2)p_0$. Since the probabilities sum to 1, we can solve for p_0:

$$p_0 = \left(1 + \frac{\lambda}{\mu_1} + \frac{\lambda}{\mu_2}\right)^{-1}$$

Now, using the results of Sections 11.1 and 11.2, we can determine the operating characteristics of the system. For example, since p_0 is the proportion of potential customers who actually enter the system, the effective average arrival rate is

$$\lambda_a = \lambda p_0$$

The average number of customers in the system is

$$L = p_1 + p_2.$$

If each customer pays \$$D$ to use the facility, the rate of return from the facility is $D \cdot \lambda_a$ (\$ per hour). The average time spent in the system per customer is obviously

$$W = \frac{1}{\mu_1} + \frac{1}{\mu_2}$$

Little's formula tells us that $L = \lambda_a \cdot W$. In this example, we can also verify this relation directly from the equations just derived. Notice, too, that

$$\begin{aligned} p_0 + p_2 &= \text{Proportion of time wash machine is idle} \\ p_0 + p_1 &= \text{Proportion of time vacuum machine is idle} \end{aligned}$$

■

Example 11.2 Enlarged Wash and Vacuum Model

To increase the rate of return on their investment, the owners of the wash-and-vacuum facility in Example 11.1 plan to increase the utilizaton of the machinery by enlarging the customer space at the facility. How can they estimate how much additional business such an expansion will generate?

Suppose that the enlarged facility will have capacity for two cars at a time: one at the washing location, and the other at the vacuuming location. Assume that there is no possibility of creating queueing space, and that the washing machinery must be used before the vacuum. In such a setup, an arriving customer can enter the system if the wash machine is free. After finishing the wash, the customer waits at that location until the vacuum is free (thereby blocking access to all potential customers during this period). Then the customer goes to the vacuum and completes the service.

In setting up an exponential model for this system, the crucial step is to choose an appropriate state space, so that the time spent in each state is an exponential random variable (it will be the *minimum* of several exponential random variable describing the arrival and service times). In this example, we need to know more than just the number of customers in the system at any time to obtain this property.

For example, if we want to describe the system at a moment when Ann is at the wash and Bob is at the vacuum, we must also know whether Ann is still using using the wash facility (in which case her remaining wash time is an exponential random variable with rate μ_1), or has finished the wash and is waiting for Bob to finish so that she can move to the vacuum location (in which case her remaining time at the wash station is an exponential random variable with rate μ_2, by the memoryless property of the vacuum time.) That is, if Ann has finished the car wash and is waiting for Bob to finish at the vacuum, it inevitably appears to her that Bob *started* using the vacuum precisely when she *finished* the wash.

We can distiguish between the two cases just described by defining the states of the system as follows:

STATE	DESCRIPTION
$(0,0)$	System empty
$(1,0)$	Car wash busy, vacuum idle
$(0,1)$	Car wash empty, vacuum busy
$(1,1)$	Car wash busy, vacuum busy
$(b,1)$	Vacuum busy, car wash idle but blocked by customer waiting for vacuum

With this choice of state space, we clearly have an exponential model, by the preceding discussion. The transition diagram for the system is shown in Figure 11.5. The arrows in the diagram are labeled with the (conditional) rates at which the transitions occur. For example, the transitions $(0,1) \rightarrow (0,0)$, $(1,1) \rightarrow (1,0)$, and $(b,1) \rightarrow (0,1)$ all have rate μ_2 because they involve the completion of service at the vacuum station. There is no transition $(1,0) \rightarrow (1,1)$ because customers must use the wash facility before they use the vacuum facility.

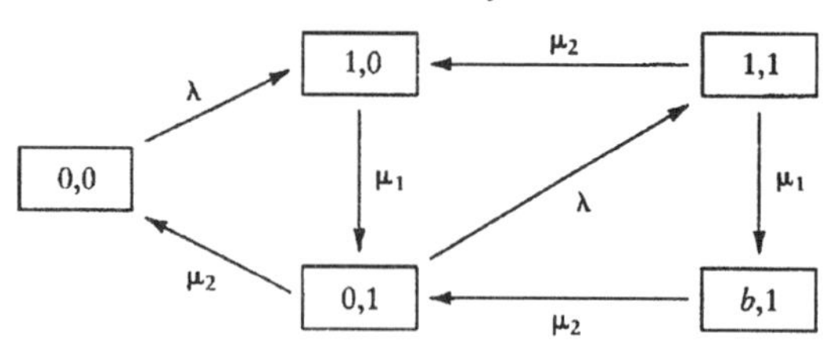

Figure 11.5: Enlarged Wash and Vacuum Model

We shall label the steady-state probabilities according to the corresponding state. From the transition diagram and the balance principle expressed in equation (11.13) we can write down the balance equations. Each rate in the diagram is multiplied by its corresponding steady-state probability (= proportion of time) to get the effective rates for entering and leaving the states of the system:

STATE	RATE ENTERING	=	RATE LEAVING
$(0,0)$	$\mu_2 p_{01}$	=	λp_{00}
$(1,0)$	$\mu_1 p_{10} + \mu_2 p_{b1}$	=	$(\mu_2 + \lambda) p_{01}$
$(0,1)$	$\lambda p_{00} + \mu_2 p_{11}$	=	$\mu_1 p_{10}$
$(1,1)$	λp_{01}	=	$(\mu_1 + \mu_2) p_{11}$
$(b,1)$	$\mu_1 p_{11}$	=	$\mu_2 p_{b1}$

This is a system of five homogeneous equations in five unknowns, but it only has rank 4 (the sum over all five equations of the coefficients of any particular p_{ij} is zero). If we add the equation

$$p_{00} + p_{10} + p_{01} + p_{11} + p_{b1} = 1$$

then we have a system of full rank 5, which can be easily solved by Gaussian elimination once the arrival and service rates are specified (see the exercises.)

We can now use this solution to analyze the operating characteristics of the enlarged facility. The average number of customers in the system is

$$L = p_{01} + p_{10} + 2(p_{11} + p_{b1}).$$

(Notice that in states p_{11} and p_{b1}, two customers are in the facility.) The proportion of time that the system is available to new customers is $p_{00} + p_{01}$, so the average arrival rate is

$$\lambda_a = (p_{00} + p_{01}) \cdot \lambda$$

(by the PASTA principle). If each customer pays $\$D$ to use the facility, then the rate of return is $D \cdot \lambda_a$ ($ per hour). The amount of time the wash machine is idle is now

$$p_{00} + p_{01} + p_{b1}$$

By taking particular values for μ_1 and μ_2, we can compare these results with those of Example 11.1 for a range of λ values, to see how much the yield has increased. (Notice that, in practice, λ might be difficult to determine, so we would like to analyze the system under varying arrival rates.) ■

11.5 Jackson's Theorem

In the two examples of service in stages in Section 11.4, the maximum queue lengths at each server were fixed. In this section we look at an example in which there is no restriction on queue length, and we find a remarkably simple description for the steady-state behavior of the system.

Example 11.3 $M/M/1$ Queues in Series

Customers arrive at a motor vehicle agency according to a Poisson process with rate λ. They queue up at server 1 to process the registration papers; after this processing is completed, the customers move to the queue at server 2 to pay for the license and pick up the vehicle plate. There is no limit to the queue lengths at either server, and the server times are exponentially distributed.

This system is described schematically by Figure 11.3. It consists of two single-server queues, in *series*, with the output from the first queue becoming the input to the second queue. We shall say that the system is in state (m, n) if there are m customers at server 1 (and so $m - 1$ in queue 1 when $m > 0$), and n customers at server 2 (and so $n - 1$ in queue 2 when $n > 0$). With this choice of state space, we have an exponential model—for the same reason as in Example 11.2. The transitions and their rates are as follows:

STATE TRANSITION			DESCRIPTION
(m, n)	$\rightarrow$	$(m+1, n)$	New customer arrives (rate λ)
(m, n)	$\rightarrow$	$(m-1, n+1)$	Customer leaves server 1, moves to queue 2 (rate μ_1 if $m > 0$)
(m, n)	$\rightarrow$	$(m, n-1)$	Customer finishes at server 2 and departs (rate μ_2 if $n > 0$)

The transition diagram and balance equations are somewhat complicated to write down, and of course we now have an infinite set of equations, one for each state. Part of the diagram is shown in Figure 11.6. ■

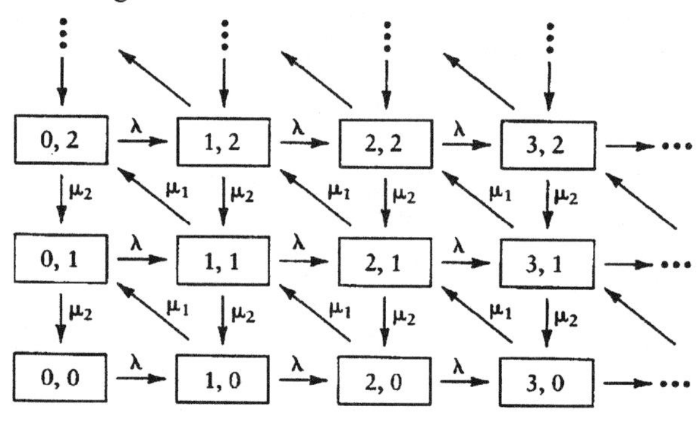

Figure 11.6: Service in Stages (Unlimited Queues)

A remarkable property of $M/M/1$ queueing systems in series is that the steady-state probabilities $p_{m,n}$ have a simple product form.

Theorem 11.1 (Jackson's Theorem). *Let $\rho_1 = \lambda/\mu_1$ and $\rho_2 = \lambda/\mu_2$. Assume that $\rho_1 < 1$ and $\rho_2 < 1$. Then*

$$p_{m,n} = (1 - \rho_1)\rho_1^m \cdot (1 - \rho_2)\rho_2^n$$

Thus, under steady-state conditions, a system with two $M/M/1$ queues in series behaves as if the two queues were independent *subsystems. In particular, the average number of customers in the system is*

$$L = L_1 + L_2 = \frac{\rho_1}{1 - \rho_1} + \frac{\rho_2}{1 - \rho_2}$$

and the average time per customer in the system is $W = L/\lambda$.

Proof When $m > 0$ and $n > 0$, the balance equation for state (m, n) is

$$(\lambda + \mu_1 + \mu_2) p_{m,n} = \lambda p_{m-1,n} + \mu_1 p_{m+1,n-1} + \mu_2 p_{m,n+1}$$

When $m = 0$ or $n = 0$, the equations have to be modified in an obvious way, since there are no transitions to states with m or n negative. The proof of the theorem consists in taking the proposed formula for $p_{m,n}$ as stated in the theorem, and then plugging it into the balance equations to see that it works. ■

Example 11.4 Steady-state Analysis

We now determine the steady-state statistics (average queue lengths, average customer times, and so on) for the system in Example 11.3. Suppose that people arrive at a rate of $\lambda = 7$ people per hour. Assume that the rate at the processing clerk is $\mu_1 = 8$ applications per hour, while the rate at the cashier is $\mu_2 = 21$ customers per hour. Thus the utilization factors at the two servers are $\rho_1 = \frac{7}{8}$ and $\rho_2 = \frac{1}{3}$. Since these are both less than 1, we can apply Jackson's Theorem.

According to Jackson's theorem, the steady-state marginal distribution of the number of people at each server is the same as for an $M/M/1$ queue with the given parameters. Thus, from Section 11.3, we find the average number of people at the processing clerk is

$$L_1 = \frac{7}{8-7} = 7$$

while the average number at the cashier is

$$L_2 = \frac{7}{21-7} = \frac{1}{2}$$

Since the arrival rate at both servers is 7 customers per hour, under steady state conditions the average amount of time (including queue time) that a customer spends at the registration clerk is

$$W_{\text{clerk}} = \left(\frac{1}{7}\right) \cdot L_1 = 1 \text{ hour}$$

The average time spent at the cashier is

$$W_{\text{cashier}} = \left(\frac{1}{7}\right) \cdot L_2 = \frac{1}{14} \text{ hour} \approx 4 \text{ minutes}$$

Clearly the bottleneck in the system is at the registration clerk! ■

11.6 $M/M/s$ Queue Statistics

In this section we make a detailed study of the multiserver $M/M/s$ queue system introduced in Example 9.4. Recall that, in this model, we assume Poisson arrivals of customers at rate λ, unlimited system capacity, and s servers ($1 \leq s < \infty$), each working independently at an exponential rate μ (the same rate for each server). The customers form a single queue, and the customer at the head of the queue goes to the first available server, as shown in Figure 11.7.

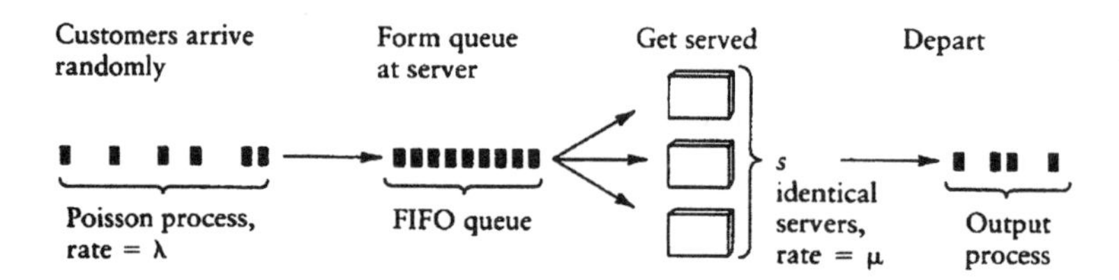

Figure 11.7: $M/M/s$ Queueing System

We worked out in Section 9.3 that the state-transition diagram for this system is the one given in Figure 11.8. Recall from the general results for birth–death processes (formulas (10.19) and (10.20) that the steady-state probability p_n of having n customers in the system is

$$p_n = \rho_n \cdot p_0 \tag{11.14}$$

where ρ_n is the product of the ratios

$$\frac{\text{Rate of entering state } k}{\text{Rate of leaving state } k}$$

for all the states $k \leq n$. From the transition diagram, we can see that these products are

$$\rho_n = \begin{cases} \dfrac{\alpha^n}{n!} & \text{if } n \leq s \\ \dfrac{\alpha^s}{s!} \cdot \rho^{n-s} & \text{if } n \geq s \end{cases} \tag{11.15}$$

Here, we have set

$$\alpha = \frac{\lambda}{\mu} = \frac{1/\mu}{1/\lambda} = \frac{\text{Average service time per customer}}{\text{Average time between arrivals}}$$

and

$$\rho = \frac{\lambda}{s\mu} = \frac{\text{Arrival rate of customers}}{\text{Maximum departure rate of customers}}$$

The parameter α is called the **traffic intensity**, and the parameter ρ is called the **utilization factor** of the system. (The justification for this terminology will become

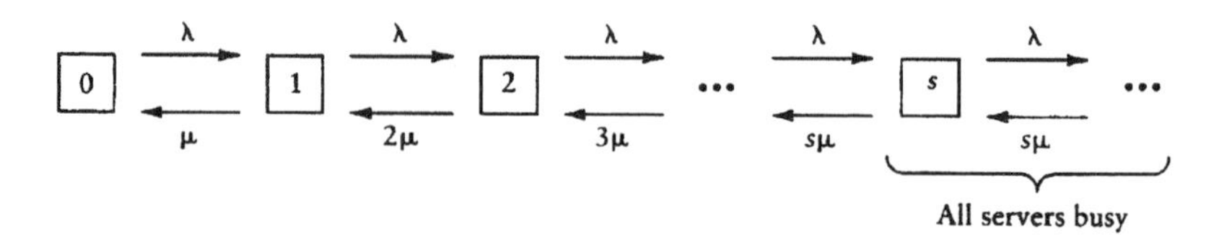

Figure 11.8: Transition Diagram for $M/M/s$ Queue

more apparent later in the section.) Of course $\alpha = s\rho$, so these parameters coincide for the single-server queue. Both parameters are *dimensionless* (they are ratios: (time)/(time)); however, it is conventional in engineering literature to measure α in *erlangs*, in honor of A. K. Erlang, a Danish mathematician who first studied queueing theory in 1909 in connection with telephone networks.

Recall that, in order for the steady-state probabilities to exist, the infinite series with terms $\{\rho_n\}$ must converge (formula (10.21)). Notice from formula (11.15) (or directly from the transition diagram) that, as soon as $n \geq s$, then

$$\frac{\rho_{n+1}}{\rho_n} = \rho$$

Thus, by the ratio test, the steady-state (stability) condition is $\rho < 1$; that is,

$$s\mu > \lambda \tag{11.16}$$

This is the requirement that the departure rate when all servers are busy must be greater than the arrival rate. From now on we shall assume that λ, s, and μ satisfy condition (11.16).

To complete the determination of the steady-state probabilities, we shall calculate p_0, as in Section 10.4, from the normalization condition

$$\frac{1}{p_0} = 1 + \sum_{n=1}^{\infty} \rho_n \tag{11.17}$$

We can express the answer in compact form by defining the **truncated exponential series** $e_m(x)$ to be the Taylor polynomial up to degree m of the function e^x:

$$e_m(x) = 1 + x + \frac{x^2}{2!} + \cdots + \frac{x^m}{m!} \tag{11.18}$$

From formulas (11.15) and (11.17), it follows that

$$\frac{1}{p_0} = e_{s-1}(\alpha) + \frac{\alpha^s}{s!}\sum_{n=s}^{\infty} \rho^{n-s}$$

The infinite geometric series on the right sums to $1/(1-\rho)$, so we obtain the formula

$$p_0 = \left(e_{s-1}(\alpha) + \frac{\alpha^s}{s!(1-\rho)}\right)^{-1} \tag{11.19}$$

This is the proportion of time during which the system is empty—that is, the proportion of time during which *all* servers are idle. (Notice that, if $s = 1$, then formula (11.19) reduces to $p_0 = 1 - \rho$, as we derived in Section 10.6). From formula (11.15), we then have

$$p_n = \begin{cases} \left(\dfrac{\alpha^n}{n!}\right) p_0 & \text{if } n \leq s \\ \rho^{n-s}\, p_s & \text{if } n > s \end{cases} \tag{11.20}$$

Observe that, for $n > s$, we are expressing p_n in terms of $p_s = (\alpha^s/s!)p_0$.

From formula (11.20), we can calculate the proportion of time during which all servers are busy:

$$\begin{aligned}\sum_{n=s}^{\infty} p_n &= \sum_{n=s}^{\infty} \rho^{n-s} p_s \\ &= \frac{1}{1-\rho} p_s = \frac{s\mu}{s\mu - \lambda} p_s \\ &= \text{Proportion of arriving customers who have to queue}\end{aligned}$$

The last equality is a consequence of the uniform distribution of conditional arrival times for a Poisson process—the PASTA principle—as we already saw in connection with the $M/M/1$ queue in Section 10.6. We thus have the important result that, under steady-state conditions,

$$P\{\text{Customer must wait in queue}\} = \frac{\alpha^s}{s!(1-\rho)} \cdot p_0 \qquad (11.21)$$

with p_0 given by formula (11.19).

The right side of formula (11.21) is a rather complicated function of the integer variable s and the traffic intensity α, with $0 \leq \alpha < s$. (Remember that $\rho = \alpha/s$.) Because of the frequent use of this probability in telephone networks and other queueing applications, it has been given the name **Erlang's C function**, and it is usually denoted by the symbol $C(s, \alpha)$. Extensive tables and graphs of this function have been published.

Steady-State Waiting Times in $M/M/s$ Queue

The next step in the analysis of this system is to calculate the long-term average queue statistics, as described in Section 11.1. Since we can obtain the average number of customers L_Q in the queue from the average queue time per customer W_Q via Little's formula $L_Q = \lambda W_Q$ (formula (11.5)), we shall determine W_Q first. To make the result statistically meaningful, we shall determine the *complete* probability distribution of a typical customer's time in the queue, under steady-state conditions. From this we can get not only the *average* time in the queue per customer, but also the variance and percentiles for the time in the queue.

Assume that the system has been operating long enough to reach steady-state conditions (see the discussion in Section 10.1). Let's follow the progress of a typical arriving customer J.DOE through the system. We define the random variable

$$T_Q = \text{Waiting time in queue for J.DOE}$$

To calculate the probability distribution of T_Q, we condition on the state of the system at the moment of J.DOE's arrival. Suppose that there are n customers already present.

Case 1: n < s In this case at least one server is free, so J.DOE doesn't have to wait in the queue. Thus $T_Q = 0$.

Case 2: n ≥ s Now all s servers are busy when J.DOE arrives, and there are $n - s$ customers in the queue. During any period when all the servers are busy, the departure rate is $s\mu$ and the departure process $D(t)$ is a Poisson process with rate $s\mu$. (We start counting departures from the moment of J.DOE's arrival.) Each departure from the system moves J.DOE up one place in the queue; so after $n - s$ departures, J.DOE will be at the head of the queue. One more departure will then make a server available to serve J.DOE. Thus, in this case,

$$T_Q = \text{Elapsed time for } n - s + 1 \text{ departures}$$

Fix a number $t \geq 0$. By the description just given, the event $\{T_Q > t\}$ is the same as the event $\{D(t) \leq n - s\}$; these inequalities on T_Q or $D(t)$ assert that there will not be enough departures by time t to make a server free for J.DOE. Thus, we can calculate the conditional distribution of J.DOE's queue time by using Poisson probabilities:

$$\begin{aligned} P\{\, T_Q > t \mid n \text{ in system at arrival}\,\} &= P\{\, D(t) \leq n - s\,\} \\ &= \sum_{k=0}^{n-s} \frac{(\mu s t)^k}{k!} e^{-\mu s t} \end{aligned}$$

We now calculate the unconditional distribution of J.DOE's waiting time by the usual conditional probability formula:

$$P\{T_Q > t\} = \sum_{n=s}^{\infty} P\{\, T_Q > t \mid n \text{ in system at arrival}\,\} \cdot p_n$$

Using formula (11.20) for p_n and the conditional probabilities just calculated, we get

$$P\{\, T_Q > t\,\} = p_s e^{-\mu s t} \cdot \sum_{n=s}^{\infty} \Big(\sum_{k=0}^{n-s} \frac{(\mu s t)^k}{k!} \Big) \rho^{n-s}$$

We substitute $j = n - s$ for the first variable of summation and interchange the order of summation (notice that $0 \leq k \leq j < \infty$):

$$P\{\, T_Q > t\,\} = p_s e^{-\mu s t} \cdot \sum_{k=0}^{\infty} \Big(\sum_{j=k}^{\infty} \rho^j \Big) \frac{(\mu s t)^k}{k!}$$

The summation involving ρ gives $\rho^k/(1 - \rho)$, so we have

$$P\{\, T_Q > t\,\} = p_s \frac{e^{-\mu s t}}{1 - \rho} \sum_{k=0}^{\infty} \frac{(\mu s \rho t)^k}{k!}$$

Finally, we observe that the infinite series in this formula is just the power series expansion of $e^{\lambda t}$, since $\mu s \rho = \lambda$. Thus, we have obtained the formula

$$P\{\, T_Q > t\,\} = p_s \frac{e^{-(s\mu - \lambda)t}}{1 - \rho}$$

for all $t \geq 0$.

As a particular case, if we set $t = 0$ in the equation, we get the formula

$$P\{T_Q > 0\} = p_s \cdot \frac{1}{1-\rho} = C(s,\alpha) \tag{11.22}$$

which we already knew from formula (11.21). For any $t > 0$, we can then write

$$P\{T_Q > t\} = P\{T_Q > 0\} \cdot e^{-(s\mu-\lambda)t} \tag{11.23}$$

Dividing by $P\{T_Q > 0\}$, we can write this formula as a conditional probability:

$$P\{T_Q > t \mid T_Q > 0\} = e^{-(s\mu-\lambda)t}$$

We have thus shown that, under steady-state conditions, *if an arriving customer has to wait in the queue, then the customer's queue time is exponentially distributed, with rate* $s\mu - \lambda$. This result is intuitively reasonable, since $s\mu$ is the departure rate from the queue when all servers are busy, and λ is the arrival rate; thus, the queue occupancy rate is just the difference of these rates.

If we denote the expected waiting time in the queue for those customers who have to queue as $\$W_Q$, the preceding calculation gives the formula

$$\$W_Q = E[T_Q \mid T_Q > 0] = \frac{1}{s\mu - \lambda} \tag{11.24}$$

To calculate the expected queue time W_Q for *all* customers (including those whose queue time is zero), we use the formula

$$E[T_Q] = E[T_Q \mid T_Q = 0] \cdot P\{T_Q = 0\} + E[T_Q \mid T_Q > 0] \cdot P\{T_Q > 0\}$$

(that is, we condition on the event $\{T_Q = 0\}$ and the complementary event $\{T_Q > 0\}$). Of course the first conditional expectation on the right is zero, so we get the formula

$$W_Q = \$W_Q \cdot P\{T_Q > 0\} = \frac{1}{s\mu - \lambda} \cdot C(s,\alpha) \tag{11.25}$$

where $C(s,\alpha)$ is calculated from formula (11.21).

Now the hard work is finished, and it only remains to apply Little's formula to get the remaining queue statistics. Thus we have

$$L_Q = \lambda W_Q \tag{11.26}$$

Since the average service time for a customer is $1/\mu$, it follows that the average time in the system per customer is

$$W = \frac{1}{\mu} + W_Q \tag{11.27}$$

just as in the single-server queue. By Little's formula again,

$$L = \lambda W = \alpha + L_Q \tag{11.28}$$

In particular, from equation (11.28) we see that

$$\text{Average number of customers being served} = \alpha$$

This gives us justification for calling α the *traffic intensity*. Of course, the average number of customers being served is the same as the average number of servers being kept busy, so the *proportion* of utilization of the system is $\rho = \alpha/s$. This explains why ρ is called the *utilization factor* of the system. (Recall that $\rho < 1$ by the steady-state hypothesis of formula (11.16).)

Returning to the queue-time distribution, we observe from formula (11.25) that the probability that an arriving customer will have to wait in the queue is the ratio $W_Q/\$W_Q$. Thus the proportion of customers who *don't* have to queue up is

$$1 - \frac{W_Q}{\$W_Q} \tag{11.29}$$

This quantity can also be calculated from Erlang's C function (formula (11.21)) as $1 - C(s,\alpha)$. Finally, the cumulative distribution function of T_Q can be obtained from formula (11.23):

$$P\{T_Q \le t\} = 1 - C(s,\alpha) \cdot e^{-(s\mu-\lambda)t} \tag{11.30}$$

Example 11.5

Suppose that a library has two copy machines. One is located in the reading room and is used by 15 people per hour. The other is located in the reserve room and is used by 5 people per hour. In both cases, the average job requires 3 minutes. Would the copy facilities function better if both copiers were in the same location?

To analyze this problem, we assume Poisson arrivals, exponential service time, and steady-state conditions. Then the existing setup consists of two independent $M/M/1$ queues, with arrival rates $\lambda_1 = 15$ per hour and $\lambda_2 = 5$ per hour, and a common service rate $\mu = 20$ per hour. Thus the utilization factors are $\rho_1 = \frac{15}{20}$ and $\rho_2 = \frac{5}{20}$. The average amount of time a user spends at the first machine is

$$W_1 = \frac{1}{\mu - \lambda_1} = \frac{1}{5} \text{ hour} = 12 \text{ minutes}$$

while the average amount of time spent at the second machine is

$$W_2 = \frac{1}{\mu - \lambda_2} = \frac{1}{15} \text{ hour} = 4 \text{ minutes}$$

(see Section 11.3).

Suppose now that we placed both machines in a central location, where the users would form a single queue for the two machines. In this case we would have an $M/M/2$ queue, with arrival rate $\lambda = \lambda_1 + \lambda_2 = 20$ per hour. (Recall that the sum of two independent Poisson processes is again a Poisson process, whose rate is the sum of the two rates.) The service rate is still 20 per hour per machine, but the utilization factor is now $\rho = \lambda/(2\mu) = \frac{20}{40}$ (the average of ρ_1 and ρ_2). Using the formulas derived in this section, we calculate that the average time a user spends at this centralized facility would be

$$W = \frac{1}{\mu \cdot (1 - \rho^2)} = \frac{1}{20} \cdot \frac{4}{3} = \frac{1}{15} \text{ hour} = 4 \text{ minutes}$$

(see Exercise 15). This is clearly a better arrangement: for the reserve-room users the system is as fast as before; while for the reading room users, it is three times faster than before. This illustrates the general principle that *a single queue feeding multiple servers is usually more efficient than independent queues for each server.*

11.7 Optimization Problems

In this section we use steady-state statistics to determine optimal values for the parameters of some queueing systems.

Consider first the $M/M/s$ queueing system studied in Section 11.6. Customers arrive at rate λ, and each server works at rate μ. We make the following assumptions about the rates of cost for servers and customers:

a. Each server costs $b\mu$ \$/hour, whether busy or not (b = constant). Thus, the service cost is proportional to the rate of service.

b. Customer time in the system costs c \$/hour per customer (c = constant). Here we consider the customer's queue time and service time as equally expensive.

The average total cost rate (\$/hour) for operating the system is then

$$\text{Service Cost Rate} + \text{Waiting Cost Rate} \;=\; b\mu s + cL$$

where L is the average number of customers in the system.

Suppose that the arrival rate λ and the number of servers s are fixed. A natural optimization problem is then to find the service rate that minimizes the cost. We calculated in Section 11.6 that

$$\begin{aligned} L &= \alpha + L_Q = \alpha + \lambda W_Q \\ &= \alpha + \frac{\alpha}{s-\alpha} \cdot C(s,\alpha) \end{aligned}$$

where $\alpha = \lambda/\mu$ is the traffic intensity and $C(s,\alpha)$ is Erlang's C function (formula (11.21)). Since we are assuming steady-state behavior, the parameter α is constrained by

$$0 < \alpha < s \tag{11.31}$$

In the formula for the total cost per hour, we can substitute λ/α for μ and use the formula for L just given to express everything in terms of the variable α. Then the function to be minimized is

$$E(\alpha) = \frac{b\lambda s}{\alpha} + c\alpha + \left(\frac{c\alpha}{s-\alpha}\right) \cdot C(s,\alpha) \tag{11.32}$$

with α constrained by condition (11.31) and λ, s fixed.

It is easy to see from formula (11.32) that $E(a) \to \infty$ when $\alpha \searrow 0$ or $\alpha \nearrow s$, since $C(s,0) = 0$ and $C(s,s) = 1$. Thus $E(\alpha)$ has a minimum on the interval $0 < \alpha < s$. The optimal value for α, minimizing $E(\alpha)$, can be found by analytic or numerical methods.

Example 11.6 Optimizing a Single-Server Queue

Suppose s = 1. Then $C(1,\alpha) = 1 - p_0 = \alpha$ (see Section 10.6). Substituting this in formula (11.32) and simplifying, we find that

$$E(\alpha) = \frac{b\lambda}{\alpha} + \frac{c\alpha}{1-\alpha}$$

Hence,

$$E'(\alpha) = -\frac{b\lambda}{\alpha^2} + \frac{c}{(1-\alpha)^2}$$

The value $\alpha = \alpha_{\min}$ that minimizes $E(\alpha)$ thus satisfies the equation $E'(\alpha) = 0$, which can be written as

$$\frac{\alpha^2}{(1-\alpha)^2} = \frac{b\lambda}{c} \tag{11.33}$$

To solve equation (11.33) for α, it is convenient to introduce the variable $z = (b\lambda/c)^{1/2}$. The equation then reads $\alpha/(1-\alpha) = z$, which has the unique solution

$$\alpha_{\min} = \frac{z}{1+z}$$

Notice that $0 < \alpha_{\min} < 1$ for any positive value of z, so the steady-state constraint expressed in formula (11.31) is satisfied. Furthermore, since $\alpha_{\min}$ is the only zero of $E'(\alpha)$ and since $E'(\alpha)$ is negative for α near 0 and positive for α near 1, it follows that the function $E(\alpha)$ must have a minimum at $\alpha_{\min}$.

We get the optimal service rate $\mu_{\min}$ for minimum cost by substituting λ/μ for α and solving for μ. The result is

$$\mu_{\min} = \lambda\left(1 + \sqrt{c/(\lambda b)}\right) \tag{11.34}$$

This equation tells us just how much faster the service rate should be than the arrival rate to operate the system at minimal cost. Thus, if the cost of providing fast service is considerably greater than the cost of customer waiting time, then $c/(\lambda b)$ will be small, and $\mu_{\min}$ will be only slightly larger than λ. Of course in this case the average customer's time W in the system will be very large, since $W = 1/(\mu - \lambda)$. ■

Erlang's Loss System

Consider a direct-mail retailing business that operates as follows. There are s telephone operators who take orders from customers. Incoming calls arrive at a Poisson rate of λ per minute, and the placing of an order requires an exponentially-distributed amount of time, with rate μ. If all operators are busy, the caller leaves the system immediately.

This system is obviously the $M/M/s$ queueing system with queue capacity zero. If we define the state of the system to be the number of telephone operators who are busy, we obtain the transition diagram shown in Figure 11.9. This model was first studied by Erlang (decades before the invention of call-waiting) in connection with telephone switching exchanges, where an incoming call is lost whenever all the switching units are busy.

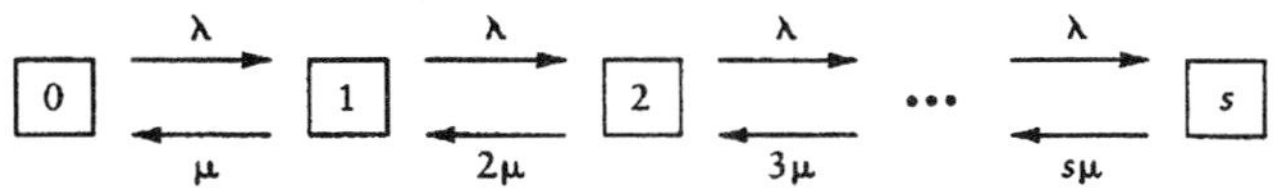

Figure 11.9: Erlang's Loss System

By calculations similar to those in Section 11.6, we find the steady-state probabilities

$$p_n = \frac{\alpha^n}{n!} \cdot p_0 \qquad \text{for } 1 \le n \le s$$

where $\alpha = \lambda/\mu$ is called the **offered load**. Since the probabilities must sum to 1, we have

$$p_0 = \left(1 + \alpha + \frac{\alpha^2}{2!} + \cdots + \frac{\alpha^s}{s!}\right)^{-1} = \frac{1}{e_s(\alpha)} \tag{11.35}$$

Here, $e_s(\alpha)$ is the Taylor polynomial of degree s for e^α, as in Section 11.6.

The probability that all phones are busy when a potential customer tries to call is

$$p_s = \frac{\alpha^s}{s! e_s(\alpha)} \tag{11.36}$$

Formula (11.36) is called **Erlang's loss formula**, since it gives the proportion of potential customers who are lost (by the PASTA principle). The function on the right-hand side in this formula is denoted by $B(s, \alpha)$ and is called **Erlang's B function**.

The average arrival rate to the system is

$$\lambda_a = (1 - B(s, \alpha))\lambda \tag{11.37}$$

We call λ_a the **carried load**. Since there is no queueing time, the average time in the system per customer is $W = 1/\mu$. Hence, by Little's formula, the average number of customers in the system is

$$L = \lambda_a W = (1 - B(s, \alpha)) \cdot \alpha \tag{11.38}$$

Example 11.7 Optimizing an Erlang Loss System

We now consider some optimization problems for an Erlang loss system. Suppose that each customer generates $\$c$ profit, on average, and each server costs $b\mu$ \$/hour. Then

$$\text{Average profit rate} = \lambda_a c - sb\mu \tag{11.39}$$

where the carried load λ_a is given by formula (11.37). Two natural optimization problems are as follows:

a. For fixed values of λ and s, find the service rate μ that maximizes the average profit rate. For example, introducing faster terminals and a faster computer system for the operators would result in an increase in μ. We want to know if the increased business would justify the capital costs of the new equipment.

b. For fixed values of λ and μ, determine the number of servers that maximizes the average profit rate.

Either of these problems can be readily solved by direct numerical calculation. In certain cases, analytic methods may be used. For example, suppose that the rate of incoming calls exactly matches the rate of serving the orders, so that the traffic intensity $\alpha = 1$. Which is more profitable: one operator or two operators?

To decide this, we calculate from Erlang's loss formula (11.36) that

$$B(1,1) = \frac{1}{1!(1+1)} = \frac{1}{2}$$

Therefore, 50% of the potential customers are lost when only one operator is employed, and the carried load is 0.5λ. On the other hand,

$$B(2,1) = \frac{1}{2!(1+1+1/2!)} = \frac{1}{5}$$

That is, customer loss would drop to 20% if we had two operators, and the carried load would be 0.8λ. Thus, by formula (11.39), the corresponding average profit rates are

$$\frac{1}{2}\lambda c - \mu b \qquad \text{when } s = 1 \tag{11.40}$$

$$\frac{4}{5}\lambda c - 2\mu b \qquad \text{when } s = 2 \tag{11.41}$$

Subtracting rate (11.40) from rate (11.41), and using our assumption that $\lambda = \mu$, we find that

$$\text{Difference in profit rates} = \frac{\lambda}{10}(3c - 10b)$$

So, when $3c > 10b$, then using two servers is more profitable than using one. Conversely, when $3c < 10b$, then one server is more profitable.

We can summarize our conclusions by using the decision diagram shown in Figure 11.10 (the slope of the line separating the two regions is $\frac{10}{3}$). It is intuitively reasonable that the decision criterion should be of this form. If the parameter c (profit per customer) is very large relative to b (server cost), then the extra business generated by having another server will easily cover the additional server costs. ■

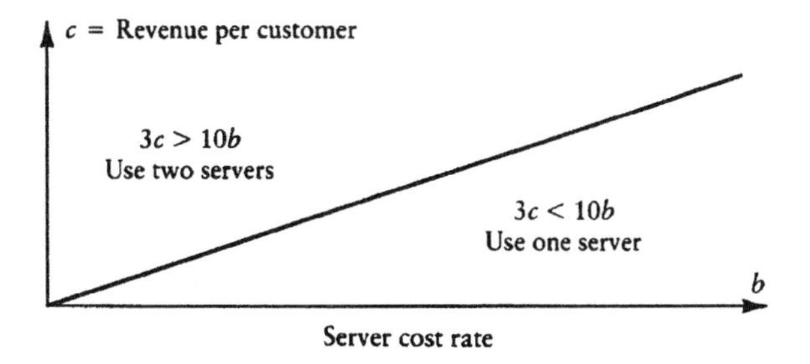

Figure 11.10: Decision Diagram

11.8 Examples of Queue Simulation

The simple exponential models that we have studied up to now are not adequate to model the queueing systems that actually arise, for example, in telecommunications networks or computer-operating systems. Fortunately, a direct experimental method is available for studying these systems by simulation. Although simulation methods tend to have limited accuracy (see Example 2.10), they at least provide a rough idea of the functioning of a system. Even for exponential models, where we can calculate the steady-state probabilities exactly, a simulation can indicate how fast the system is approaching steady-state conditions.

In this section we give some examples of queue-simulation programs and the data they generate. In Chapter 12, we will develop analytical methods for some of these examples and compare the analytic results with the simulation data.

We have already introduced the basic algorithm for queue simulation in Section 8.6. For studying the long-term behavior of a queueing system, the simulation must be coded as a computer program and run for an extended time period. From the results of the simulation, we can calculate the long-term average queue statistics described in Section 11.1.

As we showed in the simulation example in Section 8.6, quite a lot of bookkeeping is needed to keep track of the arrivals and departures of customers from the queues and service facilities. For this reason, several specialized computer languages, such as GPSS, SIMSCRIPT, SIMULA, and SIMULINK have been created to simplify the job of creating simulation programs for stochastic processes. The most widely used simulation language for queueing models is GPSS (General Purpose Simulation System). GPSS programs for queueing models have a structure similar to the flow of customers through the model, and they are quite straightforward to write. Implementations of GPSS exist for a variety of computer platforms.

In the following examples we describe some simple GPSS programs and give samples of the output they generate. This is *not* intended as a self-contained introduction to the GPSS language and its syntax; for details and for more complex examples see the references at the end of the book.

Example 11.8 Simulating a Barbershop

A barbershop with one barber has space for a maximum of two people to wait for a haircut. Potential customers arrive at a Poisson rate of 6 per hour, and get served at an exponential rate of 3 per hour. We shall describe a simulation of this system in GPSS and compare the simulation statistics with the steady-state statistics for an $M/M/1$ queue with capacity $N = 3$ (see Sections 10.6 and 11.3).

The GPSS language simulates the passage of customers (called *transactions*) through the system. A GPSS program is organized into *blocks*. Each block has a special one-word name and generally corresponds to one line of the program. The order of listing the blocks in the program corresponds to the flow of customers through the system.

Figure 11.11 gives a schematic description of the flow of customers in this example. In the figure we have indicated the GPSS block names associated with each stage of the customer's passage through the system. There are stochastic blocks, such as GENERATE (the customer arrives according to the random arrival distribution named

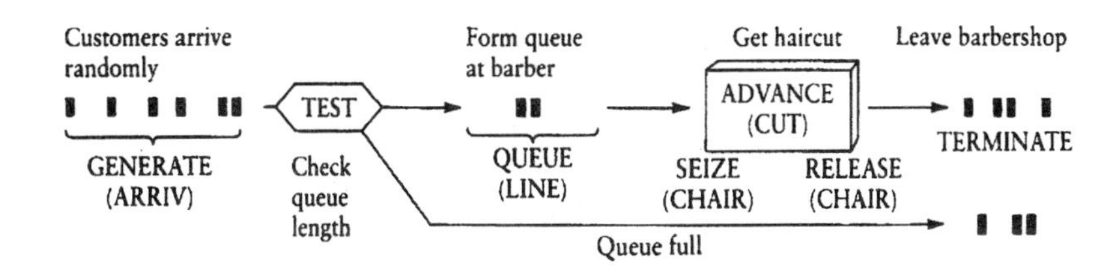

Figure 11.11: Barbershop Model

ARRIV) and ADVANCE (the customer gets served according to random service distribution named CUT). There are branching blocks, such as TEST (the arriving customer will leave if the queue is full). There are bookkeeping blocks, such as the pair QUEUE and DEPART of queue-creating blocks (the customer enters queue named LINE, moves to the front of queue as other customers are served, and then departs when the server is free) and the pair SEIZE and RELEASE of server blocks (the customer seizes the service facility named CHAIR, remains there for a random length of service time, and then releases the facility for the next customer to use). Finally, there is the TERMINATE block (the customer leaves the system).

In GPSS there are several random-number generators, called RN1, RN2, ... To generate random exponential interarrival and service times, some versions of GPSS also have an exponential function, called BE (a mnemonic for *Built-in Exponential*). To run a GPSS program simulation for a specified length of time, we create a nonrandom clock segment of the model that runs concurrently with the main segment.

The following GPSS program simulates 8 hours of this queueing system (the far right-hand column and the lines starting with * are comments which are ignored by the compiler):

```
        SIMULATE
*       Model Segment
ARRIV   FUNCTION    RN1,BE          exponential interarrival times
CUT     FUNCTION    RN2,BE          exponential service times
        GENERATE    10,FN$ARRIV     mean interarrival = 10 minutes
        TEST L      Q$LINE,2,OUT    if line < 2 then queue; else out
        QUEUE       LINE            wait for barber
        SEIZE       CHAIR           go to barber chair
        DEPART      LINE            leave waiting area
        ADVANCE     20,FN$CUT       haircut, mean service = 20 minutes
        RELEASE     CHAIR           haircut finished
OUT     TERMINATE                   leave barbershop
*       Clock segment
        GENERATE    60              one clock transaction per 60 minutes
        TERMINATE   1               count simulated hours
        START       8               simulate 8 hours operation
        END
```

Notice that each line contains an operation command, such as QUEUE, and operands (modifiers) for the operation, such as LINE (the symbolic name of the queue in this model). The branching command TEST sends customers to the location OUT.

One attractive feature of GPSS is that the results of the simulation are automatically tablulated and printed out in a form that is easy to interpret. When we run the program above, we obtain a printed output of block counts (how many customers passed through each block of the program), queue statistics, and server statistics. In this example, we can also calculate the steady-state queue statistics exactly, and compare them with the simulation results, as follows.

The states of the system are 0, 1, 2, and 3. The traffic intensity $\rho = \frac{6}{3} = 2$. By the formulas of Section 10.6, we calculate that the steady-state probabilities are

$$\begin{aligned} p_0 &= \frac{1}{15} = 0.067 = \text{Proportion of time the barber is idle} \\ p_3 &= \frac{8}{15} = 0.533 = \text{Proportion of arrivals turned away} \end{aligned}$$

and the steady-state arrival rate to the queue is

$$\lambda_a = (1 - p_3)\lambda = 2.8 \text{ customers per hour}$$

From the steady-state probabilities, we calculate

$$L = 1 \cdot p_1 + 2 \cdot p_2 + 3 \cdot p_3 = \frac{34}{15} = 2.267 \text{ customers}$$

Then by Little's formula, $W = L/\lambda_a = \frac{17}{21}$ hours, and

$$W_Q = W - \frac{1}{\mu} = \frac{10}{21} \text{ hours} = 28.57 \text{ minutes}$$

By Little's formula again, we get

$$L_Q = \lambda_a W_Q = \frac{4}{3} = 1.33 \text{ customers}$$

A particular sample run of this example generated 43 arrivals in an 8-hour period. Of this number, 14 passed the TEST block and joined the queue. This is an arrival rate of

$$\left[\lambda_a\right]_{\text{sim}} = \frac{14 \text{ customers}}{8 \text{ hours}} = 1.75 \text{ customers per hour}$$

(Compare this with the steady-state value $\lambda_a = 2.8$.) The proportion of arrivals that left because the queue was full was $\frac{29}{43} = 0.674$. (Compare this with the steady-state value $p_3 = 0.533$.)

The queue results for this sample run are given in Table 11.1. These statistics are calculated by the formulas

$$\begin{aligned} \text{Average contents} &= \frac{\Sigma \text{ (Time in queue for each customer)}}{\text{Total time of simulation}} \\ \text{Average time/trans.} &= \frac{\Sigma \text{ (Time in queue for each customer)}}{\text{Number of customers passing through queue}} \end{aligned}$$

Thus, the value for average contents yields an estimate $\left[L_Q\right]_{\text{sim}}$ for the steady-state av-

Table 11.1 Queue Statistics for Example 11.8 (8 hour simulation)

MAXIMUM CONTENTS	AVERAGE CONTENTS	ZERO ENTRIES	AVER. TIME/TRANS.
2	1.33	3	45.71

erage queue length L_Q, and the average time per transaction gives an estimate $[W_Q]_{\text{sim}}$ for the steady-state average queue time W_Q. Notice that

$$\begin{aligned}[L_Q]_{\text{sim}} &= \frac{\text{Number of customers passing through queue}}{\text{Total time of simulation}} \cdot [W_Q]_{\text{sim}} \\ &= [\lambda_a]_{\text{sim}}[W_Q]_{\text{sim}}\end{aligned}$$

This is the simulation version of Little's formula (see Section 11.2). We see that the simulated average queue length agrees very well with the steady-state value L_Q, but the simulated average queue time is not close to W_Q.

The Zero Entries listing in the queue statistics in Table 11.1 reports the number of customers who arrived when the queue was empty and were served immediately. Thus, the proportion of *all* arriving customers who found the server free was $\frac{3}{43} = 0.0699$. Since Poisson arrivals see time averages (Example 7.6), the corresponding steady-state value is $p_0 = 0.0667$.

The server results for this sample run are shown in Table 11.2. Here, average utilization refers to the proportion of time during which the server is busy. These values are not very close to the steady state values $1 - p_0 = 0.933$ for the average utilization and 20 minutes for the average service time.

Table 11.2 Server Statistics for Example 11.8 (8 hour simulation)

AVERAGE UTILIZATION	NUMBER ENTRIES	AVERAGE TIME/TRANS.
0.85	12	34

The simulation just described ran for 8 simulated hours. If we change the START statement in the program to

```
START      400      simulate 400 hours
```

and run the program, we get a much closer approximation to the steady-state values for the queue statistics. One such sample run generated 2542 arrivals, with 1161 passing the TEST block. This gave the value

$$[\lambda_a]_{\text{sim}} = \frac{1161 \text{ customers}}{400 \text{ hours}} = 2.9 \text{ customers per hour}$$

which is close to the steady-state value $\lambda_a = 2.8$. The proportion of arrivals that left because the queue was full was $1381/2542 = 0.543$, which is also close to the steady-state value $p_3 = 0.533$.

The queue results for the longer sample run were

$$[L_Q]_{\text{sim}} = 1.35 \text{ customers} \qquad [W_Q]_{\text{sim}} = 27.84 \text{ minutes}$$

both of which are much closer to the steady-state values $L_Q = 1.33$ customers and $W_Q = 28.57$ minutes than in the first simulation. The proportion of *all* arriving customers who found the server free was 0.0669, as compared to the steady-state value $p_0 = 0.0667$.

The server results for the longer sample run were

$$\text{Average Utilization} = 0.9287 \qquad \text{Average time/trans.} = 19.21$$

These values are very close to the steady state values $1 - p_0 = 0.933$ for the average utilization and 20 minutes for the average service time.

This example illustrates the general principle stated in Section 11.1: *sample time averages taken over a long time period approach steady-state probabilities in an ergodic system*. Notice, however, that the length of time needed for a simulation to approach steady-state conditions may be very long. To get some idea of the random short-term behavior of the system, we can combine simulation with the Law of Large Numbers; for example, we can make several independent simulations of an 8-hour time period and take the average of the simulation statistics. ■

Another attractive feature of GPSS is that the random aspects of the model can easily be modified while preserving the same flow of customers. In Example 11.8, we could model service times uniformly distributed between 15 and 25 minutes by changing the ADVANCE statement to

```
ADVANCE     20,5     mean service 20 ± 5 minutes
```

(This is a more realistic model for this situation). We could also modify the waiting space of the barbershop by changing the number 2 in the TEST block, or by eliminating the TEST block and allowing an unlimited queue.

Example 11.9 Simulating a Car Wash and Vacuum Facility

A car wash and vacuum facility has no queue space for arriving cars. Potential customers arrive according to a Poisson process with rate 3 cars per hour. If the wash facility is free, the customer seizes it and proceeds to wash the car. The wash time is exponentially distributed, with mean 15 minutes. If the vacuum facility is free at the moment the car wash is finished, then the customer seizes the vacuum facility and releases the wash facility for the next customer. If the vacuum facility is busy, the customer leaves without vacuuming the car. The time to vacuum a car is exponentially distributed, with mean 30 minutes.

This is a modification of Example 11.2 (there is no blocking state). We now describe how to use GPSS to simulate the operation of this service-in-stages model. (See Exercise 12 for the steady-state statistics.)

Table 11.3 Simulation Statistics for Car Wash and Vacuum

FACILITY	AVERAGE UTILIZATION	NUMBER ENTRIES	AVERAGE TIME/TRANS.
WASH	0.4167	61	16.39
VAC	0.4992	34	35.24

In this model there are two service facilities but no QUEUE blocks. We test whether a facility is busy by using the GATE block. This requires a statement of the form

GATE NU WASH,OUT

where WASH is the symbolic name of the wash facility, and OUT is the location name of the TERMINATE block. Here NU means *not in use*. If the wash facility is not in use at the moment the customer enters the gate block, customer proceeds to the next block in the program. Otherwise, the customer goes to the TERMINATE block. We use a similar statement after the RELEASE block to determine whether the vacuum facility is in use. The model segment of the GPSS program is as follows:

```
ARRIV   FUNCTION    RN1,BE          exponential interarrival times
WASH    FUNCTION    RN2,BE          exponential wash times
VAC     FUNCTION    RN3,BE          exponential vacuum times
        GENERATE    20,FN$ARRIV     mean interarrival = 20 minutes
        GATE NU     WASH,OUT        leave if wash in use
        SEIZE       WASH            start car wash
        ADVANCE     15,FN$WASH      mean wash time = 15 minutes
        RELEASE     WASH            wash finished
        GATE NU     VAC,OUT         leave if vacuum in use
        SEIZE       VAC             start vacuum
        ADVANCE     30,FN$VAC       mean vacuum time = 30 minutes
        RELEASE     VAC             vacuum finished
OUT     TERMINATE                   drive away
```

To run the program, we include a clock segment, as in Example 11.8.

A simulation run of 40 hours yielded the data shown in table 11.3. The simulated arrival rates to the facilities were

$$\left[\lambda_{\text{wash}}\right]_{\text{sim}} = \frac{61}{40} = 1.525 \text{ cars per hour}$$

$$\left[\lambda_{\text{vac}}\right]_{\text{sim}} = \frac{34}{40} = 0.85 \text{ cars per hour}$$

The block counts for this run show that 120 arrivals to the wash facility were generated. Thus, $(120 - 61)/120 = 49.2\%$ of the potential wash customers were lost because of the absence of queue space.

It is easy to modify this GPSS program it in several ways: we can change the random distribution of arrival or service times, and we can insert queues (with prescribed capacity) at either the wash or the vacuum facility. ■

We can set up a GPSS program to print out more statistics about the simulation than the ones shown in Examples 11.8 and 11.9. In particular, we can obtain the cumulative distribution of queue time. For the case of $M/M/s$ queues, this empirical distribution can be compared with the steady-state distribution that we calculated in Section 11.6. We illustrate this more extensive statistical analysis in the next example.

Example 11.10 Simulating a Ship Unloading Facility

A shipping company has a single unloading dock at a harbor. Ships randomly arrive according to a Poisson process at an exponential rate of 2 per day, anchor in the harbor until the dock is free, and then are unloaded at the dock. After being unloaded, the ships leave the harbor.

We shall use simulation to compare two models for this system. In the first model we assume that the unloading time is an exponential random variable with rate 3 per day (this is an $M/M/1$ queue). In the second model, we assume that the random unloading time is uniformly distributed between 6 hours and 10 hours. In both models the mean unloading time is 8 hours.

To compare the distributions of queue time in the two models, we put the statement

```
LINE      QTABLE      LINE,0,1,26
```

at the beginning of the GPSS program. This will create a tabulation of queue time in the line, in intervals of width 1 hour, starting at 0 and having 26 categories (the last category consisting of times greater than 24 hours).

The rest of the program is a simple modification of the program in Example 11.8, with the TEST block omitted and the following changes: in the GENERATE block, the mean interarrival time is changed to 12 hours; whereas in the first model the ADVANCE block uses an exponential function with mean 8 hours, in the second model the ADVANCE block is changed to

```
ADVANCE      8,2
```

following the format discussed after Example 11.8.

Table 11.4 Comparison of Queue Statistics for Unloading Dock

Exponential Service Time:			(226 Transactions)
MAXIMUM CONTENTS	AVERAGE CONTENTS	ZERO ENTRIES	AVER TIME/TRANS.
12	1.65	73	17.55
Uniform Service Time:			(210 Transactions)
MAXIMUM CONTENTS	AVERAGE CONTENTS	ZERO ENTRIES	AVER TIME/TRANS.
9	0.94	70	10.78

Simulation of 100 hours of these two models produced the data in Tables 11.4 and 11.5. In comparing the two models, we see that the server statistics are about the same in both. Also the percentage of ships that were able to unload without waiting was about the same (32.9% in the exponential model, and 33.5% in the uniform model). For the ships that had to queue, however, the situation in the two models was drastically different. The average queue length in the exponential model was 175% more than in the uniform model, and the average waiting time in the exponential model was 162% more than in the uniform model.

Table 11.5 Comparison of Server Statistics for Unloading Dock

Exponential Service Time:		
AVERAGE UTILIZATION	NUMBER ENTRIES	AVERAGE TIME/TRANS.
0.68	222	7.35
Uniform Service Time:		
AVERAGE UTILIZATION	NUMBER ENTRIES	AVERAGE TIME/TRANS.
0.70	209	8.00

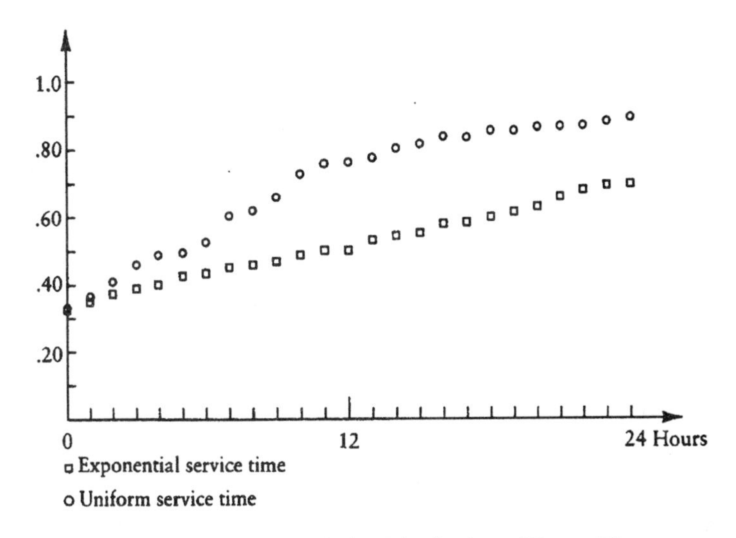

Figure 11.12: Cumulative Distribution of Queue Time

We can examine the differences between these models more closely by using the simulation data produced by the QTABLE statement in the program. This gives the plots of the cumulative distribution of simulated queue time shown in Figure 11.12. From the graphs we see that for the exponential model, the probability is only 0.5 that a ship will wait less than 12 hours in the queue. For the uniform model, this probability is 0.75. In Section 12.8, we give a theoretical explanation of this simulation data. ■

Exercises

1. Consider the simulation problem in Chapter 8, Exercise 13.

   a. Plot the graphs of $A(t)$ (number of arrivals by time t) and $D(t)$ (number of departures by time t) on the same axes for $0 \leq t \leq 20$. Use these graphs as in Section 11.2 to obtain the graph of $L(t)$ (number of customers present at time t), assuming that the fifth customer arrives after $t = 20$.

   b. From the graphs in part (a), calculate

   $$\begin{aligned} L(20) &= \text{Average number of customers in the system during } (0, 20] \\ W(20) &= \text{Average time in system per customer during } (0, 20] \\ \lambda_a(20) &= \text{Average arrival rate during } (0, 20] \end{aligned}$$

   and verify that the identity $L(20) = \lambda_a(20) \cdot W(20)$ holds (see the proof of Little's formula in Section 11.2).

2. Derive formula (11.11).

3. For the $M/M/1$ queue with finite capacity N, show that the average number of customers being served (under steady-state conditions) is approximately ρ when ρ is near zero, and is approximately $1 - \rho^{-N}$ when ρ is large.

4. A new copy machine is being installed in a library. The librarian estimates that each user will spend 3 minutes on the average with the machine, and wants the average number of users L at the facility at any moment to be at most three.

   a. Under these conditions, what is the maximum average number of users per hour that the machine can serve? Assume Poisson arrivals, exponential service times, and steady-state conditions.

   b. With the maximum allowed arrival rate from part (a), what is the average queue time for a user of the machine?

   c. Suppose more space were provided for the machine, so that the facility could accommodate twice as many users ($L = 6$), on average. What would the results of parts (a) and (b) be in this case?

5. A barber shop with one barber has space for a maximum of two people to wait for service. Potential customers arrive at a Poisson rate of 6 per hour, and get served at an exponential rate of 3 per hour. If the barber is busy at the moment of a customer's arrival, however, then—even if there is space to wait—the probability is 0.5 that the customer will leave immediately.

   a. Set up this system as a birth–death process, and draw a state-transition diagram.

   b. Calculate the steady-state probabilities for the system.

   c. Calculate the average number of customers in the shop at any moment.

   d. If the barber charges \$10 per customer, what is the average rate of income of the shop (\$ per hour)?

   e. What is the average amount of time that each customer who gets served spends at the shop?

6. In Chapter 10, Exercise 10, what is the average number of cars at the station? How long does each car spend at the station, on average?

7. Consider an ergodic birth–death process with birth rates $\{\lambda_n\}$ and death rates $\{\mu_n\}$, and assume that the process has steady-state probabilities $\{p_n\}$. Show that the steady-state average birth rate

$$\lambda_a = \sum_{n=0}^{\infty} p_n \lambda_n$$

   is equal to the steady-state average death rate

$$\mu_a = \sum_{n=1}^{\infty} p_n \mu_n$$

   (HINT: Use the balance equations.)

8. For the car-wash-and-vacuum facility in Example 11.1, assume that the washing rate is 4 cars/hour, and the vacuuming rate is 12 cars/hour. Potential customers arrive at a Poisson rate of λ cars/hour.

a. Calculate the steady-state probabilities as functions of the parameter λ.

b. Suppose that each customer pays $\$D$ to use the facility. Calculate the average gross return ($ per hour) from the business, as a function of λ.

c. Show that, if λ is large, the probability that an arriving customer will find the system free is $\approx 3/\lambda$, and business earns $\approx \$3D$ per hour.

9. For the enlarged car wash-and-vacuum facility in Example 11.2, assume that the washing rate is 4 cars/hour, and the vacuuming rate is 12 cars/hour. Potential customers arrive at a Poisson rate of λ cars/hour.

   a. Calculate the steady-state probabilities as functions of the parameter λ.

   b. Suppose that each customer pays $\$D$ to use the facility. Calculate the average gross return ($ per hour) from the business, as a function of λ.

   c. Show that if λ is large, then the probability that an arriving customer finds the system free is $\approx 48/(13\lambda)$, and the business earns $\approx \$(48/13)D$ per hour.

10. Verify that the product solution for steady-state probabilities given in Jackson's theorem satisfies the balance equations in Section 11.5.

11. Consider Example 11.3 (motor vehicle licensing agency), with the numerical values for the arrival and service rates as given at the end of that section.

    a. During what proportion of time are both servers idle?

    b. During what proportion of time is the registration clerk busy and the cashier idle?

    c. How fast would the registration clerk have to process applications in order for the average time that a customer spends at the agency to be at most $\frac{1}{2}$ hour?

12. Modify the enlarged car wash-and-vacuum model in Example 11.2, as follows. If the vacuum facility is busy when a car finishes using the wash facility, the car leaves. Assume now that the washing rate is 4 cars/hour, the vacuuming rate is 2 cars/hour, and potential customers arrive at a Poisson rate of 3 cars/hour.

    a. Define the states of this system to make it an exponential model, draw a state-transition diagram, and find the steady-state probabilities for the system.

    b. What is the average rate at which customers enter the facility?

    c. What is the average number of customers at the facility?

    d. What is the average amount of time that an entering customer spends at the facility?

13. Consider a two-stage service facility with no queue space at station A and room for at most one customer to queue at station B. Potential customers arrive at station A according to a Poisson process with rate 3 per hour. If the server at station A is free, then the customer goes there. Otherwise, the arriving customer leaves without receiving any service. The service at A is exponential, with rate 4 per hour. After finishing at station A, a customer immediately goes to station B if the queue there is empty. Otherwise, the customer leaves. The service at B

is exponential, with rate 2 per hour. Let the state of the system be (m, n) if there are m customers at station A and n customers at station B.

a. Draw a complete state-transition diagram for this system and write down the balance equations for the steady-state probabilities.

b. Solve the balance equations from part (a), and calculate the average number of customers in the system, under steady-state conditions.

14. Consider an $M/M/1$ queueing system with arrival rate 3 customers/hour, and service rate 4 customers/hour. Assume that steady-state conditions apply.

a. Find the average queue time for a customer and the average number of customers in the queue.

b. Suppose that J.Doe arrives when there are already two customers in the system (one at the server and one in the queue). What is the probability that J.Doe will wait more than $\frac{1}{2}$ hour in the queue before getting served?

15. Consider an $M/M/2$ queue with arrival rate λ and server rate μ per server. Let $\alpha = \lambda/\mu$ and $\rho = \lambda/(2\mu)$. Assume that $\rho < 1$ so that the system reaches a steady state.

a. Show that Erlang's C function is

$$C(2, \alpha) = \frac{\alpha^2}{2 + \alpha}$$

b. Show that the steady-state average waiting time in the queue (for all customers) is

$$W_Q = \frac{1}{\mu} \cdot \frac{\alpha^2}{(2 - \alpha)(2 + \alpha)}$$

c. Show that the steady-state average time in the system (for all customers) is

$$W = \frac{1}{\mu(1 - \rho^2)}$$

d. Show that the probability the system is empty is

$$p_0 = \frac{1 - \rho}{1 + \rho}$$

16. Consider a two-server queue with arrival rate $\lambda = 18$ customers/hour and service rate $\mu = 10$ customers/hour per server. Use the formulas from Section 11.6 and Exercise 15 to calculate the following queue statistics:

a. p_0 b. $P\{\text{Customer must queue}\}$ c. $\$W_Q$
d. W_Q e. L_Q f. W
g. L h. $P\{\text{Customer's queue time} \geq 1 \text{ hour}\}$

17. Consider an $M/M/1$ queue, with arrival rate λ and service rate 2μ (for the single server), and an $M/M/2$ queue, with arrival rate λ and service rate μ for each of the two servers. (Thus, the utilization factor ρ is the same at both queues,

but the traffic intensity α_2 at the two-server queue is twice the traffic intensity α_1 at the one-server queue.) Use the formulas in Section 11.6 and Exercise 15 to compare the following queue statistics for the two systems (assume $\lambda < 2\mu$ so that steady-state conditions can exist):

a. $\$W_Q$ b. W_Q c. W

(HINT: Calculate the ratios $W_Q^{(1)} / W_Q^{(2)}$ and $W^{(1)} / W^{(2)}$, where the superscript denotes the number of servers.)

18. For the copy-machine example (Example 11.5), compare the queue lengths and queue times for the two setups.

19. Show that, if $0 < \alpha < s$, then the reciprocal of Erlang's C function can be expressed as

$$C(s,\alpha)^{-1} = 1 + (s-\alpha)\frac{(s-1)!}{\alpha^s} e_{s-1}(\alpha)$$

20. Use the result of Exercise 19 to show that $\lim_{\alpha \nearrow s} C(s,\alpha) = 1$.

21. A shipping company has a single unloading dock at a harbor. Ships randomly arrive (at a Poisson rate of 1 every 12 hours), anchor in the harbor until the dock is free, and then are unloaded at an exponential rate of μ per hour (the value of μ depends on the size of the work crews that the company employs). The cost rate for unloading is $\$4800\mu$ per hour. From the time a ship arrives at the harbor, it costs the company \$100 per hour until the ship is unloaded.

    a. Find the optimal value of μ that minimizes the cost of operating the dock.

    b. For the optimal μ from part (a), what is the probability that an arriving ship will find the dock free?

    c. For the optimal μ from part (a), what is the probability that a ship will spend more than 24 hours anchored in the harbor, waiting for the dock to be free?

    (Assume that steady-state conditions apply in answering parts (a), (b), and (c).)

22. Consider Erlang's Loss System. Assume that calls come in at the rate of 20 per hour, and that each operator can fill orders at a rate of 10 per hour.

    a. What is the average number of orders taken per hour if there is one operator? What is the average number if there are two operators?.

    b. Suppose each customer's order generates \$15 profit, on average. If each operator costs $\$10b$ per hour, then for what values of b do two operators generate more profit than one operator?

23. A business has two operators to accept telephone orders. Incoming calls arrive at a Poisson rate of 3 per minute, but if both operators are busy the caller gets a busy signal and hangs up. The placing of an order requires an exponentially distributed amount of time, with rate μ per minute.

    a. Draw a state-transition diagram, and calculate the steady-state probabilities for this system.

    b. Management wants no more than one-fourth of the callers to get a busy signal. How large must μ be to achieve this?

24. Consider an Erlang's loss system with traffic intensity α and a large number s of servers.

    a. Show that the probability of having n customers in the system is approximately given by a Poisson distribution with mean α, for small values of n.

    b. Use Stirling's formula:

$$n! \sim (2\pi)^{1/2} n^{n+1/2} e^{-n}$$

    to show that Erlang's loss function is approximately

$$B(s,\alpha) \approx (2\pi s)^{-1/2} \cdot \left(\frac{e\alpha}{s}\right)^s \cdot e^{-\alpha}$$

    c. Suppose $\alpha = 10$. Use the approximation in part (b) to estimate how many servers must be provided so that fewer than 5% of the arriving customers are lost.

    (HINT: Proceed by trial-and-error. As a start, you might note that the approximation in part (b) is a monotone decreasing function of s, and that when $s = \alpha$ the value is $(2\pi s)^{-1/2}$.)

25. Verify equation (11.34).

26. For Example 11.9, compare the simulation values of the following statistics (using the simulation data given in the text) against the values calculated using steady-state probabilities (see Exercise 9):

    a. The average customer time at each facility

    b. The proportion of time that the wash facility is busy

    c. The proportion of time that the vacuum facility is busy

    d. The arrival rate to the wash facility (this refers to customers who actually use the wash facility)

    e. The arrival rate to the vacuum facility (use Little's formula to calculate the steady-state value from the steady-state average number of customers and average time at the vacuum facility)

    f. The proportion of wash customers who also used the vacuum facility

27. For Example 11.10, make the following comparisons between the simulation data for the exponential model given in the text and the steady-state probabilities (see Exercise 21):

    a. The simulated total number of arrivals versus the expected total number of arrivals in the given time period

    b. The average queue contents in the simulation versus the steady-state value L_Q

    c. The average queue time in the simulation versus the steady-state value W_Q

    d. The proportion of ships that had 0 queue time in the simulation versus the steady-state probability that an arriving ship will find the dock free

e. The simulated average unloading time versus the expected unloading time per ship

f. The proportion of time in the simulation that the dock was busy versus the steady-state value of this statistic

28. The queue statistics in the simulation of Example 11.10 showed that 30.2% of the ships waited more than 24 hours in the queue (see Fig. 11.12). Calculate the steady-state probability of this event assuming exponential service times.

29. (Simulation) Modify the model in Example 11.8 by making the service time uniformly distributed between 15 and 25 minutes. Run an 8-hour and a 400-hour simulation of this modified model, and compare the queue and server statistics you obtain with the exponential model simulation data given in the text. Compare the percentage of potential customers were turned away because the queue was full in the two models.

30. (Simulation) Customers arrive at a single-server counter with uniformly distributed interarrival times of 20 ± 10 seconds. If there are fewer than ten customers in the queue, a new arrival joins the queue; otherwise, the customer leaves (thus the maximum queue length is 10). When customers get to the server, they purchase from one to four items with the following probabilities: 50% purchase one item, 20% purchase two items, 20% purchase three items, and 10% purchase four items. It takes 15 seconds of the server's time for each item purchased (so the total service time for each customer is random and depends on the number of items purchased).

    Write a program that will simulate 1 hour of operation of this system and gather statistics on customer queue time. Using the statistics generated by running your program, answer the following questions:

    a. What proportion of arriving customers left because the queue was full?

    b. What proportion of customers waited more than 5 minutes (300 seconds) in the queue?

    Calculate the average arrival rate $[\lambda_a]_{\text{sim}}$ to the system for your simulation, and verify that Little's formula

$$[L_Q]_{\text{sim}} = [\lambda_a]_{\text{sim}}[W_Q]_{\text{sim}}$$

    is true for your simulation averages.

31. (Simulation) Modify the model in Exercise 30 as follows. The maximum allowed queue length remains 10, but now there are two servers at the counter working simultaneously (a customer at the head of the queue goes to the first free server). The distribution of service time for each server remains the same, and the arrival distribution remains the same.

    Write a program for this two-server model that will simulate 1 hour of operation and gather statistics on customer queue time. Using the statistics generated by running your program, calculate the following quantities:

    a. The proportion of arriving customers who left because the queue was full

b. The average arrival rate $[\lambda_a]_{\text{sim}}$ to the queue

c. The proportion of customers who waited more than 5 minutes (300 seconds) in the queue

32. (Simulation) Compare the efficiency of the two-server model in Exercise 31 with the that of the one-server model in Exercise 30, under the following assumptions:

    a. Each potential customer turned away because the queue is full costs \$5.

    b. Customer waiting time (for each customer who joins the queue) costs \$10 per hour spent in the queue.

    c. Each server costs $\$(20 + X)$ per hour, where X is the proportion of time the server is busy.

    (Assume that no cost is incurred for lost time when a customer is being served.) Use your simulation data to compare the total cost (customer cost + server cost) for one hour of operation of each model.

33. (Simulation) Modify the model in Exercise 30 as follows: Assume that there is a sales counter and shipping counter. Customers arrive at the sales counter with interarrival times that are *exponentially* distributed with a mean of 30 seconds. *All* customers queue up, and when they reach the sales clerk, they randomly select one, two, three, or four items with probabilities 0.5, 0.2, 0.2, and 0.1 respectively, as Exercise 30. It takes 15 seconds for the sales clerk to process each item selected. After purchasing the items, 60% of the customers leave immediately. The remaining 40% of the customers queue up at the shipping counter, where it takes 30 ± 10 seconds to arrange for the shipment of the purchases (independently of how many items were purchased). These customers then leave.

    Write a program that will simulate 1 hour of operation of this system and gather statistics on customer queue time and the number of items purchased by each customer. Using the statistics generated by running your program, make the following comparisons:

    a. The expected number of customers arriving at the sales clerk versus the simulated number of arrivals

    b. The theoretical frequencies of the various numbers of items bought with the simulated frequencies

34. (Simulation) Write a program to simulate the model in Exercise 5. Use the results of an 8-hour and a 400-hour simulation to answer parts (c), (d), and (e) of Exercise 5, and compare with the answers obtained using steady-state probabilities.

35. (Simulation) Modify Exercise 34 by assuming that the service time is uniformly distributed on $(15, 25)$. Answer parts (c), (d), and (e) of Exercise 5 using your simulation results, and compare these with the answers you calculated in Exercise 34.

Chapter 12

Renewal Processes

We have been analyzing various probability models based on exponential random variables. These models describe *completely random* (memoryless) phenomena, such as the failure of certain types of electronic equipment and the arrival of telephone calls at a switchboard, but they do not take into account any aging. The Poisson process, for example, is the continuous-time limiting version of repeated coin tosses, in which the coin never wears out and has a constant (very small) probability of showing heads.

The assumption of exponentially distributed lifetimes is clearly inappropriate when components show aging effects. In this chapter we study stochastic models of events occurring at random moments in time, with the interevent times not assumed to be exponentially distributed. This leads to **renewal processes** and a variety of models of the **renewal-reward** type. These models naturally arise when a system shows random cyclic behavior.

The basic theoretical result of the chapter is the Renewal-Reward Theorem. By suitably choosing the renewal points and the rewards, we use this theorem to calculate long-term averages of important random quantities such as operating costs. We also apply this technique to queueing models in which the service times are not exponentially distributed.

12.1 Failure Rates

The assumption of exponentially distributed lifetimes is not appropriate for many stochastic part-replacement models. For example, the lifetime of a car battery is unlikely to have the no-memory property. If a particular battery with a rated life of 48 months is currently 60 months old and is functioning, we are unlikely to accept it as equivalent to a new battery. In this section, we describe some probability models for lifetimes, expanding on the ideas introduced in Section 7.1.

Suppose a machine contains a certain type of component. We shall assume that the operating lifetime of this component is a continuous random variable $X \geq 0$ that has a probability density $f(x)$ and cumulative probability distribution $F(x)$. We denote by $\tau = E[X]$ the average lifetime, which we assume to be finite (τ has the physical dimensions of time).

If the component is working at time t, then the probability that it fails in the time interval between t and $t + \delta t$ is

$$P\{t < X < t + \delta t \mid X > t\} \tag{12.1}$$

If δt is small and the density $f(t)$ is continuous at t, then

$$P\{t < X < t + \delta t\} \approx f(t)\,\delta t$$

Assume that $F(t) < 1$, so that the event $\{X > t\}$ that we are conditioning on in formula (12.1) has positive probability. If we define the **failure rate** (also called **hazard rate) function** to be

$$h(t) = \frac{f(t)}{1 - F(t)} \tag{12.2}$$

then we can express formula (12.1) as

$$P\{\,\text{Failure occurs between } t \text{ and } t + \delta t \mid \text{Lifetime } > t\,\} \approx h(t)\,\delta t \tag{12.3}$$

When X is memoryless (exponential), then $1 - F(t) = e^{-\lambda t}$ and $f(t) = \lambda e^{-\lambda t}$, so that $h(t) = \lambda$ is constant (recall that $\lambda = 1/\tau$), as we noted in Sections 7.2 and 7.4. In general, *the probability distribution of X is completely determined by its failure rate function $h(t)$*. To verify this, notice that, since $f(x) = F'(x)$, we can write

$$h(x) = -\frac{d}{dx}\log\big(1 - F(x)\big) \tag{12.4}$$

Since $F(0) = 0$ (remember that $X \geq 0$), we can integrate the differential equation (12.4) from 0 to t, and then exponentiate, to get the formula

$$1 - F(t) = \exp\left(-\int_0^t h(x)\,dx\right) \tag{12.5}$$

The integral of the failure rate function

$$H(t) = \int_0^t h(x)\,dx$$

is called the **cumulative hazard function**, and

$$R(t) = 1 - F(t) = P\{\,\text{Lifetime } > t\,\}$$

is called the **reliability function** of the component. The relation in formula (12.5) can thus be written as

$$R(t) = e^{-H(t)} \tag{12.6}$$

For modeling purposes, it is important to notice that, starting with *any* nonnegative piecewise continuous function $h(t)$, we can define a function $F(t)$ for $t \geq 0$ by formula (12.5). Since $H(t)$ is continuous, nonnegative, and monotone increasing (it is the area under the curve $y = h(t)$ between 0 and t), it follows that $R(t)$ is *decreasing* and bounded by 1. Thus, $F(t) = 1 - R(t)$ is automatically continuous, monotone increasing, and nonnegative. The normalization condition

$$\lim_{t\to\infty} F(t) = 1$$

will be satisfied, provided that the failure rate function satisfies

$$\int_0^{\infty} h(x)\,dx = \infty \tag{12.7}$$

The divergence of the integral (12.7) means that the cumulative hazard function is unbounded as time increases.

Notice that, if condition (12.7) does *not* hold, the improper integral on the left has a finite positive value. In this case

$$P\{\text{ Component never fails }\} = \lim_{t\to\infty} R(t) = \exp\left(-\int_0^{\infty} h(x)\,dx\right) > 0$$

A probability model based on a lifetime distribution with this property is not likely to be very realistic.

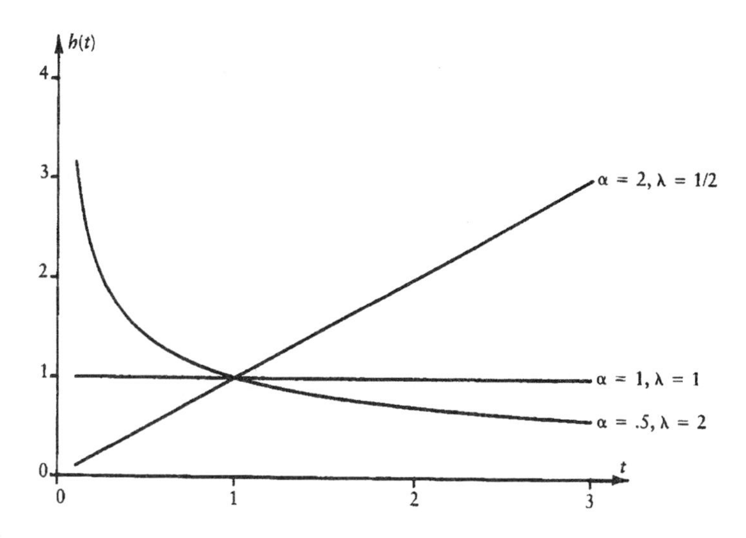

Figure 12.1: Failure Rates of Weibull Distributions

Example 12.1 Weibull Distribution

Suppose the failure rate function of a component is

$$h(x) = \lambda\alpha x^{\alpha-1}$$

where $\lambda > 0$ and $\alpha > 0$. Then the cumulative hazard function $H(t) = \lambda t^{\alpha}$, and condition (12.7) is obviously satisfied. From formula (12.6), we see that the reliability function is

$$R(t) = \exp\left(-\lambda t^{\alpha}\right)$$

This probability distribution is called the **Weibull distribution**. By adjusting the parameter α, we can model various types of aging. When $\alpha < 1$, the failure rate *decreases* with age; when $\alpha = 1$, the failure rate is *constant*, and we have an exponential distribution; when $\alpha > 1$, the failure rate *increases* with age. The parameter λ is a scaling parameter that is determined by the unit of time measurement. By adjusting the parameters α and λ, we can make a distribution of Weibull type fit a wide range of experimental data. Some examples of failure-rate functions for Weibull distributions are shown in Figure 12.1. ■

If a component has a *finite* lifetime T, so that $f(t) = 0$ when $t \geq T$, then we define the failure rate function $h(t) = +\infty$ for $t \geq T$. With this convention, formula (12.5) is still valid, since $F(t) = 1$ for $t \geq T$.

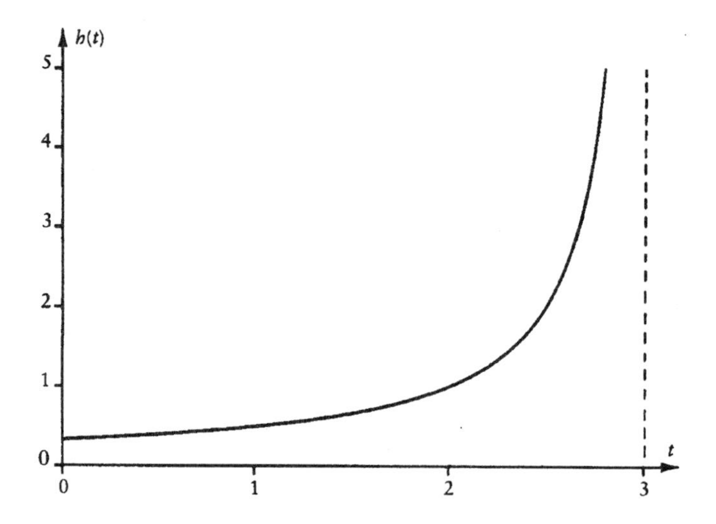

Figure 12.2: Failure Rate of Uniform $(0, 3)$ Random Variable

Example 12.2 Uniform Lifetime

Suppose that X is uniformly distributed on $(0, T)$. Then $f(t) = 1/T$ and $F(t) = t/T$ for $0 \leq t \leq T$. Thus, by formula (12.2),

$$h(t) = \frac{1/T}{1 - (t/T)} = \frac{1}{T - t} \quad \text{for } 0 \leq t < T$$

and $h(t) = +\infty$ for $t \geq T$. The graph of $h(t)$ is shown in Figure 12.2. ■

Remark Given empirical data about lifetimes, we can use statistical tests for *goodness of fit* to determine a probability distribution to model these lifetimes. (The description of such tests is outside the scope of this book but can be found in the references listed at the end of the book.)

12.2 Renewal Model

We introduce the notion of a general renewal process by using a replacement model, as we did for the Poisson process in Section 7.2. Suppose that a machine (such as a clock) contains a component (such as a battery) whose operating lifetime is a continuous positive random variable X. As in Section 12.1, we denote by $\tau = E[X]$ the average lifetime of the component, which we assume to be finite (τ has the physical dimensions of time).

Consider the following scenario. Assume that there is an unlimited supply of identical replacement components, arranged in an ordered stack. This means that the lifetimes $X_1, X_2, \ldots$ of the successive components are a sequence of mutually independent, identically distributed random variables, with common distribution function $F(x)$ and density function $f(x)$. We install the first component in the machine and turn on the machine at time $t = 0$. When this component fails, we immediately replace it with the second component, and so on.

We shall say that a **renewal** has occurred each time we replace a component. Let

$$N(t) = \text{Number of renewals up to and including time } t$$

By assumption every component has a positive lifetime, so $N(0) = 0$. We want to determine the probabilistic properties of the family of random variables $\{N(t) : t > 0\}$, just as we did for the Poisson process in Section 7.2.

Notice that $N(t)$ is a *discrete* random variable that depends on a continuous parameter $t \geq 0$. To analyze the stochastic process $\{N(t)\}$, we consider a natural complementary family of *continuous* random variables indexed by a discrete parameter n. Define

$$S_n = X_1 + X_2 + \cdots + X_n$$

Thus, S_n is the sum of the lifetimes of the first n components that are used. Recall from Section 7.2 that, for any $t \geq 0$,

$$\{S_n \leq t\} = \{N(t) \geq n\} \tag{12.8}$$

That is, if the sum of the lifetimes of the first n components is at most t, then there have been at least n renewals by time t, and vice versa.

For exponentially distributed lifetimes, it was easy to use moment-generating functions to determine the distribution of S_n (see Section 7.2). For the general renewal process, however, we use a direct probability argument. Define

$$F_n(t) = P\{\, S_n \leq t \,\}$$

for any integer $n \geq 1$. Notice that $F_1(t) = F(t)$, the cumulative lifetime distribution for a single component. We can get a recursive formula for $F_{n+1}(t)$ in terms of $F_n(t)$ by conditioning on X_1 and using the following fundamental renewal argument.

Suppose that the first renewal occurs at time x, and $0 \leq x \leq t$. Then

$$P\{\, S_{n+1} \leq t \mid X_1 = x \,\} = P\{\, S_n \leq t - x \,\} = F_n(t - x) \tag{12.9}$$

To see why formula (12.9) is true, notice that the process starts over after the first renewal. If this renewal occurs at time x, then we reset the clock, so that t becomes $t - x$. The elapsed time until n more renewals occur then has the same probability distribution as S_n.

Now we take the average of formula (12.9), relative to the probability density of X_1, to pass from a conditional probability to an absolute probability:

$$F_{n+1}(t) = \int_0^t F_n(t-x) f(x)\,dx \tag{12.10}$$

(see Section 5.3). This is the desired recursive formula for F_n. Taking the derivative relative to t of this formula and using the condition $F_n(0) = 0$, we obtain a recursive formula for the probability densities:

$$f_{n+1}(t) = \int_0^t f_n(t-x) f(x)\,dx \tag{12.11}$$

(see equation (4.13)).

Example 12.3 Constant Failure Rate

Suppose that the components in the replacement model have a constant failure rate λ. Then we know that the lifetimes are exponentially distributed, and $f(x) = \lambda e^{-\lambda x}$. Thus, from formula (12.11) with $n = 1$, we have

$$f_2(t) = \lambda^2 \int_0^t e^{-\lambda(t-x)} e^{-\lambda x}\,dx = \lambda^2 e^{-\lambda t} \int_0^t dx = \lambda^2 t e^{-\lambda t}$$

(see Example 4.11). This is a gamma density with parameters $(2, \lambda)$.

In general, S_n is the sum of n independent exponential random variables, each with rate λ. Using moment-generating functions, we found in Example 4.19 that such a sum is a **gamma random variable** with parameters (n, λ). Here, we have derived this result directly from formula (12.11) in the case $n = 2$, and a similar calculation (with induction on n) works for any value of n. (Note that the key property $e^x e^y = e^{x+y}$ of the exponential function that simplified the calculation of the integral here is the same property that makes the exponential random variable memoryless.) ■

Although we cannot explicitly calculate the integrals appearing in formula (12.10) for a general lifetime distribution $F(t)$, we can still use this formula to get information about $F_n(t)$. For example, since $F_n(t-x) \leq F_n(t)$, we can estimate $F_{n+1}(t)$ as follows:

$$F_{n+1}(t) \leq \int_0^t F_n(t) f(x)\,dx = F_n(t) \cdot F(t)$$

Hence, by induction on n, we have

$$F_n(t) \leq F(t)^n \tag{12.12}$$

Suppose that there is a positive probability that a component's lifetime is greater than t. Then $F(t) < 1$, and we see from formula (12.12) that $F_n(t) \to 0$ at a geometric rate as $n \to \infty$. (This rapid convergence to zero is true even if $F(t) = 1$, as can be proved by slightly modifying the argument leading to estimate in formula (12.12); see Exercises 9 and 10.)

Once we know the probability distribution of S_n, we can get the distribution of $N(t)$ from formula (12.8), just as we did in Section 7.2 for the Poisson process. We have

$$\begin{aligned} P\{ N(t) = n \} &= P\{ N(t) \geq n \} - P\{ N(t) \geq n + 1 \} \\ &= P\{ S_n \leq t \} - P\{ S_{n+1} \leq t \} && \text{(by equation (12.8))} \\ &= F_n(t) - F_{n+1}(t) && (12.13) \end{aligned}$$

In particular, we can calculate the expected number of renewals up to time t:

$$\begin{aligned} E[N(t)] &= \sum_{n=1}^{\infty} nP\{N(t) = n\} \\ &= \sum_{n=1}^{\infty} n\big(F_n(t) - F_{n+1}(t)\big) && \text{(by equation (12.13))} \\ &= \sum_{n=1}^{\infty} nF_n(t) - \sum_{n=1}^{\infty} nF_{n+1}(t) \end{aligned}$$

By the geometric rate of decay of $F_n(t)$ toward zero as $n \to \infty$, each of the these infinite series converges absolutely, so it is permissible to rearrange the terms. In the second series we make a shift of index $n \to n - 1$ and then recombine terms:

$$E[N(t)] = \sum_{n=1}^{\infty} nF_n(t) - \sum_{n=1}^{\infty}(n-1)F_n(t) = \sum_{n=1}^{\infty} F_n(t) < \infty \tag{12.14}$$

We have thus proved that the random variable $N(t)$ has a finite expectation. This result is plausible but not obvious from the definition of $N(t)$. Of course, for the special case of exponential lifetimes, as in Example 12.3, we already know from Section 7.2 that $N(t)$ is a Poisson random variable with finite expectation λt. The importance of this expected value in the study of renewal processes is described in Theorem 12.1.

Theorem 12.1. *For a renewal process $\{N(t)\}$, the expected number of renewals during any time interval $(0, t]$ is finite. Define the* **renewal function**

$$M(t) = E[N(t)]$$

Then the function $M(t)$ uniquely determines the interarrival time distribution $F(t)$, and hence determines the whole renewal process.

The finiteness of $M(t)$ follows from equation (12.14). The fact that $F(t)$ is completely determined by $M(t)$ will not be proved here (it uses the theory of the *renewal equation*—see Exercise 6); it implies, for example, that if $M(t) = \lambda t$, then $\{N(t)\}$ must be a Poisson process with rate λ. In working with a general renewal process, we shall not try to calculate $M(t)$ directly from equation (12.14). Rather, we shall relate the long-term behavior of $M(t)$ to the expected lifetime of a single component. This will require developing some more probabilistic tools in later sections.

We have introduced renewal processes using a particular part-replacement model. The following example illustrates a completely different model that contains a hidden renewal process.

Example 12.4 $M/G/1$ Queue with Capacity 1

Suppose that customers arrive at a single server at a Poisson rate of λ per second. Assume that there is no room to queue, so a customer gets served only if the server is idle at the moment of the customer's arrival. Let the service rate be μ customers per second, so that the average service time is $1/\mu$ seconds. We assume that successive service times are mutually independent and identically distributed, but we do *not* assume that the distribution of service time is exponential.

We shall say that a *renewal* occurs when an arriving customer finds the server idle. Define

$$X_n = \text{Service time of } n\text{th customer} + \text{Idle time of server waiting for next arrival} \tag{12.15}$$

Notice that an idle period for the server begins at the moment that a customer leaves, and ends at the moment of the next arrival. Since the service times are independent of the arrival process, we may use the shift invariance property of the Poisson process to take the moment a customer leaves as a starting point for counting the arrivals. Hence each idle period is exponentially distributed with mean $1/\lambda$. By the independent increments property of the Poisson process, successive idle periods are mutually independent. From equation (12.15), we conclude that the random variables $\{X_n\}$ are mutually independent and have the same probability distribution. Thus, we have a renewal process, with X_n being the length of the nth renewal interval.

From the definition of X_n we calculate that the average time between renewals is

$$\tau = E[X_n] = \frac{1}{\mu} + \frac{1}{\lambda}$$

We shall be able to say more about this example after determining the long-term average behavior of a renewal process in Section 12.4. ■

12.3 Random Stopping Rules and Wald's Identity

In this section, we reexamine the notion of the sum of a random number of random variables, which we studied in Section 5.4. This will lead to a very useful identity for treating renewal processes.

Suppose N is a random variable taking the values $0, 1, 2, \ldots$. We shall say that N is a **random stopping rule** for a sequence of random variables $X_1, X_2, \ldots$ if, for each integer $n \geq 0$, the event $\{N = n\}$ is independent of the random variables $X_{n+1}, X_{n+2}, \ldots$.

The intuitive meaning of a random stopping rule is most vividly illustrated in a gambling context. We play the same game repeatedly and win X_n on the nth play. Our playing strategy is to stop after the Nth game, where N is a random variable that can depend, for example, on our winnings. (We require that $N < \infty$, that is, we must stop playing in any case after a finite number of rounds. The value $N = 0$ means that we decide not to play any rounds of the game.) For any such strategy, the decision to stop after n games has to be made without knowing the outcomes of games $n+1, n+2, \ldots$. That is, the event $\{N = n\}$ cannot depend on our *future* wins or losses. This is precisely the definition of a random stopping rule.

Example 12.5

A trivial example of a random stopping rule arises when N is actually independent of *all* the random variables $\{X_j\}$. For example, N could be constant (non-random). ■

Example 12.6 Playing with a Fixed Goal

Suppose that we are playing a game in which we have a constant probability p of winning in each round of the game, and successive rounds are mutually independent. Before starting to play, we fix an integer $r \geq 1$ and agree to stop playing as soon as we win r times. The random variable

$$N = \text{Total number of rounds played to obtain } r \text{ wins}$$

is clearly a random stopping rule. We know from Example 3.5 that N has a *negative-binomial* distribution. In particular,

$$P\{N < \infty\} = 1$$

So with this strategy we are certain to stop playing after a finite number of rounds, even though that is not explicitly assumed. ■

The fundamental result relating to random stopping rules is the following formula for the expectation of the sum of a random number of random variables (compare with formula (5.15)).

Lemma 12.1 (Wald's Identity). *Suppose that the random variables $X_1, X_2, \ldots$ all have the same expectation τ. If N is a random stopping rule for the sequence $\{X_j\}$, then*

$$E\Big[\sum_{j=1}^{N} X_j\Big] = \tau \cdot E[N] \tag{12.16}$$

(When N takes on the value 0 then the sum inside the expectation on the left in equation (12.16) is defined to be zero.)

Proof Define the cutoff random variable I_j by

$$I_j = \begin{cases} 1 & \text{if } N \geq j \\ 0 & \text{if } N < j \end{cases}$$

We can then replace N by ∞ in the upper limit of summation, if we multiply each X_j by I_j:

$$\sum_{j=1}^{N} X_j = \sum_{j=1}^{\infty} X_j \cdot I_j \tag{12.17}$$

This follows because the event $\{N = n\}$ occurs if and only if $I_j = 1$ for $j \leq n$ and $I_j = 0$ for $j > n$. Thus, the factors I_j cut off the sum beyond n. (Of course, this is a *random* cutoff.)

The crucial point is that, for each integer j, the random variable I_j is *independent* of X_j. This is true because the event $\{I_j = 0\} = \{N \leq j - 1\}$ is independent of $X_j, X_{j+1}, \ldots$, by definition of a stopping rule. Thus, taking expectations in equation (12.17), we get

$$E\Big[\sum_{j=1}^{N} X_j\Big] = \sum_{j=1}^{\infty} E[X_j \cdot I_j] = \sum_{j=1}^{\infty} E[X_j] \cdot E[I_j]$$

(The interchange of summations and expectations can be justified by general results from integration theory.) Now $E[I_j] = P\{\, N \geq j \,\}$, while $E[X_j] = \tau$ doesn't depend on j. Thus, the left side of equation (12.16) is

$$\tau \sum_{j=1}^{\infty} P\{\, N \geq j \,\}.$$

To evaluate this sum, we denote $P\{\, N = j \,\}$ by p_j. Then

$$\begin{array}{rcl} P\{\, N \geq 1 \,\} & = & p_1 + \ p_2 + \ p_3 + \cdots \\ P\{\, N \geq 2 \,\} & = & \qquad\quad p_2 + \ p_3 + \cdots \\ P\{\, N \geq 3 \,\} & = & \qquad\qquad\qquad p_3 + \cdots \\ & \vdots & \end{array}$$

Adding all of these equations, we find that

$$\begin{aligned} \sum_{j=1}^{\infty} P\{\, N \geq j \,\} &= p_1 + 2p_2 + 3p_3 + \cdots \\ &= E[N] \end{aligned} \qquad (12.18)$$

This completes the proof of Wald's identity. (Formula (12.18) is valid for any nonnegative integer-valued random variable—see Exercise 26 in Chapter 3.) ■

Example 12.7 Expected Winnings with Fixed Goal

Return to the situation of Example 12.6. Suppose that, in each round of the game, we win \$1 with probability p and \$0 with probability q, where $0 < p \leq 1$ and $q = 1 - p$. If X_i is the amount we win in the ith round, then $E[X_i] = \$p$ for all i.

Suppose that our strategy is to stop playing as soon as we have won twice. This strategy means that we are certain to win \$2 (assume that the game costs nothing to play); in particular, our expected gain is \$2. On the other hand, the number of rounds N that we must play to win twice is a negative-binomial random variable with parameters $2, p$. Thus, $E[N] = 2/p$ (see Example 4.17). Wald's identity (equation (12.16) asserts that our expected gain is

$$E[N] \cdot E[X_i] = \frac{2}{p} \cdot p = 2$$

which of course agrees with the answer obtained directly from the definition of N. ■

Example 12.8 Expected Winnings with Stopping Rule

Consider the situation of Example 12.7, but suppose that, in each round of the game, we lose \$1 with probability q and win \$1 with probability p. If X_i is the amount we win in the ith round, then

$$E[X_i] = \$(p - q)$$

for all i.

Let N be any random stopping rule for the game. Then, by Wald's identity, our expected winnings when we play the game using the rule N are

$$\$(p - q) \cdot E[N]$$

In particular, *if we are playing a fair game* $(p = q = \frac{1}{2})$, *our expected winnings when we use any stopping rule are always zero.* ■

12.4 Renewals with Rewards

We return now to studying a renewal process $\{N(t) \,:\, t \geq 0\}$. As in the case of birth–death models, we shall analyze the process in terms of its long-term average behavior. There are many interesting stochastic models associated with renewal processes that can be described as follows.

Suppose that at the end of the nth renewal interval, a random **reward** R_n is paid. We allow R_n to be positive or negative; negative values correspond to costs incurred in a renewal cycle. Then

$$Y(t) = \sum_{n=1}^{N(t)} R_n \qquad (12.19)$$

is the total reward paid, by time t. A positive value of $Y(t)$ corresponds to a net profit up to time t, while a negative value corresponds to a net loss. We assume that the rewards $R_1, R_2, \ldots, R_n$ are mutually independent and identically distributed random variables for all n. It is important for the applications to allow the reward R_n to depend on the length X_n of the nth renewal interval. However, we will insist on the following condition:

> For every n, the reward R_n is independent of the earlier renewal intervals X_j for $j < n$.

When this condition is satisfied, we shall say that $\{N(t),\ R_n\}$ is a **renewal–reward process.**

Remark The independence assumptions about the rewards are crucial for the renewal property of the model: whenever a renewal occurs and the reward is paid, we can reset the clock to zero and have a system which is probabilistically identical to the original one. This is true because the rewards in future cycles depend neither on the length of the cycle just completed nor on the reward just paid. To set up such a model in realistic situations, however, we must choose the cycles carefully so that this independence (no-memory) property holds. We shall consider several examples of renewal–reward processes throughout this chapter.

Example 12.9 Insurance Claims

An insurance company pays out claims on its automobile accident policies in accordance with a Poisson process that has rate λ per week. Assume that there is no relation between the amounts of successive claims or between the amount of a claim and the number of claims received in a week. (Thus, we ignore the situation of a major disaster simultaneously affecting many cars insured by this company.) If the average amount paid on a claim is $\$r$, what is the expected amount paid out in one year?

Solution Let R_n be the amount paid to settle the nth claim, and let $N(t)$ be the number of claims paid up to time t (where $t = 0$ corresponds to the beginning of the year). The description of the situation makes it reasonable to *assume* that the sequence $\{R_n\}$ is mutually independent and identically distributed, and that each R_n is *independent* of $N(t)$ for all t. Since the amount paid out by time t is given by formula (12.19), the results of Section 5.4 on the sum of a random number of random variables imply that

$$E[Y(t)] = E[R_1] \cdot E[N(t)] = \$r \cdot \lambda t \tag{12.20}$$

Thus setting $t = 52$, we find that the expected total claims paid in a year amount to $\$52r\lambda$. The expected time τ between occurrences of claims is $1/\lambda$, so we can also write equation (12.20) in the form

$$\frac{E[Y(t)]}{t} = \frac{r}{\tau} = \frac{\text{Expected cost per claim}}{\text{Expected time between claims}} \tag{12.21}$$

The left side of equation (12.21) is the average **cost rate** (\$/week) to the company for paying claims. Notice that there are two kinds of time measurements involved in this equation: on the left side is the nonrandom calendar time t (one week or one year, for example); on the right side is expectation of the *random* time interval between occurrences of claims. ∎

Example 12.10 Part Replacement Model

Consider the part-replacement model of Section 12.2 and the associated renewal process $\{N(t)\}$. Assume that a part costs \$10 to replace, but when it fails at age X, however, it has salvage value $\$5e^{-X}$ (notice that the salvage value starts at \$5 but goes to zero rapidly as X gets large). Thus, the net cost of replacing the nth part is

$$R_n = 10 - 5e^{-X_n}$$

Clearly $\{N(t),\ R_n\}$ is a renewal-reward process, because R_n is determined by X_n, and hence it is independent of the lifetimes of the earlier components. ∎

We now shall turn to the fundamental theoretical result of this chapter: *on a long-term basis, equation* (12.21) *is valid for any renewal-reward model.* The argument in Example 12.9 that we used to obtain this formula does not suffice to prove the general result, since it assumed the mutual independence of the rewards and the process $\{N(t)\}$. This independence assumption is certainly *not* true in Example 12.10, since the rewards $\{R_n\}$ are completely determined by the process $\{N(t)\}$ in this case. On the other hand, the validity of Wald's identity would suffice to get equation (12.20). So the key property to establish is Lemma 12.2

Lemma 12.2. *Let $\{N(t),\ R_n\}$ be a renewal–reward process. Then, for any value of $t > 0$, the random variable $N = N(t) + 1$ is a random stopping rule for the sequence $R_1, R_2, \ldots$ of rewards.*

Proof First notice that the event $\{N = n\} = \{N(t) = n - 1\}$ means that, at time t, there have already been $n - 1$ renewals but the nth renewal has not yet occurred. In terms of the random variables $S_k = X_1 + \cdots + X_k$, we can express this event as

$$\{ N = n \} = \{ S_{n-1} \leq t \} \cap \{ S_n > t \}$$

Thus, this event is completely determined by the lengths of the first n renewal cycles. But by definition of a renewal–reward process, the rewards $R_{n+1}, R_{n+2}, \ldots$ are independent of the first n renewal cycles. Hence, the event $\{ N = n \}$ is independent of the random variables $\{R_j\ :\ j > n\}$. ■

We can now state and prove the general limiting version of equation (12.21).

Theorem 12.2 (Renewal–Reward Theorem). *Let $\{N(t),\ R_n\}$ be a renewal–reward process. Let τ be the average time between renewals, and let r be the average reward paid at each renewal. Then, for t large, the expected total reward up to time t is approximately rt/τ. That is,*

$$\lim_{t\to\infty} \frac{E[Y(t)]}{t} = \frac{r}{\tau} \tag{12.22}$$

As a special case of the Renewal–Reward Theorem, we can let the reward be 1 at each renewal. Then $Y(t) = N(t)$, and the average reward $r = 1$. Hence we get the following information about the renewal function $M(t)$ for large values of t.

Theorem 12.3 (Elementary Renewal Theorem). *Over a long time interval of length t, the expected number of renewals is approximately t/τ, so that the long-term* **renewal rate** *is $1/\tau$:*

$$\lim_{t\to\infty} \frac{M(t)}{t} = \frac{1}{\tau} \tag{12.23}$$

where $M(t) = E[N(t)]$ is the renewal function.

Proof of the Renewal Theorems By Lemma 12.2 and Wald's identity,

$$E\Big[\sum_{n=1}^{N(t)+1} R_n \Big] = r \cdot M(t) + r \tag{12.24}$$

The left side of equation (12.24) is $E\big[Y(t) + R_{N(t)+1}\big]$, so we conclude that

$$E[Y(t)] = r \cdot M(t) + r - E\big[R_{N(t)+1}\big] \tag{12.25}$$

Assume that R_n and X_n are bounded by some constant c. Then $E[R_{N(t)+1}]$ is bounded by c. Thus, with this assumption, the Renewal–Reward Theorem will follow from (12.25), once we prove the Elementary Renewal Theorem.

To prove the Elementary Renewal Theorem, we first apply equation (12.25) to the renewal–reward process with rewards $R_n = X_n$. In this case, $r = \tau$; therefore, from equation (12.25), we get another formula for the renewal function $M(t)$:

$$E\Big[\sum_{n=1}^{N(t)} X_n\Big] = \tau \cdot M(t) + \tau - E\big[X_{N(t)+1}\big] \tag{12.26}$$

But we can make a direct estimate of the random sum inside the expectation in equation (12.26) (just as we did in Example 12.7). The event $\{\, N(t) = n \,\}$ can be described by the pair of inequalities

$$X_1 + \cdots + X_n \le t \quad \text{and} \quad X_1 + \cdots + X_{n+1} > t$$

(The sum of the first n renewal intervals is at most t, and the $(n+1)$st renewal occurs after time t.) Thus, for any value of t, we have

$$t - X_{N(t)+1} \le \sum_{j=1}^{N(t)} X_j \le t \tag{12.27}$$

Taking expectations in estimate (12.27) and using equation (12.26) and the bound $E[X_{N(t)+1}] \le c$, we get the lower and upper bounds

$$t - \tau \le \tau \cdot M(t) \le t - \tau + c \tag{12.28}$$

Dividing by $t\tau$ and letting $t \to \infty$ in formula (12.28), we obtain the Elementary Renewal Theorem, and hence the Renewal–Reward Theorem. (The proof when the renewal cycles and rewards are not bounded uses a truncation argument whose justification requires integration theory.) ■

Remark Let us say that a **cycle** has been completed each time a renewal occurs. Then the Renewal–Reward Theorem asserts that

$$E[\,\text{Long-term reward rate}\,] = \frac{E[\,\text{Reward in cycle}\,]}{E[\,\text{Length of cycle}\,]} \tag{12.29}$$

Example 12.11 Renewals in $M/G/1$ Queue with Capacity 1

We can now finish our discussion of the $M/G/1$ Queue with capacity 1 (see Example 12.4). Recall that the average time between renewals in that example was

$$\tau = \frac{1}{\mu} + \frac{1}{\lambda}$$

where μ is the service rate and λ is the (Poisson) arrival rate. By the Elementary Renewal Theorem, the rate at which customers enter the system is

$$\lambda_a = \frac{1}{\tau} = \frac{\lambda\mu}{\lambda + \mu} \tag{12.30}$$

(In this model, only one customer gets served in each renewal cycle.) Since there is no queue, the average time an entering customer spends in the system is $W = 1/\mu$. Hence, by Little's formula, the average number of customers in the system is

$$L = \frac{\lambda_a}{\mu} = \frac{\lambda}{\lambda + \mu}$$

By the now-familiar PASTA principle—see Section 7.4—we know that the proportion of potential customers who actually get served is λ_a/λ. From formula (12.30), we thus find that

$$\text{Proportion of potential customers served} = \frac{\mu}{\lambda + \mu}$$

■

12.5 Age-Replacement Model

The Renewal–Reward Theorem of Section 12.4 has many applications. In this section we use it to analyze a modified version of the replacement model of Section 12.2 that takes into account the costs associated with replacements.

Example 12.5 Flashing Light Buoy

The Port Authority maintains a flashing light buoy at the entrance to a harbor. Let Y be the (random) lifetime of the flashing light unit in the buoy. Assume that on the basis of past experience, the cumulative probability distribution $G(y)$ of Y is known. The goal is to determine a minimal-cost strategy for keeping the light buoy in working order at all times, using the following **age-replacement policy**. A fixed maximum service period T (days) is specified for the light unit. Every day a patrol boat makes the rounds of the harbor. Whenever the light unit in the buoy has already been in service for T days, then the patrol boat routinely replaces it at a cost of C_1, even though it is still working. However, if the light unit fails at a random time before age T, then a boat is immediately dispatched to the buoy to replace the light at a cost of $C_1 + C_2$. Here C_2 is the extra cost for the special trip. The costs C_1 and C_2 are fixed, but the length T of the service period can be adjusted by the managers. ■

To study the costs associated with the stochastic model in Example 12.12, we set up a renewal process, as follows. Let the events of the process be the replacements of the light units, so that a *cycle* consists of the interval between successive renewals of the units. From the description of the age-replacement policy, it follows that the length of the nth cycle is

$$X_n = \min\{Y_n, T\} \tag{12.31}$$

where Y_n is the random lifetime of the n flashing light unit installed. (Notice that this renewal process has shorter cycles, in general, than does the replacement model process of Section 12.2, whose cycles have length Y_n.)

Suppose that the random variable Y_n has probability density $g(y)$ and cumulative distribution function $G(y)$. By equation (12.31) and the Law of the Unconscious

Statistician (Section 3.5), we can calculate

$$\begin{aligned} E[\text{Length of cycle}] &= \int_0^\infty \min\{y, T\}\, g(y)\, dy \\ &= \int_0^T y g(y)\, dy + \int_T^\infty T g(y)\, dy \\ &= \int_0^T y g(y)\, dy + T \cdot \big(1 - G(T)\big) \end{aligned} \tag{12.32}$$

This formula can be written more neatly, as follows. We can integrate by parts in the first integral, since $g(y) = G'(y)$:

$$\begin{aligned} \int_0^T y g(y)\, dy &= [yG(y)]\big|_{y=0}^{y=T} - \int_0^T G(y)\, dy \\ &= T \cdot G(T) - \int_0^T G(y)\, dy \end{aligned}$$

Substituting this in equation (12.32) gives

$$E[\text{Length of cycle}] = \int_0^T \big(1 - G(y)\big)\, dy = \int_0^T R(y)\, dy \tag{12.33}$$

where $R(y) = 1 - G(y)$ is the reliability function for the flasher (see Section 12.1). We denote by

$$L(T) = \int_0^T R(y) dy$$

the expected length of the replacement cycle from equation (12.33).

Next, we observe that

$$\begin{aligned} P\{\text{Light unit routinely replaced at age T}\} &= P\{Y > T\} \\ P\{\text{Light unit replaced because of random failure}\} &= P\{Y \le T\} \end{aligned}$$

Thus, from the description of the costs of making renewals, we calculate that

$$\begin{aligned} E[\text{Cost of cycle}] &= C_1 \cdot P\{Y > T\} + (C_1 + C_2) \cdot P\{Y \le T\} \\ &= C_1 \cdot (1 - G(T)) + (C_1 + C_2) \cdot G(T) \\ &= C_1 + C_2 \cdot G(T) \end{aligned} \tag{12.34}$$

Hence by substituting equations (12.33) and (12.34) into the basic renewal–reward formula (12.29), we find that

$$E[\text{Long-term cost rate}] = \frac{C_1 + C_2 \cdot G(T)}{L(T)}$$

Optimization Problem Denote by

$$C(T) = \frac{C_1 + C_2 \cdot G(T)}{L(T)} \tag{12.35}$$

the expected long-term cost rate. What choice of T will minimize $C(T)$?

Before solving this problem in a specific example, let's examine the behavior of $C(T)$ in the extreme cases $T \searrow 0$ and $T \to \infty$. Since the reliability function $R(y)$ is continuous and monotone decreasing, with $R(0) = 1$, we can estimate the integral in equation (12.33) as follows:

$$R(T) \cdot T = \int_0^T R(T)\, dy \le \int_0^T R(y)\, dy \le \int_0^T R(0)\, dy = T$$

Hence, we have the bounds

$$R(T) \cdot T \le L(T) \le T$$

In particular, $L(T) \approx T$ for T near 0. Since $G(0) = 0$, we conclude that

$$T \cdot C(T) \approx C_1 \quad \text{as } T \searrow 0$$

Because $C_1 > 0$ is fixed, the cost must become very large when the scheduled replacement interval T is small. On the other hand, from formula (12.32), it is also clear that, when $T \to \infty$, then $L(T) \approx \tau_G$ (the mean lifetime of the light unit). Since $G(T) \approx 1$ for large T, we see from formula (12.35) that

$$C(T) \approx \frac{C_1 + C_2}{\tau_G} \quad \text{as } T \to \infty \tag{12.36}$$

If the lifetime is exponentially distributed, then $C(T)$ turns out to be a monotone decreasing function of T, and the only solution to the optimization problem is $T = \infty$; that is, we make all replacements at the random time of failure (see Exercise 16). This is intuitively reasonable, by the no-memory property of the exponential random variable. On the other hand, for lifetime distributions that show aging, there may be a choice of T for which the cost rate is less than the limiting value in formula (12.36). Example 12.13 is such a case.

Example 12.13 Optimal Maintenance Policy

In the situation of Example 12.12, suppose that the lifetime Y of the light unit is uniformly distributed over $(0, 1)$ (with a suitable choice of time scale). If we choose a fixed age $T \ge 1$ in our replacement policy, then *all* replacements will be made at random times because of the failure of the light unit. Thus, the average length of a replacement cycle will simply be the average lifetime of the light unit in this case, so $L(T) = \frac{1}{2}$ with this policy. Since $G(T) = 1$ for $T \ge 1$, we conclude from formula (12.35) that

$$C(T) = 2(C_1 + C_2) \quad \text{if } T \ge 1$$

Suppose that we choose T in the range $0 < T < 1$ in our replacement policy. Since $R(y) = 1 - y$ for $0 \le y \le 1$, we calculate from formula (12.33) that

$$L(T) = \int_0^T (1 - y)\, dy = T - \frac{1}{2}T^2$$

Also $G(T) = T$ in this range, so formula (12.35) becomes

$$C(T) = \frac{C_1 + C_2 T}{T - \frac{1}{2}T^2} \quad \text{if } 0 < T \leq 1$$

In particular, note that the cost rate becomes infinite when $T \searrow 0$, as we previously observed.

To find the value $0 < T_{\min} < 1$ that minimizes the cost, we first set

$$r = \frac{C_1}{C_2}$$

Then the formula for $C(T)$ can be written as

$$C(T) = 2C_2 \cdot \frac{r + T}{2T - T^2}$$

Treating T as a continuous variable and calculating the derivative of the cost rate relative to T, we find that

$$C'(T) = 2C_2 \cdot \frac{T^2 + 2rT - 2r}{\left(2T - T^2\right)^2}$$

Since $r > 0$, we see from the numerator that $C'(T) < 0$ when T is near 0, while $C'(T) \approx 2C_2 > 0$ when T is near 1. Thus, $C(T)$ has a minimum in $(0, 1)$, which occurs at the unique root of $C'(T)$:

$$T_{\min} = \left(r^2 + 2r\right)^{1/2} - r \tag{12.37}$$

Numerical Examples Suppose that a scheduled replacement of a flasher unit costs \$200, while the added cost of dispatching a boat for random replacement is \$50. Then $r = 4$, and (from formula (12.37)) the optimal time interval for scheduled replacement is $T_{\min} = (24)^{1/2} - 4 = 0.90$. This is quite near the end of the maximum lifetime of the unit. (Since the units are expensive relative to the cost of the boat trip, it is better to get more use out of each flasher unit.) On the other hand, if each flasher unit costs \$10, while the extra trip for random replacement costs \$200, then $r = 0.05$, and the optimal value for T is $(0.1025)^{1/2} - 0.05 = 0.27$. Thus, when the flasher units are cheap relative to the extra cost for nonscheduled replacement, they should be replaced quite early. ■

12.6 Regeneration Points

The methods of renewal theory can be applied to any system that exhibits random *cyclic* behavior as time goes on, provided that the successive cycles are probabilistically independent but identically distributed.

Example 12.14 Renewal Cycles in $G/G/1$ Queue

Customers arrive at a single-server queue according to a renewal process (that is, the times between successive arrivals are mutually independent and identically distributed continuous random variables). We assume that the service times for each customer are mutually independent and identically distributed. Define

$$X(t) = \text{Number of customers in the system at time } t$$

Suppose that we observe the system, with time $t = 0$ being the moment when the first customer arrives. We record the special moments $T_1, T_2, \ldots$ at which a customer arrives when the queue is empty and the server is free. The significance of these (randomly distributed) moments is that from a probabilistic point of view, each of the shifted processes $\{X(t + T_j)\}_{t \geq 0}$ has the same distributions as the original process $\{X(t)\}_{t \geq 0}$. (This follows from the definition of the queueing system, since the past history of the system can be forgotten at each of these special arrival times.)

We say that each such random moment T_j is a **regeneration point** (or **renewal point**) for the system, and that the behavior of the system between successive regeneration points constitutes a **renewal cycle**. Clearly the lengths of the time intervals between successive regeneration points are a sequence of independent, identically distributed random variables. Thus, we can define a renewal process $\{N(t)\}_{t \geq 0}$ that counts the number of regeneration points up to time t. Notice that random quantities defined (such as customer waiting times) will be mutually independent if they occur in different cycles and are completely determined by the behavior of the system in one cycle. ■

In general, suppose we have a continuous-time stochastic process $\{X(t)\}_{t \geq 0}$. The random moments in time $T_1, T_2, \ldots$ are said to be *regeneration points* (also called *renewal points*) for the process if the starting-over and independence properties in Example 12.14 are true. In particular, the regeneration points define a renewal process; that is, the sequence of random variables $X_j = T_{j+1} - T_j$, for $j = 1, 2, \ldots$, is independent and identically distributed.

Once we have identified a sequence of regeneration points for a stochastic process, we can try to determine the steady-state behavior of the system, using the renewal–reward technique of Section 12.4.

For example, in the queueing system of Example 12.14, suppose that we pay a reward

$$R = \$1 \cdot \text{Length of busy period for the server}$$

in each cycle. Notice that, by our choice of regeneration points, a cycle consists of a busy period for the server followed by an idle period. (The idle period could have length 0 if a new customer happens to arrive at the precise moment when the queue is empty and the current service is completed.) Thus,

$$E[\text{Length of cycle}] = E[\text{Length of busy period}] + E[\text{Length of idle period}] \tag{12.38}$$

We shall assume that these expectations are finite. The total reward paid up to time t is the total amount of time that the server has been busy in all the cycles that have been completed by time t. If we define

$$p_B = \text{Long-term proportion of time the server is busy}$$

then the Renewal–Reward Theorem asserts that

$$p_B = \frac{E[\text{Length of busy period}]}{E[\text{Length of cycle}]} \tag{12.39}$$

To analyze the behavior of the queueing system, we would now like to calculate the numerator and denominator in formula (12.39) in terms of the arrival and service time parameters. This can be quite complicated in general, since the cycles are defined in a way that involves both the arrival times and the service times. So we restrict our attention to the following special case.

Example 12.15 $M/G/1$ Queue

Consider a single-server queue with Poisson arrivals at rate λ and general random service time S with $E[S] = 1/\mu$ (where μ is the service rate). We assume unlimited queue space.

This is a good model for a variety of applications. Arrivals to a queue are often uncontrolled and completely random (and hence comprise a Poisson process). On the other hand, the random properties of the service, whether it be performed by a person or a computer, vary considerably from one application to another; so the service time is often not exponentially distributed. Furthermore, in the design of queueing systems, such as computer networks, it is typically the service mechanism that can be adjusted to achieve a desired level of performance of the system.

The mean time between arrivals is $1/\lambda$, so the utilization factor ρ is λ/μ (see Section 11.6). From our study of the $M/M/1$ queue, it is reasonable to require that

$$\rho < 1$$

as a necessary condition for steady-state behavior. Because the arrival process is Poisson, the idle periods for the server are exponentially distributed, with mean $1/\lambda$. (What properties of the Poisson process does this use?) Thus, to determine p_B from formula (12.39), we only need to calculate the expected length of the busy period. We do this by applying the following ingenious renewal argument, due to L. Takács.

Consider a randomly chosen busy period and the following random variables:

B = Length of the busy period

S = Service time of the first customer J.DOE in the busy period

N = Number of customers arriving during J.DOE's service time

We shall calculate the conditional expectation of B, given both N and S. For example, if $N = 0$, then it is obvious that

$$E[B \mid N = 0,\ S = s] = s$$

In analyzing the situation when $N > 0$, we shall call a customer *special* if the queue is empty at the moment the customer's service begins. Of course, every customer who arrives when the server is idle is a special customer. But a customer can also be special even if he has to wait in line, provided that no other customers arrive during his queue time.

If we mark the t-axis at the moments that the special customers begin service, then the no-memory property of the Poisson arrival process implies that the distances between successive marks have the *same* probability distribution as B, the busy period. That is, we can forget about the special customer's waiting time, and say that a new busy period commences whenever a special customer's service begins.

For example, suppose that $N = 1$. In this case, the single customer C_1 who arrived during J.DOE's service is a special customer, and a new busy period begins at the moment J.DOE leaves, since the queue becomes empty at this moment. Furthermore, the original busy period B will have length S plus the length of this new busy period. Since all busy periods have the same probability distribution, we can write

$$E[B \mid N = 1,\ S = s] = s + E[B]$$

What is the situation if we know that $N = n > 1$? Suppose that customers $C_1, \ldots, C_n$ are in the queue at the moment J.DOE finishes service. During the remainder of this busy period, more customers may arrive after C_n; we shall call any such arrivals *regular* customers. We observe that the length of the busy period will be the same, regardless of the order in which $C_1, \ldots, C_n$ and the regular customers are served. Suppose we move $C_2, \ldots, C_n$ into a special customer line and only allow them to be served, one at a time, when there is no regular customer in the queue at the moment the server becomes available. This means that C_1 is served immediately after J.DOE, but C_2 has to wait until the regular line is empty and the only customers in the system are $C_2, C_3, \ldots, C_n$. This priority arrangement thus forces C_2 to be a special customer (even though C_2 has to wait while C_1 and any regular customers who arrived after C_n are served—see Figure 12.3).

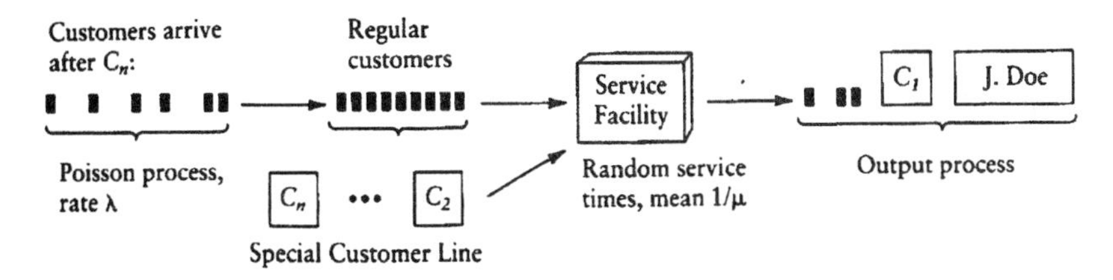

Figure 12.3: Analysis of Busy Period Using Special Customer Line

Since the arrival process has no memory, C_2 has to wait through a full busy period after C_1 begins service (until the regular customer queue is empty) before he is allowed to be served. Similarly, C_3 has to wait through a full busy period after C_2 begins service before he can start getting served, and so on. It will take a total of n busy periods before $C_1, \ldots, C_n$ leave the system and the original busy period B is finished.

Although the duration of each of these n busy periods is a different random variable, these random variables all have the same probability distribution and hence the same expectation as the random variable B. Therefore,

$$E[B \mid N = n,\ S = s] = s + n \cdot E[B]$$

Considering the conditional expectation as a random variable, we can write this relation as

$$E[B \mid N,\ S] = S + N \cdot E[B].$$

Hence, taking expectations, we get

$$E[B] = \frac{1}{\mu} + E[N] \cdot E[B] \tag{12.40}$$

Since the customers arrive at a Poisson rate λ and the arrival process is independent of the service time S, we have $E[N \mid S] = \lambda S$. Taking expectations once again, we get

$$E[N] = E[E[N \mid S]] = \lambda E[S] = \rho$$

Substituting this in formula (12.40) and solving for $E[B]$, we find that

$$E[B] = \frac{1}{\mu(1-\rho)} = \frac{1}{\mu - \lambda} \tag{12.41}$$

Now we return to formulas (12.38) and (12.39) at the beginning of this section. From formulas (12.38) and (12.41), we can calculate

$$E[\text{Length of cycle}] = \frac{1}{\mu - \lambda} + \frac{1}{\lambda} = \frac{1}{\rho(\mu - \lambda)} \tag{12.42}$$

In particular, this shows that the steady-state condition $\lambda < \mu$ (which we assumed to hold) does ensure that the regeneration cycles have finite expected length. However, this expected length becomes infinite as the arrival rate increases toward the service rate. This agrees with our intuition, since when the mean time between arrivals is only slightly more than the mean service time, the occurrence of an empty queue is a rare event.

Taking the ratio of formulas (12.41) and (12.42), we find from formula (12.39) that

$$p_B = \frac{E[B]}{E[\text{Length of cycle}]} \tag{12.43}$$

Since $\rho = \lambda/\mu = (1/\mu)/(1/\lambda)$, we can also write formula (12.43) as

$$p_B = \frac{E[\text{Service time}]}{E[\text{Time between arrivals}]}$$

This is the same value we previously derived for the $M/M/1$ queue, using steady-state probabilities. Thus the traffic intensity ρ measures the average utilization of the server, just as it did for the $M/M/1$ queue. ■

Example 12.16 Dispatching Shuttle Buses

An amusement park runs a fleet of shuttle buses to transport customers from the parking lot to the main entrance. The buses are lined up at the bus stop. Customers arrive at the bus stop according to a renewal process with mean interarrival time τ. They board the bus at the head of the line, which departs as soon as it has k passengers (where k is a fixed, nonrandom number). Assume that

a. The cost to the park operators for lost revenue due to customer waiting time at the bus stop is $\$c$ per hour per customer.

b. The cost to operate each bus is $a per trip.

What is the long-term rate of cost of the shuttle bus system?

Solution In this example, we may take the moments that the buses depart as the regeneration points, so a cycle consists of k successive customer arrivals. (The renewal process $\{N(t)\}$ defined by these regeneration points simply counts the number of buses that have departed up to time t.) Thus

$$E[\text{Length of cycle}] = k\tau \tag{12.44}$$

Let $X_1, X_2, \dots X_{k-1}$ be the time intervals between the successive customer arrivals in a particular cycle. When the bus arrives, the k passengers will have waited the following lengths of time at the bus stop:

$$\begin{array}{rccccc}
\text{1st customer:} & X_1 + & X_2 + & X_3 + \cdots + & X_{k-1} \\
\text{2nd customer:} & & X_2 + & X_3 + \cdots + & X_{k-1} \\
\text{3rd customer:} & & & X_3 + \cdots + & X_{k-1} \\
 & & \cdots & & \\
(k-1)\text{st customer:} & & & & X_{k-1} \\
k\text{th customer:} & & & \text{No waiting time} &
\end{array}$$

Thus, the total customer waiting time in the cycle is

$$X_1 + 2X_2 + 3X_3 + \cdots + (k-1)X_{k-1}$$

By assumptions (a) and (b), it follows that

$$\begin{aligned}
E[\text{Cost of cycle}] &= a + c\tau(1 + 2 + 3 + \cdots + (k-1)) \\
&= a + c\tau \cdot \frac{k(k-1)}{2}
\end{aligned} \tag{12.45}$$

We can now apply the Renewal-Reward Theorem from Section 12.4, using formulas (12.44) and (12.45):

$$\text{Average long-term cost rate} = \frac{a}{k\tau} + \frac{c(k-1)}{2} \tag{12.46}$$

(Notice that the cost of a particular cycle is independent of the lengths of all the other cycles, so the hypotheses of the Renewal–Reward Theorem are satisfied.)

Optimization Problem For fixed values of a, c, τ, choose the bus load value k to minimize the cost in formula (12.46). If we denote the right side of this formula as $F(k)$ and treat k as a continuous variable, then, by elementary calculus, we find that $F(k)$ is a concave function on $(0, \infty)$ with a global minimum at

$$k_{\min} = \left(\frac{2a}{\tau c}\right)^{1/2} \tag{12.47}$$

For the actual minimization with integer k, we can take either $k = [k_{\min}]$ or $k = [k_{\min}] + 1$—whichever gives the lower cost (where $[x] =$ largest integer $\leq x$).

Formula (12.47) does give some interesting information about the behavior of the optimal policy as a function of the parameters. Notice, for example, that if the arrival rate goes up by a factor of 4 ($\tau \to \tau/4$), then k_{min} only doubles, because of the square root in the formula.

In a realistic analysis, there is another constraint on k: the bus capacity is finite. Thus if k_{min} exceeds this capacity, the optimal strategy is to fill up the bus (see Exercise 21). ■

12.7 Lifetime Sampling and the Inspection Paradox

We return to our original model for a renewal process, as described in Section 12.2: a machine contains a certain type of component whose operating lifetime is a random variable X. We let $X_1, X_2, \ldots$ be the lifetimes of the successive components installed in the machine, and we let $S_n = X_1 + \cdots + X_n$ be the sum of the lifetimes of the first n components used.

In this section, we consider the following statistical sampling problem: Suppose that we periodically inspect the machine and record the current age $A(t)$ of the component in use at the time t of inspection. We can then use these data (generated from many observations over a long time period) to estimate the steady-state distribution of the **current age** A of the component in use. That is, A is the random variable whose distribution function is

$$P\{A \le c\} = \lim_{T\to\infty} \{\text{Proportion of interval } (0, T) \text{ on which } A(t) \le c\} \qquad (12.48)$$

In this formula, we are assuming that the limit exists and is the same for all sample observations of $A(t)$, just as in Section 10.5. (This is the *ergodic hypothesis*, which can be proved using the Law of Large Numbers, as in Section 6.9.)

Using our inspection data, we also tabulate the remaining life $Z(t)$ of the component in use at time t. (Notice that $Z(t)$ can't be determined empirically by the data up to time t; we must continue collecting inspection data and record when the component in use at time t finally fails.) From this data, we can estimate the steady-state distribution of the **remaining life** Z of the component in use. That is, Z is the random variable whose distribution function is

$$P\{Z \le c\} = \lim_{T\to\infty} \{\text{Proportion of interval } (0, T) \text{ on which } Z(t) \le c\} \qquad (12.49)$$

In this formula we are assuming that the limit exists and is the same for all sample observations of $Z(t)$; this will be true under the same ergodic hypothesis as for formula (12.48).

We shall use the Renewal–Reward Theorem to determine the probability distributions expressed in formulas (12.48) and (12.49). The answer turns out to be somewhat surprising—even paradoxical. This result will also play a key role when we return to the $M/G/1$ queue model in Section 12.8.

We start by recalling that the number of components that have failed up to time t is $N(t)$, so the component that we are inspecting at time t has total lifetime $X_{N(t)+1}$. Thus,

$$X_{N(t)+1} = A(t) + Z(t)$$

The random function $A(t)$ has a sawtooth graph, since $A(t) = t - S_{N(t)}$ when $S_{N(t)} \leq t < S_{N(t)+1}$. A typical graph of a sample function for the stochastic process $\{A(t)\}$ in an interval $(0, T)$ is shown in Figure 12.4.

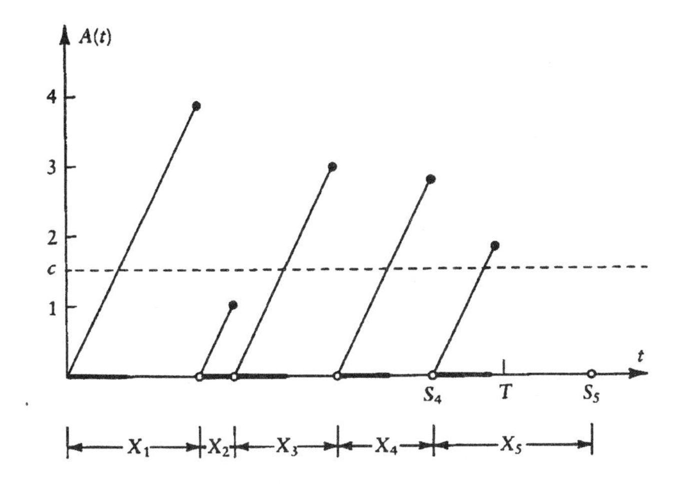

Figure 12.4: Sample Observation of Current Age at time t ($N(T) = 4$)

The graph of the remaining life $Z(t)$ is also a sawtooth, with teeth of the same height but facing the opposite way, since $Z(t)$ is constantly decreasing from the moment a component is installed. For the sample graph of $A(t)$ in Figure 12.4, the corresponding sample graph of $Z(t)$ is shown in Figure 12.5. In each graph, the segments on the t axis on which $A(t) \leq c$ (respectively, $Z(t) \leq c$) have been emphasized.

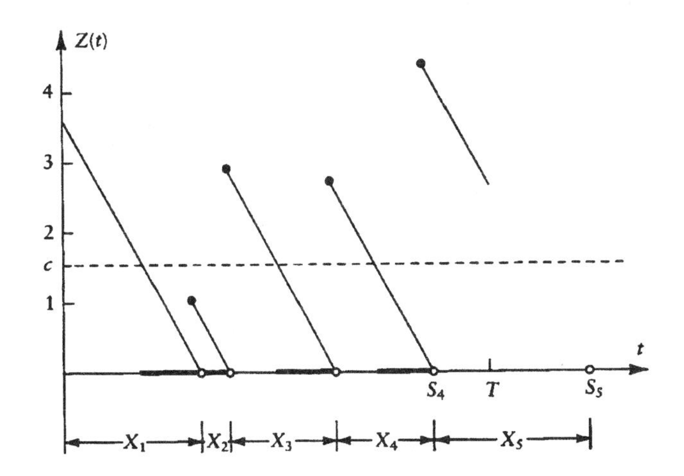

Figure 12.5: Sample Observation of Remaining Life at time t ($N(T) = 4$)

Consider the sample graphs of $A(t)$ and $Z(t)$ up to a fixed time T, and let N

be the number of renewals up to time T. It is clear from the graphs that, in each of the N renewal intervals, the subinterval on which $A(t) \leq c$ has the same length as the subinterval on which $Z(t) \leq c$. (It is only in the final interval between the Nth renewal point and T that the subintervals for $A(t)$ and $Z(t)$ don't have the same length.) Adding up the lengths of all these subintervals up to time T and dividing by T, we see that

$$\begin{aligned}&\{\text{Proportion of time current age } \leq c\} \approx \\ &\qquad\qquad \{\text{Proportion of time remaining life } \leq c\}\end{aligned} \tag{12.50}$$

for any constant $c > 0$. Notice that the two proportions in relation (12.50) are only approximately equal because of the final small subintervals in the two graphs (see Figures 12.4 and 12.5); however, it is clear from these graphs that the proportions differ by less than $(X_{N+1})/T$.

The relation expressed in formula (12.50) holds for any particular sample function over any finite time interval $(0, T)$, and the discrepancy between the two sides of (12.50) goes to zero as $T \to \infty$. Thus, identifying long-term time averages with probabilities (as in Section 10.5), we conclude that

$$P\{A \leq c\} = P\{Z \leq c\} \tag{12.51}$$

by formulas (12.48) and (12.49). This shows that *the steady-state current age and remaining life are identically distributed.*

Now we determine the common probability distribution of A and Z using the renewal–reward technique. Fix $c > 0$, and set up a reward process by paying a reward

$$R_k = \min\{X_k, c\}$$

when the kth component is replaced. From the graph of $A(t)$, it is clear that the total reward $Y(T)$ paid up to time T is essentially the same as the sum of the lengths of the time intervals up to T on which $A(t) \leq c$, if we ignore the final renewal interval containing T. Thus,

$$\lim_{T\to\infty} \frac{E[Y(T)]}{T} = \{\text{Long-term proportion of time that } A(t) \leq c\} \tag{12.52}$$

(The contribution of the ignored interval is at most c/T, which can be omitted in the limit as $T \to \infty$.)

Identifying long-term time averages with probabilities, we can write limit (12.52) as

$$\lim_{T\to\infty} \frac{E[Y(T)]}{T} = P\{A \leq c\} \tag{12.53}$$

by definition of the random variable A. The Renewal–Reward Theorem gives a formula for calculating the left side of (12.53)—namely,

$$\frac{E[\text{Reward per renewal}]}{E[\text{Length of renewal interval}]} \tag{12.54}$$

Let $F(x)$ be the cumulative probability distribution of the component lifetime, and let $R(x) = 1 - F(x)$ be the reliability function. We assume that $F(x)$ has a density

$f(x)$ and that the mean lifetime is τ. We already calculated in formula (12.33) that

$$E[\min\{X, c\}] = \int_0^c R(x)\,dx \tag{12.55}$$

Thus, combining formulas (12.53), (12.54) and (12.55), we obtain the cumulative probability distribution of the steady-state current-age random variable A:

$$P\{A \le c\} = \frac{1}{\tau}\int_0^c R(x)\,dx \tag{12.56}$$

From formula (12.56), we see that the random variable A has probability density

$$f_A(a) = \frac{R(a)}{\tau} = \frac{1}{\tau}\int_a^\infty f(x)\,dx \tag{12.57}$$

We can calculate the expectation of A from formula (12.57) by interchanging orders of integration (we assume that the integrals converge):

$$\begin{aligned} E[A] &= \frac{1}{\tau}\int_0^\infty a\left(\int_a^\infty f(x)\,dx\right)da \\ &= \frac{1}{\tau}\int_0^\infty \left(\int_0^x a\,da\right) f(x)\,dx \\ &= \frac{1}{2\tau}\int_0^\infty x^2 f(x)\,dx = \frac{E[X^2]}{2E[X]} \end{aligned} \tag{12.58}$$

Recall that, for any random variable X with finite mean τ and variance σ^2, we have

$$\sigma^2 = E[X^2] - \tau^2$$

Since A and Z are identically distributed, we find from formula (12.58) and this formula for the variance that

$$E[A] = E[Z] = \frac{\tau}{2} + \frac{\sigma^2}{2\tau} \tag{12.59}$$

Example 12.17 Current Age for Poisson Process

Suppose the component lifetime is an exponential random variable with mean $\tau = 1/\lambda$ (thus, the replacement process is a Poisson process with rate λ). Then $1 - F(x) = \exp(-\lambda x)$; so, from formula (12.57) we see that A and Z are exponentially distributed with mean $1/\lambda$ in this case. (Of course, this also follows directly from the memoryless property of the exponential distribution without taking long-term limits, since, for every t, $Z(t)$ is exponentially distributed with mean $1/\lambda$.) ■

By formula (12.59), the long-term average lifetime of the component currently in use is

$$\tau + \frac{\sigma^2}{\tau} \tag{12.60}$$

since it is the sum of the long-term average current age and remaining life. But if X is not constant, then $\sigma > 0$. Hence from formula (12.60) we arrive at the **inspection paradox**:

Expected lifetime of the component currently in use >
Expected lifetime of a typical component

Furthermore, the more random the lifetime (giving larger σ), the more long-lived the component currently in use will be, compared to an *average* component.

Example 12.18 Inspection Paradox for Uniform Lifetimes

Suppose the lifetime X is uniformly distributed on $(0, 1)$. In this case $\tau = \frac{1}{2}$ and $\sigma^2 = \frac{1}{12}$. Thus, the expected life of a randomly selected component is $\frac{1}{2}$, but the expected life of the component currently in use is $\frac{1}{2} + \frac{1}{6} = \frac{2}{3}$. ■

One explanation of the inspection paradox is that, if we wanted to determine the average lifetime of *all* the components by statistical sampling, then we should put a large (theoretically infinite) quantity of them in an urn, pick one at random from the urn, and use it until it fails. Then we would repeat the test on another component picked at random from the urn, and so on. After testing a certain number of components to failure, we would calculate the average of the lifetimes of all the components tested.

Of course, the amount of time that this procedure would take is random, since we would have to wait until all of the components tested failed before we could make the final calculation. On the other hand, by making our sampling consist of inspections of the machine at *fixed* times and by observing the component currently in use, we are introducing a bias: the longer-lived components will be inspected more often, and thus will skew our calculation of average lifetimes in the direction of longer-than-average life. We may describe this in terms of the sampling procedure by saying that the component "currently in use" is not a component that has been picked at random from an urn; the fact that it is currently functioning makes it special. (For a more analytical explanation of the inspection paradox, see Exercises 27 and 28.)

12.8 Waiting Times for the $M/G/1$ Queue

For another application of the lifetime sampling results of Section 12.7, we return to the $M/G/1$ queue (see Example 12.15). We shall apply virtually all the ideas developed in the chapter to derive an important formula of Pollaczek–Khintchine–Cramer, which vividly shows the effect of randomness in the service times.

As in Section 12.6, we assume that the arrivals are Poisson of rate λ, service is provided at rate μ (mean service time $\tau = 1/\mu$), and the utilization factor $\rho = \lambda/\mu < 1$. By this assumption about ρ, we know that the system approaches steady-state behavior and that ρ is the proportion of time during which the server is busy. Furthermore,

$$\text{Average length of busy period for server} = \frac{\tau}{1-\rho}$$

(see formulas (12.41) and (12.43) of Section 12.6). Thus from the server's point of view, any service-time distribution—whether it be exponential, gamma, or even constant—with a given mean τ produces the same results in the system. We shall now see that the situation is very different from the customer's point of view.

Suppose that the queueing system has been in operation long enough so that the random variables associated with the system can be treated using steady-state distributions. Let T_Q be the waiting time in the queue (excluding service time) for a typical customer J.DOE, and let N_Q be the number of customers in the queue at the moment J.DOE arrives (by the completely random nature of Poisson arrivals, this is a random moment in time). Of course, $T_Q = 0$ if J.DOE arrives when the server is idle. By the PASTA principle we have

$$P\{T_Q > 0\} = P\{\text{Server busy}\} = \rho$$

Thus,

$$E[T_Q] = \rho \cdot E[T_Q \mid T_Q > 0] \tag{12.61}$$

$$E[N_Q] = \rho \cdot E[N_Q \mid T_Q > 0] \tag{12.62}$$

The random variables N_Q and T_Q are related by Little's formula of Section 11.2: $E[N_Q] = \lambda E[T_Q]$. Substituting this relation in formula (12.62), we get

$$E[N_Q \mid T_Q > 0] = \frac{E[T_Q]}{\tau} \tag{12.63}$$

Consider now the situation at the moment of J.DOE's arrival. If $T_Q > 0$, then there are N_Q customers already in the queue, and 1 customer being served. Let Z be the remaining service time of this customer. Then

$$T_Q = X_1 + \cdots + X_{N_Q} + Z$$

where $\{X_j\}$ are the service times of the customers ahead of J.DOE. Since these service times are independent of N_Q, and each has expectation τ, we can calculate the expectation of this sum of a random number of random variables by the method of Section 5.4. This gives

$$E[T_Q \mid T_Q > 0] = \tau \cdot E[N_Q \mid T_Q > 0] + E[Z]$$

But by formula (12.63), we can write this result as

$$E[T_Q \mid T_Q > 0] = E[T_Q] + E[Z]$$

Now we use formula (12.61) to get

$$E[T_Q] = \frac{\rho}{1-\rho} \cdot E[Z] \tag{12.64}$$

The final step is to use formula (12.59) for $E[Z]$. (Notice that we are taking the successive service times as the renewal intervals; that is, the *components* are the customers, and the *lifetimes* are the service times. The random variable Z is then the remaining life of the component currently in use.) Substituting the value for $E[Z]$

given by this formula into formula (12.64) and simplifying, we get the **Pollaczek–Khintchine–Cramer formula** for the steady-state average queue waiting time:

$$W_Q = \frac{\rho}{1-\rho} \cdot \left(\frac{\tau}{2} + \frac{\sigma^2}{2\tau}\right) \tag{12.65}$$

(where τ and σ^2 are the mean and variance of the service-time distribution).

Example 12.19 Analytic Explanation of Simulation Data

Consider the two models for a shipping dock that were studied by simulation in Example 11.10. These are single-server queue models with arrival rate $\lambda = \frac{1}{12}$ customer per hour and service rate $\mu = \frac{1}{8}$ customer per hour. Thus the traffic intensity is $\rho = \frac{2}{3}$, and the mean service time is $\tau = 8$ hours. The steady-state utilization of the server is $\rho = 0.67$ in both models, which confirms the values determined by simulation. The steady-state average length of a busy period for the unloading crew at the dock is

$$\frac{\tau}{1-\rho} = 24 \text{ hours}$$

in both models.

When the service times are exponentially distributed, the service-time variance is $\sigma^2 = \tau^2 = 64$. By formula (12.65) we thus have

$$[W_Q]_{\text{exponential}} = \frac{\rho}{1-\rho} \cdot \tau = 16 \text{ hours}$$

(see formula (11.9). By Little's formula, the steady-state average queue length is

$$[L_Q]_{\text{exponential}} = \frac{16}{12} = 1.33$$

When the service times are uniform on (6, 10), then the service-time has variance $\sigma^2 = 4^2/12 = \frac{4}{3}$. By formula (12.65) we thus have

$$[W_Q]_{\text{uniform}} = 2 \cdot \left(\frac{8}{2} + \frac{4/3}{2 \cdot 8}\right) = 8.17 \text{ hours}$$

By Little's formula, the steady-state average queue length is

$$[L_Q]_{\text{uniform}} = \frac{8.17}{12} = 0.68$$

These are about half the corresponding values for the $M/M/1$ model.

Thus, the Pollaczek–Khintchine–Cramer formula furnishes a theoretical explanation for the large differences in mean queue time and queue length that we observed in the simulation data for the two models. ■

Example 12.20 Exponential versus Uniform Service Times

Let's compare an $M/M/1$ queue with mean service time 1 with a $M/G/1$ queue in which the service time is uniformly distributed on a small interval $(1-\epsilon, 1+\epsilon)$ around 1. In both cases, we assume an arrival rate $\lambda = 0.8$, so that the utilization factor $\rho = 80\%$ and the mean service time $\tau = 1$ for both systems. For the exponential service times, $\sigma^2 = \tau = 1$; so, from formula (12.65), we calculate that

$$W_Q = \left(\frac{0.8}{0.2}\right) \cdot \left(\frac{1}{2} + \frac{1}{2}\right) = 4$$

On the other hand, for the uniformly distributed service times, $\sigma^2 = \epsilon^2/3$, which yields the value

$$W_Q = 4 \cdot \left(\frac{1}{2} + \frac{\epsilon^2}{6}\right) = 2 + \delta$$

where $\delta = (\frac{2}{3})\epsilon^2$. Thus, if the service times are approximately constant, the average queue waiting time is cut in half, compared to the exponential service times. ■

Exercises

1. The random lifetime X (in years) of a certain type of electric light bulb has the failure rate function $h(t) = t$ for $t > 0$.

   a. Calculate $P\{X > 2 \text{ years}\}$.

   b. What is the *probability density* function of X?

   c. Calculate $E[X]$ and $\text{Var}(X)$.

   (HINT: X is related to the gamma random variable by a change of variable.)

2. The failure rate of a certain type of component is assumed to fit a Weibull distribution with $\alpha = 2$. Let $R(t)$ be the reliability function.

   a. Show that the first difference

   $$\log R(t + \Delta t) - \log R(t)$$

   is a *linear* function of t, for any choice of time step Δt. (This gives an empirical test for the reliability data to fit the assumed distribution.)

   b. After testing a large number of components, researchers observed that 15% of the components that lasted 90 hours failed before 100 hours. Use this data to determine the failure rate function.

3. Suppose that an electrical component is subjected to overloads (such as voltage surges) that occur at random times according to a Poisson process with rate λ. Assume that the component can withstand one overload, but fails when it receives a second overload.

   a. Show that the lifetime of the component has a gamma distribution, with parameters $2, \lambda$.

   b. Determine the failure rate function $h(t)$, plot it, and show that $h(t) \approx \lambda$ for large t.

4. The failure rate functions of Weibull type all satisfy $h(0) = 0$. Consider a modified version of the Weibull distribution, with failure rate function

$$h(t) = \beta + \lambda t^{\alpha-1}$$

with α, β, and λ positive parameters. Calculate the reliability function $R(t)$ for this distribution.

5. Consider Example 11.1. Show that this example can be modeled as an $M/G/1$ queue with capacity 1, where the service time is the sum of two independent exponential random variables with rates μ_1 and μ_2. What is the average time between successive renewals? (Say that a *renewal* occurs every time a car enters the wash facility.)

6. Let $N(t)$ be a renewal process with mean-value function $M(t)$. Let $f(x)$ be the probability density and $F(x)$ be the cumulative distribution function for the intervals $\{X_j\}$ between renewals.

   a. Suppose the first renewal occurs at time x. Show that

$$E[N(t) \mid X_1 = x] = \begin{cases} 0 & \text{if } x > t \\ 1 + M(t-x) & \text{if } x \le t \end{cases}$$

   b. Use part (a) to show that $M(t)$ satisfies the *renewal equation*

$$M(t) = F(t) + \int_0^t M(t-x) f(x)\, dx$$

   c. Let $m(t) = M'(t)$. By differentiating the renewal equation in part (b), show that $m(t)$ satisfies the equation

$$m(t) = f(t) + \int_0^t m(t-x) f(x)\, dx$$

   (HINT: Recall that $N(0) = 0$, so $M(0) = 0$.)

   d. Suppose that $m(t) = \lambda$ is constant. Use part (c) to find $f(x)$.

   (HINT: Differentiate the equation in part (c).)

7. In the $M/G/1$ queue with capacity 1, arrival rate λ, and service time cumulative distribution $G(t)$, show that the cumulative probability distribution for the length of a cycle (busy period + idle period) is given by

$$F(t) = \lambda \int_0^t G(t-x) e^{-\lambda x}\, dx$$

   (HINT: Condition on the length of the idle period.)

8. (Continuation of Exercise 7) Consider an $M/M/1$ queue with capacity 1, arrival rate λ, and service rate μ. Assume $\mu \neq \lambda$.

a. Use the formula given in Exercise 7 to show that the probability density for the length of a cycle is

$$f(t) = \lambda\mu \left(\frac{e^{-\lambda t} - e^{-\mu t}}{\mu - \lambda} \right)$$

b. Show (by calculus) that the limit as $\mu \to \lambda$ in part (a) gives a gamma density, with parameters $(2, \lambda)$. Then give a direct probability argument for this result.

9. Consider a renewal process with renewal intervals $X_1, X_2, \ldots$. Let $\tau = E[X_i]$ and $\sigma^2 = \text{Var}(X_i)$, which we assume to be finite. Let S_n be the sum of the lengths of the first n renewal intervals, as in Section 12.2. Let $F_n(t)$ be the cumulative distribution function of S_n, and let $f_n(t)$ be the probability density of S_n. Use Chebyshev's inequality to prove that

$$F_n(t) \le \frac{n\sigma^2}{(n\tau - t)^2} \qquad \text{for } n > t/\tau$$

10. Let $F_n(t)$ and $f_n(t)$ be defined as in Exercise 9.

a. Use the renewal argument that yielded equation (12.10) to show that, for any positive integers k and m,

$$F_{k+m}(t) = \int_0^t F_k(t-x) f_m(x)\, dx$$

b. Use the result of part (a) to show that

$$F_{k+m}(t) \le F_k(t) \cdot F_m(t)$$

c. Fix $t > 0$. Use the results of part (b) and Exercise 9 to find constants $\rho < 1$ and $C > 0$ such that

$$F_n(t) \le C\rho^n \quad \text{for } n = 1, 2, \ldots$$

(HINT: Use the estimate in Exercise 9 to find an integer r such that $F_r(t) < 1$. With r fixed, write any positive integer n in the form $rq + k$, with $0 < k \le r$, and apply the inequality in part (b).)

11. Suppose that $X_1, X_2, \ldots$ is a sequence of independent Bernoulli random variables with probability p of success.

a. Define $N = \min\{n : X_1 + \cdots + X_n = 5\}$. Show that N is a random stopping rule for this sequence of random variables.

b. Define $N' = 2$ if $X_2 = 1$ and $N' = 3$ if $X_2 = 0$. Is N' a random stopping rule for this sequence of random variables?

c. Define $N'' = 2$ if $X_2 = 1$ and $N'' = 1$ if $X_2 = 0$. Is N'' a random stopping rule for this sequence of random variables?

(HINT: Imagine that this sequence describes a coin-tossing game, in which you win $\$X_n$ on the nth toss. Interpret N, N', and N'' in terms of stopping strategies for the game. What information must you know in order to decide to stop on the nth toss, according to each of these rules?)

12. Let $X_1, X_2, \ldots$ and N be as in Exercise 11. Calculate

$$E\Big[\sum_{j=1}^{N} X_j\Big]$$

in two ways:

a. By using Wald's identity

(HINT: N has a negative-binomial distribution.)

b. Directly from the definition of N

13. You are betting on the outcomes of a sequence of Bernoulli trials. On each bet, you win \$1 with probability p or lose \$1 with probability q. Suppose that you adopt the strategy that you will stop betting *just before* your first loss.

a. Does this strategy satisfy the definition of a random stopping rule?

b. Let X_n be the amount you win on trial n and N the total number of trials that you bet on when you use this strategy. What are your expected winnings? Is Wald's identity satisfied?

14. (Continuation of Exercise 5) Consider Example 11.1.

a. Use the Elementary Renewal Theorem to calculate the average rate λ_a at which customers enter the car wash.

b. Use the Renewal–Reward Theorem to calculate the proportion of time during which the facility is busy. (Imagine that each customer pays \$1 per hour at the facility.)

c. Show that your answers in parts (a) and (b) agree with the values found in Example 11.1, using an exponential model.

15. Customers arrive at a single service counter at a Poisson rate $\lambda = \frac{1}{6}$ per minute. There is no room to queue, so a customer leaves without being served if the server is busy at the moment of arrival. If the server is free, a customer randomly orders one, two, or three items with probabilities 0.5, 0.4, and 0.1, respectively. Each item ordered requires exactly 2 minutes of service time.

a. Calculate the average service time of the customers who place orders.

b. Calculate $E[X]$, where the random variable X is the time between successive arrivals of the customers who get served (ignore the potential customers who leave because the server is busy).

c. Calculate λ_a, the long-term arrival rate of customers.

16. In Example 12.12, assume that the lifetime of the flasher unit is exponentially distributed. Show that the cost-rate function $C(T)$ defined by equation (12.35) is a monotone decreasing function of T in this case.

17. In Example 12.12, assume that the flasher unit has a lifetime that is uniformly distributed between 1 and 2 years. A scheduled replacement costs \$200, and the added cost for a replacement due to random failure is \$100.

a. Calculate the long-term average cost rate $C(T)$ ($ per year) when the scheduled replacements are made at age T.

b. Find the value of T, with $1 \leq T \leq 2$, that minimizes $C(T)$.

c. With T as in part (b), what is the average total cost per year for this policy?

d. Suppose that $T = 2$ (that is, the flasher units are only replaced when they randomly fail). What is the long-term average total cost per year for this policy?

18. Carry out the calculations in Exercise 17, with the added assumption that a working flasher unit that is T years old has a salvage value of $\$10 \cdot (2 - T)$, when $T < 2$. A non-working unit has zero salvage value.

    (HINT: Be careful in setting up the formula for $C(T)$. The salvage value is subtracted from the replacement cost, but it is random, since it depends on whether the flasher is working at the time of replacement.)

19. Telephone calls arrive at a ticket reservation office at a Poisson rate of 2 per minute. The time required for the single operator to take an order is uniformly distributed between 15 and 25 seconds. Incoming calls are put on hold if the operator is busy.

    a. What proportion of time is the operator busy, under steady-state conditions?

    b. What proportion of incoming calls are *not* put on hold?

    (HINT: Use the completely random (Poisson) property of the arrival times.)

20. Consider the $M/G/1$ queue with arrival rate λ and traffic intensity $\rho < 1$, as in Example 12.15. Assume steady-state conditions. Let the random variable X be the length of a cycle (= Length of busy period + Length of idle period). Let the random variable C be the number of customers served in a cycle.

    a. Set up a renewal–reward process by letting the reward in a cycle be the number of customers served in the cycle. What is the long-term reward rate of this process?

    (HINT: The queue-length is unlimited, so all arriving customers eventually get served.)

    b. Use the Renewal–Reward Theorem and part (a) to calculate $E[C]$.

    (HINT: Use the value for $E[X]$ that was calculated in Section 12.6.)

21. In Example 12.16, assume that the bus cost is $10 per trip and that the customer waiting cost is $20 per customer-hour. Suppose that the buses have a maximum capacity of 10 people. For what values of the mean interarrival time τ will the long-term average cost of operating the system be minimized by filling up each bus?

22. Derive formula (12.47).

23. A shuttle-bus system uses vehicles with a capacity of 10 passengers. Customers arrive at the bus terminal according to a Poisson process with rate λ and board the bus that is waiting there. As soon a bus is full, it leaves and an empty bus moves up to the loading area.

a. Let $X_1, X_2, \ldots$ be the length of time between successive bus departures. What is the probability distribution of these random variables?

b. Suppose that passenger waiting time at the terminal costs \$10 per hour per passenger and that each shuttle bus trip costs \$20. Use the Renewal–Reward Theorem to estimate the average cost of operating this system for 1,000 hours.

24. Suppose that light bulbs of the type described in Exercise 1 are installed one at a time at a particular location, and replaced whenever they fail. Let A be the age of the bulb currently in use. Calculate $E[A]$.

25. Suppose that you always replace your car battery by a new one of the same type when the old one fails. In the long run, what percentage of time is the battery currently in use in your car less than one year old, if the lifetime of this type of battery is

a. uniformly distributed between 0 and 2 years?

b. exponentially distributed, with mean 1 year?

(Notice that the expected lifetime of the battery is 1 year in both cases.)

26. In Exercise 25(a), what is the expected remaining lifetime of the battery currently in use? What is the expected total lifetime of the battery currently in use?

27. In Figures 12.4 and 12.5, the lifetimes of the first five components are $x_1 = 3.41$, $x_2 = 0.88$, $x_3 = 2.59$, $x_4 = 2.47$, and $x_5 = 3.94$. The length of the sampling interval is $T = 11$, and $N(T) = 4$.

a. Calculate the average of the lifetimes of these components.

b. Calculate the *average current age*

$$\frac{1}{T}\int_0^T A(t)\,dt$$

for the sample function in Figure 12.4, with $T = 11$.

(HINT: Interpret the integral as the area under the graph of $A(t)$.)

c. Calculate the *average remaining life*

$$\frac{1}{T}\int_0^T Z(t)\,dt$$

for the sample function in Figure 12.5, with $T = 11$.

(HINT: Interpret the integral as the area under the graph of $Z(t)$.)

d. Compare the sample mean in part (a) with the sum of the time averages in parts (b) and (c). (This is a numerical illustration of the inspection paradox.)

28. Let $A(t)$ be the current age and $Z(t)$ the remaining life at time t of the component currently in use in a renewal process. Fix a non-random time T, and let $N = N(T)$ be the number of components replaced up to time T. Let S_N be the sum of the lifetimes of these components. This exercise gives another way

to obtain formula (12.59), as well as an analytical explanation of the inspection paradox.

a. Show that the time average of $A(t)$ can be expressed in terms of the component lifetimes $\{X_j\}$ as

$$\frac{1}{T}\int_0^T A(t)\,dt = \frac{1}{2T}(X_1^2 + \cdots + X_N^2 + U_N)$$

where $U_N = (T - S_N)^2$.

b. Show that the time average of $Z(t)$ can be expressed in terms of the component lifetimes $\{X_j\}$ as

$$\frac{1}{T}\int_0^T Z(t)\,dt = \frac{1}{2T}(X_1^2 + \cdots + X_N^2 + V_N)$$

where $V_N = (T - S_N)(X_{N+1} + S_{N+1} - T)$.

c. Assume that the component lifetime X is bounded. Use the Renewal–Reward Theorem to show that the expected value of each of these time averages converges to $E[X^2]/(2E[X])$ as $T \to \infty$.

(HINT: Let the rewards $R_n = X_n^2$. Then, by parts (a) and (b), the time averages of $A(t)$ and $Z(t)$ are the same as the reward paid up to time T, except for the remainder terms U_N and V_N. If the lifetimes are bounded by a constant c, then $U_N \leq (X_{N+1})^2 \leq c^2$, so $E[U_N] \leq c^2$, and the same estimate holds for V_N. Now let $T \to \infty$. The restriction that X be bounded can be removed, when X has finite variance, by using results from integration theory.)

29. For the $M/G/1$ queue with fixed arrival rate of 1 customer per minute, calculate the average queue waiting time W_Q as a function of the utilization factor ρ, with $0 < \rho < 1$, for the following service-time distributions:

    a. Exponential with mean ρ

    b. Uniformly distributed on $(0, 2\rho)$

    c. Gamma with parameters $n = 2$, $\lambda = 2/\rho$

    (Notice that all three service times have mean ρ.)

30. Consider an $M/G/1$ queue.

    a. Starting from formula (12.65) for W_Q, calculate L_Q, W, and L, using Little's formula.

    b. Verify that, when the service times are exponentially distributed, your results agree with those found in Section 11.3.

31. Consider a single-server queue with Poisson arrivals at rate $\lambda = 2$ per hour. Suppose the service time is uniformly distributed on the interval $(0, \epsilon)$, where $\epsilon < 1$ hour. Assume that queue space is unlimited.

    a. Calculate the expected duration of a busy period for the server.

    b. Calculate the steady-state average queue time.

    c. Calculate the steady-state average number of customers in the queue.

Tables

Table 1 Discrete Probability Distributions

RANDOM VARIABLE	PROBABILITY MASS FUNCTION $p_X(k)$	MOM. GEN. FUNCTION $\phi_X(t)$	MEAN μ_X	VAR. σ_X^2
Bernoulli, parameter p	$\begin{cases} p & \text{if } k = 1 \\ q & \text{if } k = 0 \end{cases}$	$q + pe^t$	p	pq
Binomial, parameters n, p	$\binom{n}{k} p^k q^{n-k}$ $k = 0, 1, \ldots, n$	$(q + pe^t)^n$	np	npq
Geometric, parameter p	$q^{k-1} p$ $k = 1, 2, \ldots$	$\dfrac{pe^t}{1 - qe^t}$	$\dfrac{1}{p}$	$\dfrac{q}{p^2}$
Negative binomial, parameters r, p $r = 1, 2, \ldots$	$\binom{k-1}{r-1} q^{k-r} p^r$ $k = r, r+1, \ldots$	$\left(\dfrac{pe^t}{1 - qe^t}\right)^r$	$\dfrac{r}{p}$	$\dfrac{rq}{p^2}$
Poisson, parameter $\lambda > 0$	$\dfrac{\lambda^k}{k!} e^{-\lambda}$ $k = 0, 1, 2, \ldots$	$\exp\left(\lambda(e^t - 1)\right)$	λ	λ

NOTE: $\binom{n}{k} = \dfrac{n!}{k!(n-k)!}$ is the binomial coefficient, $0 < p \leq 1$, and $q = 1 - p$.

Table 2 Continuous Probability Distributions

RANDOM VARIABLE X	PROBABILITY DENSITY $f_X(x)$	MOM. GEN. FUNCTION $\phi_X(t)$	MEAN μ_X	VAR. σ_X^2
Uniform on (a, b)	$\dfrac{1}{b-a}$ $a < x < b$	$\dfrac{e^{bt} - e^{at}}{t(b-a)}$	$\dfrac{a+b}{2}$	$\dfrac{(b-a)^2}{12}$
Exponential, parameter $\lambda > 0$	$\lambda e^{\lambda x}$ $x > 0$	$\dfrac{\lambda}{\lambda - t}$	$\dfrac{1}{\lambda}$	$\dfrac{1}{\lambda^2}$
Gamma, parameters $\lambda > 0$ and $r > 0$	$\lambda e^{-\lambda x} \cdot \dfrac{(\lambda x)^{r-1}}{\Gamma(r)}$ $x > 0$	$\left(\dfrac{\lambda}{\lambda - t}\right)^r$	$\dfrac{r}{\lambda}$	$\dfrac{r}{\lambda^2}$
Normal, parameters $\mu > 0$ and $\sigma > 0$	$\dfrac{1}{\sigma\sqrt{2\pi}} \exp\left(-\dfrac{(x-\mu)^2}{2\sigma^2}\right)$	$\exp\left(\mu t + \dfrac{\sigma^2 t^2}{2}\right)$	μ	σ^2

NOTE: The density function $f_X(x) = 0$ for x outside the indicated range.

The Gamma function is $\Gamma(r) = \int_0^\infty x^{r-1} e^{-x}\, dx$ for $r > 0$. Special values: $\Gamma(n) = (n-1)!$ for $n = 1, 2, \ldots$ and $\Gamma\left(\frac{1}{2}\right) = \sqrt{\pi}$. Recursion: $\Gamma(r+1) = r\Gamma(r)$.

Table 3 Cumulative Normal Distribution Function

$$\Phi(a) = \frac{1}{\sqrt{2\pi}} \int_{-\infty}^{a} e^{-\frac{x^2}{2}} dx \text{ FOR } 0{\cdot}00 \leq a \leq 4{\cdot}99.$$

a	·00	·01	·02	·03	·04	·05	·06	·07	·08	·09
·0	·5000	·5040	·5080	·5120	·5160	·5199	·5239	·5279	·5319	·5359
·1	·5398	·5438	·5478	·5517	·5557	·5596	·5636	·5675	·5714	·5753
·2	·5793	·5832	·5871	·5910	·5948	·5987	·6026	·6064	·6103	·6141
·3	·6179	·6217	·6255	·6293	·6331	·6368	·6406	·6443	·6480	·6517
·4	·6554	·6591	·6628	·6664	·6700	·6736	·6772	·6808	·6844	·6879
·5	·6915	·6950	·6985	·7019	·7054	·7088	·7123	·7157	·7190	·7224
·6	·7257	·7291	·7324	·7357	·7389	·7422	·7454	·7486	·7517	·7549
·7	·7580	·7611	·7642	·7673	·7703	·7734	·7764	·7794	·7823	·7852
·8	·7881	·7910	·7939	·7967	·7995	·8023	·8051	·8078	·8106	·8133
·9	·8159	·8186	·8212	·8238	·8264	·8289	·8315	·8340	·8365	·8389
1·0	·8413	·8438	·8461	·8485	·8508	·8531	·8554	·8577	·8599	·8621
1·1	·8643	·8665	·8686	·8708	·8729	·8749	·8770	·8790	·8810	·8830
1·2	·8849	·8869	·8888	·8907	·8925	·8944	·8962	·8980	·8997	·90147
1·3	·90320	·90490	·90658	·90824	·90988	·91149	·91309	·91466	·91621	·91774
1·4	·91924	·92073	·92220	·92364	·92507	·92647	·92785	·92922	·93056	·93189
1·5	·93319	·93448	·93574	·93699	·93822	·93943	·94062	·94179	·94295	·94408
1·6	·94520	·94630	·94738	·94845	·94950	·95053	·95154	·95254	·95352	·95449
1·7	·95543	·95637	·95728	·95818	·95907	·95994	·96080	·96164	·96246	·96327
1·8	·96407	·96485	·96562	·96638	·96712	·96784	·96856	·96926	·96995	·97062
1·9	·97128	·97193	·97257	·97320	·97381	·97441	·97500	·97558	·97615	·97670
2·0	·97725	·97778	·97831	·97882	·97932	·97982	·98030	·98077	·98124	·98169
2·1	·98214	·98257	·98300	·98341	·98382	·98422	·98461	·98500	·98537	·98574
2·2	·98610	·98645	·98679	·98713	·98745	·98778	·98809	·98840	·98870	·98899
2·3	·98928	·98956	·98983	$\cdot9^{2}0097$	$\cdot9^{2}0358$	$\cdot9^{2}0613$	$\cdot9^{2}0863$	$\cdot9^{2}1106$	$\cdot9^{2}1344$	$\cdot9^{2}1576$
2·4	$\cdot9^{2}1802$	$\cdot9^{2}2024$	$\cdot9^{2}2240$	$\cdot9^{2}2451$	$\cdot9^{2}2656$	$\cdot9^{2}2857$	$\cdot9^{2}3053$	$\cdot9^{2}3244$	$\cdot9^{2}3431$	$\cdot9^{2}3613$
2·5	$\cdot9^{2}3790$	$\cdot9^{2}3963$	$\cdot9^{2}4132$	$\cdot9^{2}4297$	$\cdot9^{2}4457$	$\cdot9^{2}4614$	$\cdot9^{2}4766$	$\cdot9^{2}4915$	$\cdot9^{2}5060$	$\cdot9^{2}5201$
2·6	$\cdot9^{2}5339$	$\cdot9^{2}5473$	$\cdot9^{2}5604$	$\cdot9^{2}5731$	$\cdot9^{2}5855$	$\cdot9^{2}5975$	$\cdot9^{2}6093$	$\cdot9^{2}6207$	$\cdot9^{2}6319$	$\cdot9^{2}6427$
2·7	$\cdot9^{2}6533$	$\cdot9^{2}6636$	$\cdot9^{2}6736$	$\cdot9^{2}6833$	$\cdot9^{2}6928$	$\cdot9^{2}7020$	$\cdot9^{2}7110$	$\cdot9^{2}7197$	$\cdot9^{2}7282$	$\cdot9^{2}7365$
2·8	$\cdot9^{2}7445$	$\cdot9^{2}7523$	$\cdot9^{2}7599$	$\cdot9^{2}7673$	$\cdot9^{2}7744$	$\cdot9^{2}7814$	$\cdot9^{2}7882$	$\cdot9^{2}7948$	$\cdot9^{2}8012$	$\cdot9^{2}8074$
2·9	$\cdot9^{2}8134$	$\cdot9^{2}8193$	$\cdot9^{2}8250$	$\cdot9^{2}8305$	$\cdot9^{2}8359$	$\cdot9^{2}8411$	$\cdot9^{2}8462$	$\cdot9^{2}8511$	$\cdot9^{2}8559$	$\cdot9^{2}8605$
3·0	$\cdot9^{2}8650$	$\cdot9^{2}8694$	$\cdot9^{2}8736$	$\cdot9^{2}8777$	$\cdot9^{2}8817$	$\cdot9^{2}8856$	$\cdot9^{2}8893$	$\cdot9^{2}8930$	$\cdot9^{2}8965$	$\cdot9^{2}8999$
3·1	$\cdot9^{3}0324$	$\cdot9^{3}0646$	$\cdot9^{3}0957$	$\cdot9^{3}1260$	$\cdot9^{3}1553$	$\cdot9^{3}1836$	$\cdot9^{3}2112$	$\cdot9^{3}2378$	$\cdot9^{3}2636$	$\cdot9^{3}2886$
3·2	$\cdot9^{3}3129$	$\cdot9^{3}3363$	$\cdot9^{3}3590$	$\cdot9^{3}3810$	$\cdot9^{3}4024$	$\cdot9^{3}4230$	$\cdot9^{3}4429$	$\cdot9^{3}4623$	$\cdot9^{3}4810$	$\cdot9^{3}4991$
3·3	$\cdot9^{3}5166$	$\cdot9^{3}5335$	$\cdot9^{3}5499$	$\cdot9^{3}5658$	$\cdot9^{3}5811$	$\cdot9^{3}5959$	$\cdot9^{3}6103$	$\cdot9^{3}6242$	$\cdot9^{3}6376$	$\cdot9^{3}6505$
3·4	$\cdot9^{3}6631$	$\cdot9^{3}6752$	$\cdot9^{3}6869$	$\cdot9^{3}6982$	$\cdot9^{3}7091$	$\cdot9^{3}7197$	$\cdot9^{3}7299$	$\cdot9^{3}7398$	$\cdot9^{3}7493$	$\cdot9^{3}7585$
3·5	$\cdot9^{3}7674$	$\cdot9^{3}7759$	$\cdot9^{3}7842$	$\cdot9^{3}7922$	$\cdot9^{3}7999$	$\cdot9^{3}8074$	$\cdot9^{3}8146$	$\cdot9^{3}8215$	$\cdot9^{3}8282$	$\cdot9^{3}8347$
3·6	$\cdot9^{3}8409$	$\cdot9^{3}8469$	$\cdot9^{3}8527$	$\cdot9^{3}8583$	$\cdot9^{3}8637$	$\cdot9^{3}8689$	$\cdot9^{3}8739$	$\cdot9^{3}8787$	$\cdot9^{3}8834$	$\cdot9^{3}8879$
3·7	$\cdot9^{3}8922$	$\cdot9^{3}8964$	$\cdot9^{4}0039$	$\cdot9^{4}0426$	$\cdot9^{4}0799$	$\cdot9^{4}1158$	$\cdot9^{4}1504$	$\cdot9^{4}1838$	$\cdot9^{4}2159$	$\cdot9^{4}2468$
3·8	$\cdot9^{4}2765$	$\cdot9^{4}3052$	$\cdot9^{4}3327$	$\cdot9^{4}3593$	$\cdot9^{4}3848$	$\cdot9^{4}4094$	$\cdot9^{4}4331$	$\cdot9^{4}4558$	$\cdot9^{4}4777$	$\cdot9^{4}4988$
3·9	$\cdot9^{4}5190$	$\cdot9^{4}5385$	$\cdot9^{4}5573$	$\cdot9^{4}5753$	$\cdot9^{4}5926$	$\cdot9^{4}6092$	$\cdot9^{4}6253$	$\cdot9^{4}6406$	$\cdot9^{4}6554$	$\cdot9^{4}6696$
4·0	$\cdot9^{4}6833$	$\cdot9^{4}6964$	$\cdot9^{4}7090$	$\cdot9^{4}7211$	$\cdot9^{4}7327$	$\cdot9^{4}7439$	$\cdot9^{4}7546$	$\cdot9^{4}7649$	$\cdot9^{4}7748$	$\cdot9^{4}7843$
4·1	$\cdot9^{4}7934$	$\cdot9^{4}8022$	$\cdot9^{4}8106$	$\cdot9^{4}8186$	$\cdot9^{4}8263$	$\cdot9^{4}8338$	$\cdot9^{4}8409$	$\cdot9^{4}8477$	$\cdot9^{4}8542$	$\cdot9^{4}8605$
4·2	$\cdot9^{4}8665$	$\cdot9^{4}8723$	$\cdot9^{4}8778$	$\cdot9^{4}8832$	$\cdot9^{4}8882$	$\cdot9^{4}8931$	$\cdot9^{4}8978$	$\cdot9^{5}0226$	$\cdot9^{5}0655$	$\cdot9^{5}1066$
4·3	$\cdot9^{5}1460$	$\cdot9^{5}1837$	$\cdot9^{5}2199$	$\cdot9^{5}2545$	$\cdot9^{5}2876$	$\cdot9^{5}3193$	$\cdot9^{5}3497$	$\cdot9^{5}3788$	$\cdot9^{5}4066$	$\cdot9^{5}4332$
4·4	$\cdot9^{5}4587$	$\cdot9^{5}4831$	$\cdot9^{5}5065$	$\cdot9^{5}5288$	$\cdot9^{5}5502$	$\cdot9^{5}5706$	$\cdot9^{5}5902$	$\cdot9^{5}6089$	$\cdot9^{5}6268$	$\cdot9^{5}6439$
4·5	$\cdot9^{5}6602$	$\cdot9^{5}6759$	$\cdot9^{5}6908$	$\cdot9^{5}7051$	$\cdot9^{5}7187$	$\cdot9^{5}7318$	$\cdot9^{5}7442$	$\cdot9^{5}7561$	$\cdot9^{5}7675$	$\cdot9^{5}7784$
4·6	$\cdot9^{5}7888$	$\cdot9^{5}7987$	$\cdot9^{5}8081$	$\cdot9^{5}8172$	$\cdot9^{5}8258$	$\cdot9^{5}8340$	$\cdot9^{5}8419$	$\cdot9^{5}8494$	$\cdot9^{5}8566$	$\cdot9^{5}8634$
4·7	$\cdot9^{5}8699$	$\cdot9^{5}8761$	$\cdot9^{5}8821$	$\cdot9^{5}8877$	$\cdot9^{5}8931$	$\cdot9^{5}8983$	$\cdot9^{6}0320$	$\cdot9^{6}0789$	$\cdot9^{6}1235$	$\cdot9^{6}1661$
4·8	$\cdot9^{6}2067$	$\cdot9^{6}2453$	$\cdot9^{6}2822$	$\cdot9^{6}3173$	$\cdot9^{6}3508$	$\cdot9^{6}3827$	$\cdot9^{6}4131$	$\cdot9^{6}4420$	$\cdot9^{6}4696$	$\cdot9^{6}4958$
4·9	$\cdot9^{6}5208$	$\cdot9^{6}5446$	$\cdot9^{6}5673$	$\cdot9^{6}5889$	$\cdot9^{6}6094$	$\cdot9^{6}6289$	$\cdot9^{6}6475$	$\cdot9^{6}6652$	$\cdot9^{6}6821$	$\cdot9^{6}6981$

Example: $\Phi(3{\cdot}57) = \cdot9^{3}8215 = 0{\cdot}9998215.$

SOURCE: Reprinted, with permission, from A. Hald, *Statistical Tables and Formulas*, p. 35. Copyright 1952, John Wiley & Sons, New York.

Table 4 Random Numbers

Line/Col.	(1)	(2)	(3)	(4)	(5)	(6)	(7)	(8)	(9)	(10)	(11)	(12)	(13)	(14)
1	10480	15011	01536	02011	81647	91646	69179	14194	62590	36207	20969	99570	91291	90700
2	22368	46573	25595	85393	30995	89198	27982	53402	93965	34095	52666	19174	39615	99505
3	24130	48360	22527	97265	76393	64809	15179	24830	49340	32081	30680	19655	63348	58629
4	42167	93093	06243	61680	07856	16376	39440	53537	71341	57004	00849	74917	97758	16379
5	37570	39975	81837	16656	06121	91782	60468	81305	49684	60672	14110	06927	01263	54613
6	77921	06907	11008	42751	27756	53498	18602	70659	90655	15053	21916	81825	44394	42880
7	99562	72905	56420	69994	98872	31016	71194	18738	44013	48840	63213	21069	10634	12952
8	96301	91977	05463	07972	18876	20922	94595	56869	69014	60045	18425	84903	42508	32307
9	89579	14342	63661	10281	17453	18103	57740	84378	25331	12566	58678	44947	05585	56941
10	85475	36857	43342	53988	53060	59533	38867	62300	08158	17983	16439	11458	18593	64952
11	28918	69578	88231	33276	70997	79936	56865	05859	90106	31595	01547	85590	91610	78188
12	63553	40961	48235	03427	49626	69445	18663	72695	52180	20847	12234	90511	33703	90322
13	09429	93969	52636	92737	88974	33488	36320	17617	30015	08272	84115	27156	30613	74952
14	10365	61129	87529	85689	48237	52267	67689	93394	01511	26358	85104	20285	29975	89868
15	07119	97336	71048	08178	77233	13916	47564	81056	97735	85977	29372	74461	28551	90707
16	51085	12765	51821	51259	77452	16308	60756	92144	49442	53900	70960	63990	75601	40719
17	02368	21382	52404	60268	89368	19885	55322	44819	01188	65255	64835	44919	05944	55157
18	01011	54092	33362	94904	31273	04146	18594	29852	71585	85030	51132	01915	92747	64951
19	52162	53916	46369	58586	23216	14513	83149	98736	23495	64350	94738	17752	35156	35749
20	07056	97628	33787	09998	42698	06691	76988	13602	51851	46104	88916	19509	25625	58104
21	48663	91245	85828	14346	09172	30168	90229	04734	59193	22178	30421	61666	99904	32812
22	54164	58492	22421	74103	47070	25306	76468	26384	58151	06646	21524	15227	96909	44592
23	32639	32363	05597	24200	13363	38005	94342	28728	35806	06912	17012	64161	18296	22851
24	29334	27001	87637	87308	58731	00256	45834	15398	46557	41135	10367	07684	36188	18510
25	02488	33062	28834	07351	19731	92420	60952	61280	50001	67658	32586	86679	50720	94953
26	81525	72295	04839	96423	24878	82651	66566	14778	76797	14780	13300	87074	79666	95725
27	29676	20591	68086	26432	46901	20849	89768	81536	86645	12659	92259	57102	80428	25280
28	00742	57392	39064	66432	84673	40027	32832	61362	98947	96067	64760	64584	96096	98253
29	05366	04213	25669	26422	44407	44048	37937	63904	45766	66134	75470	66520	34693	90449
30	91921	26418	64117	94305	26766	25940	39972	22209	71500	64568	91402	42416	07844	69618
31	00582	04711	87917	77341	42206	35126	74087	99547	81817	42607	43808	76655	62028	76630
32	00725	69884	62797	56170	86324	88072	76222	36086	84637	93161	76038	65855	77919	88006
33	69011	65797	95876	55293	18988	27354	26575	08625	40801	59920	29841	80150	12777	48501
34	25976	57948	29888	88604	67917	48708	18912	82271	65424	69774	33611	54262	85963	03547
35	09763	83473	73577	12908	30883	18317	28290	35797	05998	41688	34952	37888	38917	88050
36	91567	42595	27958	30134	04024	86385	29880	99730	55536	84855	29080	09250	79656	73211
37	17955	56349	90999	49127	20044	59931	06115	20542	18059	02008	73708	83517	36103	42791
38	46503	18584	18845	49618	02304	51038	20655	58727	28168	15475	56942	53389	20562	87338
39	92157	89634	94824	78171	84610	82834	09922	25417	44137	48413	25555	21246	35509	20468
40	14577	62765	35605	81263	39667	47358	56873	56307	61607	49518	89656	20103	77490	18062
41	98427	07523	33362	64270	01638	92477	66969	98420	04880	45585	46565	04102	46880	45709
42	34914	63976	88720	82765	34476	17032	87589	40836	32427	70002	70663	88863	77775	69348
43	70060	28277	39475	46473	23219	53416	94970	25832	69975	94884	19661	72828	00102	66794
44	53976	54914	06990	67245	68350	82948	11398	42878	80287	88267	47363	46634	06541	97809
45	76072	29515	40980	07391	58745	25774	22987	80059	39911	96189	41151	14222	60697	59583
46	90725	52210	83974	29992	65831	38857	50490	83765	55657	14361	31720	57375	56228	41546
47	64364	67412	33339	31926	14883	24413	59744	92351	97473	89286	35931	04110	23726	51900
48	08962	00358	31662	25388	61642	34072	81249	35648	56891	69352	48373	45578	78547	81788
49	95012	68379	93526	70765	10593	04542	76463	54328	02349	17247	28865	14777	62730	92277
50	15664	10493	20492	38391	91132	21999	59516	81652	27195	48223	46751	22923	32261	85653

SOURCE: Reprinted, with permission, from *CRC Handbook of Mathematical Sciences*, 5th ed., p. 797. Copyright 1978 CRC Press, Inc., Boca Raton, Fla.

Suggestions for Further Reading

Historical Background

David, F. N. *Games, Gods and Gambling: A History of Probability and Statistical Ideas.*. London: Charles Griffin, 1962 (reprinted by Dover Publications, 1998).

Introductory Textbooks on Probability and Stochastic Processes

Chung, K. L. *Elementary Probability Theory with Stochastic Processes.* 3d ed. New York: Springer-Verlag, 1979.

Clarke, A. B., and Disney, R. L. *Probability and Random Processes.* 2d ed. New York: John Wiley & Sons, 1985.

Feller, W. *An Introduction to Probability Theory and Its Applications.* Vol. 1, 3d ed. New York: John Wiley & Sons, 1968.

Haight, F. A. *Applied Probability.* New York: Plenum Press, 1981.

Karlin, S., and Taylor, H. M. *An Introduction to Stochastic Modeling*. Orlando, Fla.: Academic Press, 1984.

Ross, S. *A First Course in Probability*. 2d ed. New York: Macmillan Publishing Co, 1984.

Ross, S. *Introduction to Probability Models.* 3d ed. Orlando, Fla.: Academic Press, 1985.

Random Numbers and Simulation

Dunning, K. A. *Getting Started in GPSS.* San Jose: Engineering Press, 1981.

Knuth, D. E. *The Art of Computer Programming, Vol. 2: Seminumerical Algorithms.* 2d ed. Reading, Mass.: Addison-Wesley, 1982.

Ripley, B. D. *Stochastic Simulation*. New York: John Wiley & Sons, 1987.

Solomon, S. L. *Simulation of Waiting-Line Systems*. Englewood Cliffs, N.J.: Prentice-Hall, 1983.

Markov Chains

Isaacson, D. L., and Madsen, R. *Markov Chains: Theory and Applications*. New York: John Wiley & Sons, 1976.

Kemeny, J. G., and Snell, J. L. *Finite Markov Chains.* 2d printing. New York: Springer-Verlag, 1976.

Queueing Theory and Computer Science Applications

Allen, A. O. *Probability, Statistics, and Queueing Theory, with Computer Science Applications*. New York: Academic Press, 1978.

Cooper, R. B. *Introduction to Queueing Theory.* 2d ed. New York: Elsevier North Holland, 1981.

Heyman, D. P., and Sobel, M. J. *Stochastic Models in Operations Research, Vol. 1: Stochastic Processes and Operating Characteristics.* New York: McGraw-Hill, 1982 (reprinted by Dover Publications, 2003).

Kleinrock, L. *Queueing Systems, Vol. 1: Theory*. New York: John Wiley & Sons, 1975.

Trivedi, K. S. *Probability and Statistics with Reliability, Queuing, and Computer Science Applications*. Englewood Cliffs, N.J.: Prentice-Hall, 1982.

More Advanced Textbooks on Probability and Stochastic Processes

Billingsley, P. *Probability and Measure.* 2d ed. New York: John Wiley & Sons, 1986.

Chung, K. L. *A Course in Probability Theory.* 2d ed. New York: Academic Press, 1974.

Feller, W. *An Introduction to Probability Theory and Its Applications.* Vol. 2, 2d ed. New York: John Wiley & Sons, 1971.

Karlin, S., and Taylor, H. M. *A First Course in Stochastic Processes.* 2d ed. New York: Academic Press, 1975.

Numerical Solutions to Exercises

Chapter 1

1. (a) $|S| = 36$ (b) $|E \cap F| = 18$ $|E^c| = 9$

2. (a) $E \cup F \cup G$ (b) $E^c \cap F^c \cap G^c$
   (c) $(E \cap F^c \cap G^c) \cup (E^c \cap F \cap G^c) \cup (E^c \cap F^c \cap G)$
   (d) the union of the sets in parts (b) and (c)

5. (a) $(N \cap T^c) \cup (M \cap T^c)$ (b) $(T \cap N \cap M^c) \cup (T \cap M \cap N^c)$
   (c) $(T \cap M \cap N) \cup (M^c N^c)$

6. (a) 12 (b) (2, 5), (2, 7), (1, 4), (3, 4), (1, 6), (3, 6)
   (c) (2, 4), (2, 6) (d) (1, 5), (1, 7), (3, 5), (3, 7)

Chapter 2

1. (a) 0.4226 (b) 0.0475 (c) 0.0211 (d) 0.00024

2. (a) 3.993×10^{-2} (b) 9.235×10^{-6}

3. (a) 0.1348 (b) 0.2135

4. (a) $|S| = n^n$ (b) $n!/n^n$ (c) $e^{-n}\sqrt{2\pi n}$

5. (a) For $N = 8$, exact probability $= 3.79 \times 10^{-3}$
   (b) Probability $\approx e^{-N}\sqrt{(N-1)2\pi e} = 3.67 \times 10^{-3}$ for $N = 8$

6. (a) 0.2167 (b) 0.2109

7. 0.07473

8. (a) 0.06667 (b) 0.1644

9. (a) 0.01429 (b) 0.7714

10. (a) 0.6 (b) 0.496 (c) 0.44

11. (a) 0.28 (b) 0.252 (c) 0.972

12. 0.2777

13. (a) $\frac{1}{2}$ (b) No

15. $N = 2:$ $p = 0.5$ $N = 4:$ $p = 0.375$ $N = 6:$ $p = 0.3681$
    Need $N \geq 6$ for $p = 0.368$ to 3 decimals

16. (a) $(n-k)!/n!$ (b) $\frac{(n-k)!}{n!}\, e_{n-k}(-1)$, with $e_j(x) = 1+x+\cdots+\frac{x^j}{j!}$
    (c) $\frac{1}{k!}\, e_{n-k}(-1)$

17. 0.0511

19. $\frac{1}{2}$

20. 0.5143

21. 0.0483

22. $\frac{1}{10}$

23. 0.4545

24. (a) $\frac{1}{3}$ (b) $\frac{1}{5}$ (c) $n \geq 7$

25. (a) 0.03030 (b) 0.05882

26. (b) 0.2727

27. (a) 0.01001 (about 1 in 100) (b) 0.0009 (less than 1 in 1000)

28. $\delta < 9 \times 10^{-4}$

33. (a) 0.1875 (b) 0.3125 (c) 0.01831 (d) 0.2727

34. (a) 0.6667 (b) 0.8889

35. (a) 0.6826 (b) 0.6480 (c) 0.6

36. 0.5551

37. (b) $6p^2 - 8p^3 + 3p^4$

38. Alice (win probability 0.6651) over Barbara (win probability 0.6187)

39. (b) q^2

40. (a) 0.4843 (b) 0.5338 (c) 0.2584

Chapter 3

1. $\mu_X = 2$ $\sigma_X = 1.84$

2. (a) 0.3125 (b) 0.125

3. Range of X: 1, 2, 3, 4, 5, 6, 7, 8
   Probabilities: $\frac{3}{10}, \frac{7}{30}, \frac{7}{40}, \frac{1}{8}, \frac{1}{12}, \frac{1}{20}, \frac{1}{40}, \frac{1}{120}$

4. $P\{X=0\} = \frac{1}{12}$ $P\{X=1\} = \frac{7}{12}$ $P\{X=2\} = \frac{1}{3}$

5. (a) $P\{X=0\} = \frac{5}{6}$ $P\{X=10\} = \frac{5}{36}$ $P\{X=40\} = \frac{1}{36}$
   (b) $E[X] = \$2.50$

7. Most likely value $= 2$ $E[X] = 1.8$

8. (a) $\binom{6}{3}p^3q^3$ (b) p^3q^3 (c) $\binom{4}{2}p^3q^2$

9. (a) $c = 6$ (b) $\mu_X = 0.5$ $\sigma_X = (0.05)^{1/2}$

10. $E[X^n] = \frac{1}{n+1}$ $\mathrm{Var}(X^n) = \frac{n^2}{(2n+1)(n+1)^2}$

11. (a) 0.2835 (b) 0.2031 (c) 0.2835

14. (a) 0.9817 (b) e^{-y}

19. (a) $X = 0, 1, 2$ with probabilities 0.6367, 0.3265, 0.0367
(b) $E[X] = 0.40$ (c) Most likely value 0

20. $X = 3, 4, 5$ with probabilities 0.2593, 0.3749, 0.3658

21. (a) 0.4232 (b) 0.8428

22. $38

23. (a) $P\{X = 2\} = 0.2852$ $P\{Y = 2\} = 0.2707$
(b) 0.5940 (Poisson approx.) 0.5943 (exact binomial)

24. (a) $2^{-19} \approx 1.9 \times 10^{-6}$ (b) No: $E[X] = \infty$

27. (a) $\frac{1}{2}$ (b) $\frac{1}{4}$

28. (a) 0.1587 (b) 0.3174

29. (a) 0.7475 (b) 0.3694 (c) 0.5889 (d) 0.3781

31. Median $= (\log 2)/\lambda \approx 0.6931/\lambda$

32. (a) $E[X] = 2$ (b) mode $= 1$

34. (a) $\exp(-3^\beta + 1)$ (c) $\sqrt{\pi}$

35. (a) $Y = 0, 1, 4$ with probabilities $\frac{1}{4}, \frac{1}{2}, \frac{1}{4}$
(b) $E[Y] = 1.5$ $E[Y^2] = 4.5$

36. (b) $E[Y^n] = (e^n - 1)/n$

37. (c) Minimum $e(1/2) = 1/12$

38. (b) Sample X values 0, 1, 1, 2, 1

39. (b) Sample X values 7, 2, 3, 5, 1

40. (b) Sample X values 203.2, 215.2, 227.4, 112.4, 99.9

41. (b) Sample X values 17.5, 22.4, 23.0, 28.0, 26.8

Chapter 4

1. (b) $E[X] = 0.8 \quad E[Y] = 1.1$ (d) $E[Z] = -0.3$

2. (b) $E[X] = \frac{3}{2} \quad E[Y] = 2\log 2$ (c) $\text{Cov}(X, Y) = 2 - 3\log 2$

3. (b) $E[M] = n/(n+1)$

5. $E[N] = 12 \quad \text{Var}(N) = 36$

6. (a) $\text{Cov}(I_A, I_B) = -\frac{1}{27}$ (b) $\text{Cov}(X, Y) = -n/27$

7. $E[X] = (1 - q^N)/p$, where $p = \frac{1}{365}$ and $q = 1 - p$
   $\text{Var}(X) = (q^N - q^{2N})/p + q((1 - 2p)^N - q^{2N})/p^2$

8. $\frac{7}{8}$

9. (a) $\frac{1}{3}$ (b) z^2 for $0 \le z \le 1$ and $2z - z^2$ for $1 \le z \le 2$

11. (a) 0.6321 (b) 3.437

12. (a) 3 (b) 4.333

14. (a) $(e^{-t} + 1 + e^t)/3$ (b) $E[X] = 0 \quad E[X^2] = \frac{2}{3}$

15. $E[X] = 1 \quad \text{Var}(X) = \frac{1}{3}$

16. (a) $P\{49 < X < 51\} \ge 0.75$ (b) $P\{49 < X < 51\} \approx 0.9544$

17. $P\{X > 55\} \approx 0.1357$

18. $P\{X \le 100\} \approx 0.9803$

19. (a) $\frac{1}{4}$ (b) 0 (c) 0.01 (d) 0.0455

20. (c) 2.493×10^{-4}

22. (c) Z takes values 0, 1, 2, 3, 4, 5 with probabilities $\frac{3}{18}, \frac{5}{18}, \frac{4}{18}, \frac{3}{18}, \frac{2}{18}, \frac{1}{18}$

23. (b) $n(1 - y)^{n-1}$ for $0 \le y \le 1$ (c) $1/(n+1)$

24. (b) $n(n-1)(x - y)^{n-2}$ for $0 \le y \le x \le 1$
    (c) $\text{Var}(X) = \text{Var}(Y) = n/\big((n+2)(n+1)^2\big)$
    $\text{Cov}(X, Y) = 1/\big((n+2)(n+1)^2\big)$
    (d) $f(z) = n(n-1)z^{n-1}(1 - z)$ for $0 \le z \le 1$

25. (a) 0.3846
    (b) $\text{Var}(X_i) = 0.2367 \quad \text{Cov}(X_i, X_j) = -1.7 \times 10^{-3}$ for $i \ne j$
    (c) $E[X] = 0.3846 \quad \text{Var}(X) = 0.9264$

30. (a) $\sqrt{1-p} + 1/\sqrt{1-p}$ (b) $\big(\sqrt{1-p} + 1/\sqrt{1-p}\big)/\sqrt{r}$
    (c) 2 (d) $2/\sqrt{r}$

31. (a) $p^2/(1-p)+(1-p)^2/p$ (b) $(6+q+q^2+q^3)/q^2$ where $q=1-p$
    (c) 9 (d) 3

33. (b) $P\{70 < X < 130\} \geq 0.9743$ (using (a))
    (c) $P\{70 < X < 130\} \geq 0.9074$ (using Chebyshev)
    (d) $P\{70 < X < 130\} \approx 0.999$ (using Central Limit Theorem)

Chapter 5

1. (a) 6 (b) 7

2. (a) 0.026
   (b) $P\{C=4 \mid C+T=4\}=0.198$ $E[C \mid C+T=4]=2.667$

3. (a) $E[X]=E[Y]=1$

4. (a) 2 (b) 8.667

5. (a) $\frac{3}{4}$ hour (b) Tom gets \$6

6. (a) exponential, mean 1 (b) Uniform on $(0, y)$ (c) $(\frac{1}{3})Y^2$

7. $E[S]=35$ $\mathrm{Var}(S)=110.8$

8. (a) $(y/2)e^{-xy}$ for $0<x<\infty$ and $0<y<2$ (b) e^{-5y}
   (c) $(1-e^{-10})/10$

9. (a) $\mathrm{Cov}(X,Y)=\alpha$ (b) $\rho=\alpha/(\alpha^2+\sigma^2)^{1/2}$

10. (a) Normal with mean $\rho^2(y-\beta)/\alpha$ and variance $(\sigma\rho/\alpha)^2$
    (b) $\rho^2(Y-\beta)/\alpha$

Chapter 6

2. (a) 0.512 (b) 0.818

3. 0.559

4. (b) 0.50018

5. (b) p^3+p^5q (c) $p^3/(1-p^2q)$

8. (c) $\frac{11}{16}$

9. (a) 3.48, 2.33, 1.71, 1.36, 1.14

14. (a) 8 (b) 14 (c) 6 (d) 12

16. (a) $\frac{2}{3}$ (b) 5 generations

17. (c) $(1+p+p^2+qp^2)/(1-p^2q^2)$ (d) $p^3(1+q)/(1-p^2q^2)$

18. (c) for $p = 0.3$, win probability $= 0.0195$ and average duration $= 2.255$
for $p = 0.7$, win probability $= 0.580$ and average duration $= 4.747$

19. (a) for $p = 3:$ 2.46 for $p = 7:$ 0.805

20. 6

21. (b) [0.5530, 0.1014, 0.3456], good/poor $= 5.45$, good/broken $= 1.60$

22. (a) 2.89 (b) 5.5 (c) 5.21 (d) 3.54

23. (b) $\begin{bmatrix} \frac{2}{13} & \frac{6}{13} & \frac{5}{13} \end{bmatrix}$ (c) State 2 shortest, State 1 longest

24. State 3

Chapter 7

1. (a) 0.039 (b) 0.095

2. 259 hours

3. (a) 0.9502 (b) 0.6335

6. (a) 0.112 (b) 40 min. (c) 0.264 (d) $\frac{1}{3}$ (e) 0.393

7. $s = 2:$ 0.90 $s = 10:$ 0.61 $s = 20:$ 0.37

8. $s = 2:$ 0.99 $s = 10:$ 0.91 $s = 20:$ 0.74

9. $\lambda \min(s, t)$

10. Binomial, with parameters n and $p = s/t$

11. (a) λt (b) uniform on $(0, \lambda)$ (c) $\lambda/2$ (d) $(1 - e^{-\lambda})/\lambda$

Chapter 8

2. (a) 0.052 (b) 8

3. (a) 0.9958 (b) 0.9998

5. (a) 0.264 (b) 6 (c) 0.302

8. Mean $= \$40,000$ Var $= 1.6 \times 10^8$

10. (a) 0.56 (b) 0.130

12. (a) 1 (b) 6 (c) $3t^2 \exp(-t^3)$

13. (a) Arrivals at $t = 0.85, 10.11, 10.64, 11.10$
    (b) Service times are 8.87, 6.22, 9.09, 9.00
    (c) average queue length = 0.730 average queue time = 7.30 sec.

14. (a) States visited: 0, 1, 2, 1, 2, ...
    (b) Sojourn times: 1.43, 5.87, 3.00, 3.93, ...
    (c) server idle 9.55% of time

Chapter 9

1. (a) $P\{B \geq 4\} = e^{-2n}$

3. States visited: 0, 1, 2, 1, 2, ...
   Sojourn times: 0.072, 0.352, 0.150, 0.236, ...

4. (a) 0.2231 (b) 0.2667

6. (a) $ne^{\lambda t}$ (b) $E[X(0)] \cdot e^{\lambda t}$

Chapter 10

2. (a) $P\{X(1) = 0\} = 0.8426$ and $P\{X(10) = 0\} = 0.6027$
   (b) $P\{X(1) = 0\} = 0.2361$ and $P\{X(10) = 0\} = 0.5960$

3. $t \geq 3.79$ min.

4. $T \geq 13.33$ min.

10. (a) 0.90 (b) 0.54 (c) 5.46 more customers

11. (a) $C_{\text{machines}} = 0.112$ mean number of machines broken = 2.03
    (b) $C_{\text{repairmen}} = 0.211$ both repairmen busy 66% of time

12. (a) $C_{\text{machines}} = 0.165$ mean number of machines broken = 1.07
    (b) $C_{\text{repairmen}} = 0.258$

13. (a) 12.9 customers/hour (b) 0.38

15. (b) λ

Chapter 11

1. (b) $L(20) = 1.67$ $W(20) = 8.34$ sec. $\lambda_a(20) = 0.2$ per sec.

4. (a) $\lambda \leq 15$ per hour (b) 9 min. (c) 18.2 min.

5. (c) $L = 1.71$ customers (d) \$25.71/hour (e) 40 min.

6. $L = 2.1$ customers $W = 11.6$ min.

11. (a) $\frac{1}{12}$ (b) $\frac{7}{12}$ (c) $\mu_1 \geq 9.33$ customers/hour

12. (b) 1.71 cars/hour (c) $L = \frac{33}{35}$ (d) 33 minutes

13. (b) average number of customers = 1.5

14. (a) $W_Q = \frac{3}{4}$ hour $L_Q = 2.25$ customers (b) 0.406

16. (a) 0.053 (b) 0.853 (c) 30 min. (d) 26 min.
(e) 7.67 (f) 32 min. (g) 9.47 (h) 0.115

17. (a) $\$W_Q^{(1)} = \$W_Q^{(2)}$ (b) $W_Q^{(1)} > W_Q^{(2)}$ (c) $W^{(1)} < W^{(2)}$

18. Reading room: $L_Q = 2.25$ $W_Q = 9$ min.
Reserve room: $L_Q = 0.083$ $W_Q = 1$ min.
Combined facility: $L_Q = 0.33$ $W_Q = 1$ min.

21. (a) $\frac{1}{8}$ per hour (b) $\frac{1}{3}$ (c) 0.245

22. (a) 6.67 per hour (one operator) 12 per hour (two operators)
(b) $b < 8$

23. (b) $\mu > 2.47$

24. (c) 15 servers

28. 0.245

Chapter 12

1. (a) 0.1353 (b) $x \exp(-x^2/2)$
(c) $E[X] = \sqrt{\pi/2}$ $\text{Var}(X) = 2 - (\pi/2)$

2. (b) $\lambda = 8.55 \times 10^{-5}$

13. (b) p/q

15. (a) 3.2 min. (b) 9.2 min. (c) 0.109 min.

17. (b) $T_{\min} = 1.45$ years (c) cost = \$181 per year

18. (b) $T_{\min} = 1.37$ years (c) cost = \$179 per year

19. (a) $\frac{2}{3}$ (b) $\frac{1}{3}$

20. (c) $1/(1 - \rho)$

21. $\tau < \frac{2}{3}$ min. (arrival rate > 90 customers/hour)

25. (a) 75% (b) 63%

26. Expected remaining lifetime = $\frac{2}{3}$ year

Expected total lifetime = $\frac{4}{3}$ year

27. (a) 2.65 (b) 1.27 (c) 1.61

29. (a) $\rho/(\mu(1-\rho))$ (b) $\frac{2}{3}\times$ value in part (a)

(c) $\frac{3}{4}\times$ value in part (a)

31. (b) $W_Q = \epsilon^2/(3(1-\epsilon))$

Index